AF781553

NANOTECHNOLOGY SCIENCE AND TECHNOLOGY

Additional books in this series can be found on Nova's website under the Series tab.

Additional e-books in this series can be found on Nova's website under the e-book tab.

NANOTECHNOLOGY SCIENCE AND TECHNOLOGY

ADVANCED NANOCOMPOSITES

TYPES, PROPERTIES AND APPLICATIONS

NANOTECHNOLOGY SCIENCE AND TECHNOLOGY

ADVANCED NANOCOMPOSITES

TYPES, PROPERTIES AND APPLICATIONS

ABEL MAHARRAMOV
MAHAMMAD ALI RAMAZANOV
AND
MOHAMMAD REZA SABOKTAKIN

OLGUN GUVEN
EDITOR

nova publishers
New York

Copyright © 2013 by Nova Science Publishers, Inc

All rights reserved. No part of this book may be reproduced, stored in a retrieval system or transmitted in any form or by any means: electronic, electrostatic, magnetic, tape, mechanical photocopying, recording or otherwise without the written permission of the Publisher.

For permission to use material from this book please contact us:
Telephone 631-231-7269; Fax 631-231-8175
Web Site: http://www.novapublishers.com

NOTICE TO THE READER

The Publisher has taken reasonable care in the preparation of this book, but makes no expressed or implied warranty of any kind and assumes no responsibility for any errors or omissions. No liability is assumed for incidental or consequential damages in connection with or arising out of information contained in this book. The Publisher shall not be liable for any special, consequential, or exemplary damages resulting, in whole or in part, from the readers' use of, or reliance upon, this material. Any parts of this book based on government reports are so indicated and copyright is claimed for those parts to the extent applicable to compilations of such works.

Independent verification should be sought for any data, advice or recommendations contained in this book. In addition, no responsibility is assumed by the publisher for any injury and/or damage to persons or property arising from any methods, products, instructions, ideas or otherwise contained in this publication.

This publication is designed to provide accurate and authoritative information with regard to the subject matter covered herein. It is sold with the clear understanding that the Publisher is not engaged in rendering legal or any other professional services. If legal or any other expert assistance is required, the services of a competent person should be sought. FROM A DECLARATION OF PARTICIPANTS JOINTLY ADOPTED BY A COMMITTEE OF THE AMERICAN BAR ASSOCIATION AND A COMMITTEE OF PUBLISHERS.

Additional color graphics may be available in the e-book version of this book.

Library of Congress Cataloging-in-Publication Data

Saboktakin, M. R.
Advanced nanocomposites : types, properties and applications / Abel Maharramov, Mahammadali Ramazanov and Mohammad Reza Saboktakin.
pages cm
Includes bibliographical references and index.
ISBN: 978-1-62808-308-8 (hardcover)
1. Nanocomposites (Materials) 2. Nanocomposites (Materials)--Research. I. Mdhdrrdmov, A. M. (Abel Mdmmdddli oglu) II. Ramazanov, Mahammadali. III. Title.
TA418.9.N35S23 2013
620.1'18--dc23 2013029300

Published by Nova Science Publishers, Inc. † New York

CONTENTS

PREFACE

The field of nanotechnology is one of the most popular areas for current research and development in virtually all technical disciplines. This includes polymer science and technology and investigations in this field cover a broad range of topics. This includes microelectronics (which is now being referred to as nanoelectronics) as the critical dimension scale for modern devices now below 100 nm. Other areas include polymer-based biomaterials, nanoparticle drug delivery, miniemulsion particles, fuel cell electrode polymer bound catalysts, layer-by-layer self-assembled polymer films, electrospun nanofibers, imprint lithography, polymer blends and nanocomposites. In the field of nanocomposites, there are many diverse topics including composite reinforcement, barrier properties, flame resistance, electro-optical properties, cosmetic applications, and bactericidal properties. Nanotechnology is not new to polymer science as prior studies before the age of nanotechnology involved nanoscale dimensions but were not specifically referred to as nanotechnology until recently. Nanoparticles are defined as particulate dispersions or solid particles with a size in the range of 10-1000nm. The drug is dissolved, entrapped, encapsulated or attached to a nanoparticle matrix. Depending upon the method of preparation, nanoparticles, nanospheres or nanocapsules can be obtained. Nanocapsules are systems in which the drug is confined to a cavity surrounded by a unique polymer membrane, while nanospheres are matrix systems in which the drug is physically and uniformly dispersed. Nanotechnology offers unique approaches to probe and control a variety of biological and medical processes that occur at nanometer length scales, and is expected to have a revolutionary impact on biology and medicine. Among the approaches for exploiting nanotechnology in medicine, nanoparticles offer some unique advantages as sensing, image enhancement, and delivery agents. Several varieties of nanoparticles with biomedical relevance are available including, polymeric nanoparticles, metal nanoparticles, liposomes, micelles, quantum dots, dendrimers, and nanoassemblies. To further the application of nanoparticles in disease diagnosis and therapy, it is important that the systems are biocompatible and capable of being functionalized for recognition of specific target sites in the body after systemic administration. We will discuss the novel nanocomposites and nanoparticles from various applications and drug delivery systems for cancer therapy.

Chapter 1

WHAT ARE NANOCOMPOSITES?

1.1. INTRODUCTION

The field of nanotechnology is one of the most popular areas for current research and development in many technical disciplines. This includes polymer science and technology. Investigations in this field cover a broad range of topics. This includes microelectronics (which is now referred to as nanoelectronics) since the critical dimension scale for modern devices are now below 100 nm. Other areas of research include polymer-based biomaterials, nanoparticle drug delivery, miniemulsion particles, fuel cell electrode polymer bound catalysts, layer-by-layer self-assembled polymer films, electrospun nanofibers, imprint lithography, polymer blends and nanocomposites. Even in the field of nanocomposites, many diverse topics exist including composite reinforcement, barrier properties, flame resistance, electro-optical properties, cosmetic applications and bactericidal properties. Nanotechnology is not new to polymer science as prior studies before the age of nanotechnology involved nanoscale dimensions but were not specifically referred to as nanotechnology until recently.

Phase separated polymer blends often achieve nanoscale phase dimensions; block copolymer domain morphology is usually at the nanoscale level; asymmetric membranes often have nanoscale void structure, miniemulsion particles are below 100 nm; and interfacial phenomena in blends and composites involve nanoscale dimensions. Even with nanocomposites, carbon black reinforcement of elastomers, colloidal silica modification and even naturally occurring fiber (e.g., asbestos-nanoscale fiber diameter) reinforcement are subjects that have been investigated for decades. Almost lost in the present nanocomposite discussions are the organic– inorganic nanocomposites based on sol–gel chemistry which have been investigated for several decades [1–3]. In essence, the nanoscale of dimensions is the transition zone between the macro level and the molecular level. Recent interest in polymer matrix based nanocomposites has emerged initially with interesting observations involving exfoliated clay and more recent studies with carbon nanotubes, carbon nanofibers, exfoliated graphite (graphene), anocrystalline metals and a host of additional nanoscale inorganic filler or fiber modifications. This review will discuss polymer matrix based nanocomposites with exfoliated clay being one of the key modifications. While the reinforcement aspects of nanocomposites are the primary area of interest, a number of other properties and potential applications are important including barrier properties, flammability resistance, electrical/electronic properties, membrane properties and polymer blend compatibilization. An important consideration involves the comparison of properties of nanoscale dimensions relative to larger scale dimensions. The synergistic advantage of nanoscale dimensions (''nano-effect'') relative to larger scale modification is an important consideration. Understanding the property changes as the particle (or fiber) dimensions decrease to the nanoscale level is important to optimize the resultant nanocomposite. As will be noted, many nanocomposite systems noted in the literature can still be modeled using continuum models where the absolute size is not important since only shape and volume fraction loading are necessary to predict properties. Nanoscale is considered where the dimensions of the particle, platelet or fiber modification are in the range of 1–100 nm. With the platelet or fiber, the smallest dimension is considered for that range (platelet thickness or fiber diameter).

1.2. Fundamental Considerations

In the area of nanotechnology, polymer matrix based nanocomposites have generated a significant amount of attention in the recent literature. This area emerged with the recognition that exfoliated clays could yield significant mechanical property advantages as a modification of polymeric systems [4–6]. The results achieved were at least initially viewed as unexpected (''nano-effect'') offering improved properties over that expected from continuum mechanics predictions. More recent results have, however, indicated that while the property profile is interesting, the clay-based nanocomposites often obey continuum mechanics predictions. There are situations where nanocomposites can exhibit properties not expected with larger scale particulate reinforcements. It is now well-recognized that the crystallization rate and degree of crystallinity can be influenced by crystallization in confined spaces. In these cases, the dimensions available for spherulitic growth are confined so that primary nuclei are not present for heterogeneous crystallization and homogeneous nucleation thus results. This results in the value of n in the Avrami equation approaching one and often leads to a reduced crystallization rate, degree of crystallinity and melting point. This has been observed in phase separated block copolymers [7, 8] and polymer blends [9]. Confined crystallization of linear polyethylene in nanoporous alumina showed homogeneous nucleation with pore diameters of 62–110 nm but heterogeneous nucleation for 15–48 nm pores [10]. Linear polyethylene [11] and syndiotactic polystyrene [12] in nanoporous alumina both showed decreased crystallinity versus bulk crystallization. With nanoparticle incorporation in a polymer matrix, similarities to confined crystallinity (as noted above for crystallization in nanopores) exist as well as nucleation effects and disruption of attainable spherulite size. With inorganic particle and nanoparticle inclusions, nucleation of crystallization can occur. At the nanodimension scale, the nanoparticle can substitute for the absence of primary nuclei thus competing with the confined crystallization. At higher nanoparticle contents, the increased viscosity (decreased chain diffusion rate) can lead to decreased crystallization kinetics. Thus, the crystallization process is complex and influenced by several competing factors. Nucleation of crystallization (at low levels of addition) evidenced by the onset temperature of crystallization (Tc) and crystallization half-time has been observed in various nanocomposites (poly- (3-caprolactone)–nanoclay [13], polyamide 66–nanoclay [14, 15], polylactide–nanoclay [16], polyamide 6–nanoclay [17], polyamide 66–multi-walled carbon nanotube [18], polyester–nanoclay [19], poly(butylene terephthalate)–nanoclay [20], polypropylene–nanoclay (sepiolite) [21] and polypropylene/multi-walled carbon nanotube [22]). At higher levels of nanoparticle addition, retardation of the crystallization rate has been observed even in those systems where nucleation was observed at low levels of nanoparticle incorporation [15, 18, 20, 22–24]. The higher level of nanoparticle inclusion was noted to yield retardation of crystallization due to diffusion constraints. This was also apparent in a study where unmodified and organically modified clay were incorporated in maleic anhydride grafted polypropylene [25]. Nucleation was observed with unmodified clay, whereas the exfoliated clay yielded a reduced crystallization rate. The crystallization behavior of layered silicate clay nanocomposites noted that while nucleation is observed in many systems, the overall crystallization rate is generally reduced particularly at higher levels of nanoclay addition [26].

Another ''nano-effect'' noted in the literature has been the change in the Tg of the polymer matrix with the addition of nanosized particles. Both increases and decreases in the Tg have been reported dependent upon the interaction between the matrix and the particle. In essence, if the addition of a particle to an amorphous polymer leads to a change in the Tg, the resultant effect on the composite properties would be considered a ''nano-effect'' and not predictable employing continuum mechanics relationships unless the Tg changes were properly accounted for or were quite minor. The glass transition of a polymer will be affected by its environment when the chain is within several nanometers of another phase. An extreme case of this is where the other environment is air (or vacuum). It has been well-recognized in the literature that the Tg of a polymer at the air–polymer surface or thin films (<100 nm) may be lower than that in bulk [27]. This can also be considered to be a confinement effect. A specific experimental example was reported where poly (2-vinyl pyridine) showed an increase in Tg, poly (methyl methacrylate) (PMMA) showed a decrease in Tg and polystyrene showed no change with silica nanosphere incorporation. These differences were ascribed to surface wetting [28]. The Tg decrease for PMMA was ascribed to free volume existing at the polymer surface interface due to poor wetting. In most literature examples where Tg values have been obtained, usually only modest changes are reported (<10 _C) as noted in various examples tabulated in Table 1. In some cases, the organic modification of clay can result in a decrease in Tg due to plasticization [29]. It should be noted that the values noted in Table 1 involved relatively low levels of nanoparticle incorporation (<0.10 wt fraction and even lower volume fraction) and larger changes in Tg could be expected at much higher volume fraction loadings. For crosslinked polymers, another consideration is necessary as the presence of nanoparticles could yield a crosslink density change over the unmodified composite. This could be due to preferential interactions of the crosslinking agent with the nanoparticle surface or interruption of the crosslink density due to confinement effects. A theoretical model has been developed to predict the glass transition temperature of nanocomposites [30]. The model predicts both increases and decreases in Tg dependent upon specific interactions and shows good agreement with the experimental data noted above [28].

A situation does exist where significant increases in the glass transition temperature have been noted involving polyhedral oligomeric silsesquioxane (POSS) cage structures chemically reacted into the polymeric network [31–34]. These cage structures with a particle diameter in the range of 1–3 nm can be functionalized to provide chemical reactivity with various polymer systems. Examples include octavinyl (R¼ vinyl group) incorporation for copolymerization with PMMA [31], amine groups for incorporation into polyamides [32] and polyimides [33]. This parallels the glass transition increase often noted in the sol–gel inorganic–organic networks.

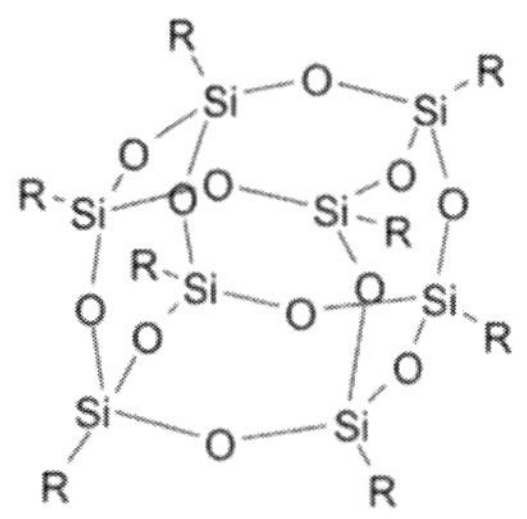

POSS

R = alkyl, aryl, cycloaliphatic, vinyl, amino, nitrile halogen. alcohol, ester, isocyanate, glycidyl etc.

While the glass transition temperature and crystallinity are the major property changes of interest of the nanocomposite polymer matrix, other ''nano-effects'' or property improvements over larger scale dimensions can be observed. Disruption of packing of rigid chain polymers resulting in higher free volume has been observed in permeability studies [44], surface area effects in photovoltaic applications involving conjugated polymers, surface area effects for catalysts incorporated in polymers, polymer chain dimensions where the radius of gyration is greater than the distance between adjacent nanoparticles, optical properties and nanofiber scaffolds for tissue engineering are additional areas. The ''aging'' of polymers is a thickness dependent property with rapid change at nanoscale dimensions [45, 46]. This property is due to the ability of free volume to diffuse out of the sample and the diffusion coefficient (although very low) becomes important in the time scale associated with polymer utility (days to years) at nanoscale thicknesses. Surface area effects including catalysts and bioactivity often require nanolevel dimensions to achieve optimum performance.

Table 1. Glass transition changes with nanofiller incorporation

Polymer	Nanofiller	T_g change (°C)	Reference
Polystyrene	SWCNT	3	[34]
Polycarbonate	SiC (0.5–1.5 wt%) (20–60 nm particles)	No change	[35]
Poly(vinyl chloride)	Exfoliated clay (MMT) (<10 wt%)	−1 to −3	[36]
Poly(dimethyl siloxane)	Silica (2–3 nm)	10	[37]
Poly(propylene carbonate)	Nanoclay (4 wt%)	13	[38]
Poly(methyl methacrylate)	Nanoclay (2.5–15.1 wt%)	4–13	[39]
Polyimide	MWCNT (0.25–6.98 wt%)	−4 to 8	[40]
Polystyrene	Nanoclay (5 wt%)	6.7	[41]
Natural rubber	Nanoclay (5 wt%)	3	[42]
Poly(butylene terephthalate)	Mica (3 wt%)	6	[43]
Polylactide	Nanoclay (3 wt%)	−1 to −4	[29]

SWCNT = single-walled carbon nanotubes; MMT = montmorillonite; MWCNT = multi-walled carbon nanotubes.

1.3. Clay-Based Polymer Nanocomposites

1.3.1. Structure of Montmorillonite

The clay known as montmorillonite consists of platelets with an inner octahedral layer sandwiched between two silicate tetrahedral layers [47] as illustrated in Figure 1.

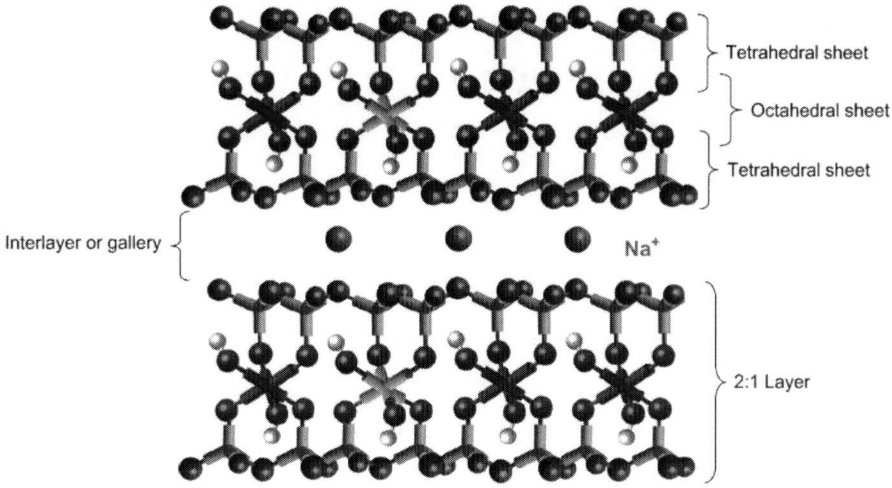

Figure 1. Structure of sodium montmorillonite. Courtesy of Southern Clay Products, Inc.

The octahedral layer may be thought of as an aluminum oxide sheet where some of the aluminum atoms have been replaced with magnesium. The difference in valences of Al and Mg creates negative charges distributed within the plane of the platelets that are balanced by positive counterions, typically sodium ions, located between the platelets or in the galleries as shown in Figure 1. In its natural state, this clay exists as stacks of many platelets. Hydration of the sodium ions causes the galleries to expand and the clay to swell. These platelets can be fully dispersed in water. The sodium ions can be exchanged with organic cations, such as those from an ammonium salt, to form an organoclay [48–57]. The ammonium cation may have hydrocarbon tails and other groups attached and is referred to as a ''surfactant'' owing to its amphiphilic nature. The extent of the negative charge of the clay is characterized by the cation exchange capacity, i.e., CEC. The X-ray d-spacing of completely dry sodium montmorillonite is 0.96 nm while the platelet itself is about 0.94 nm thick [47, 58]. When the sodium is replaced with much larger organic surfactants, the gallery expands and the X-ray d-spacing may increase by as much as 2 to 3-fold [59, 60]. While the thickness of montmorillonite platelets is a well-defined crystallographic dimension, the lateral dimensions of the platelets are not.

They depend on how the platelets grew from solution in the geological process that formed them. Many authors grossly exaggerate the lateral size with dimensions quoted of the order of microns or even tens of microns.

A commonly used montmorillonite was accurately characterized recently by depositing platelets on a mica surface from a very dilute suspension and then measuring the lateral dimensions by atomic force microscopy [61].

Since the platelets are not uniform or regular in lateral size or shape, the platelet area, A, was measured and its square-root was normalized by platelet thickness, t, to calculate an ''aspect ratio''. The distribution of the aspect ratios that were found is shown in Figure 2. If each platelet were circular with a diameter D, then $\sqrt{A}/t$, $\sqrt{\pi/4}(D/t) = 0.98(D/t)$ would be. Since t is approximately 1 nm, Figure 2 shows that the most probable lateral dimension is in the range of 100–200 nm.

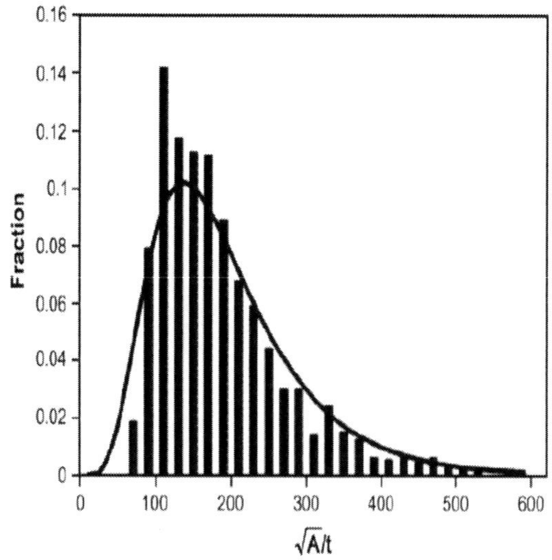

Figure 2. Aspect ratio distribution of native sodium montmorillonite platelets. Reproduced with permission of American Chemical Society.

1.3.2. Nanocomposite Formation: Exfoliation

Nanocomposites can, in principle, be formed from clays and organoclays in a number of ways including various in situ polymerization [4, 6, 62–68], solution [51, 53], and latex [69, 70] methods.

However, the greatest interest has involved melt processing [71– 139] because this is generally considered more economical, more flexible for formulation, and involves compounding and fabrication facilities commonly used in commercial practice. For most purposes, complete exfoliation of the clay platelets, i.e., separation of platelets from one another and dispersed individually in the polymer matrix, is the desired goal of the formation process. However, this ideal morphology is frequently not achieved and varying degrees of dispersion are more common. While far from a completely accurate or descriptive nomenclature, the literature commonly refers to three types of morphology: immiscible (conventional or microcomposite), intercalated, and miscible or exfoliated. These are illustrated schematically in Figure 3 along with sample transmission electron microscopic, TEM, images and the expected wide angle X-ray scans [48–53, 83]. For the case called ‘‘immiscible’’ in Figure 3, the organoclay platelets exist in particles comprised of tactoids or aggregates of tactoids more or less as they were in the organoclay powder, i.e., no separation of platelets. Thus, the wide angle X-ray scan of the polymer composite is expected to look essentially the same as that obtained for the organoclay powder; there is no shifting of the X-ray d-spacing.

Generally, such scans are made over a low range of angles, 2q, such that any peaks from a crystalline polymer matrix are not seen since they occur at higher angles. For completely exfoliated organoclay, no wide angle X-ray peak is expected for the nanocomposite since there is no regular spacing of the platelets and the distances between platelets would, in any case, be larger than what wide angle X-ray scattering can detect.

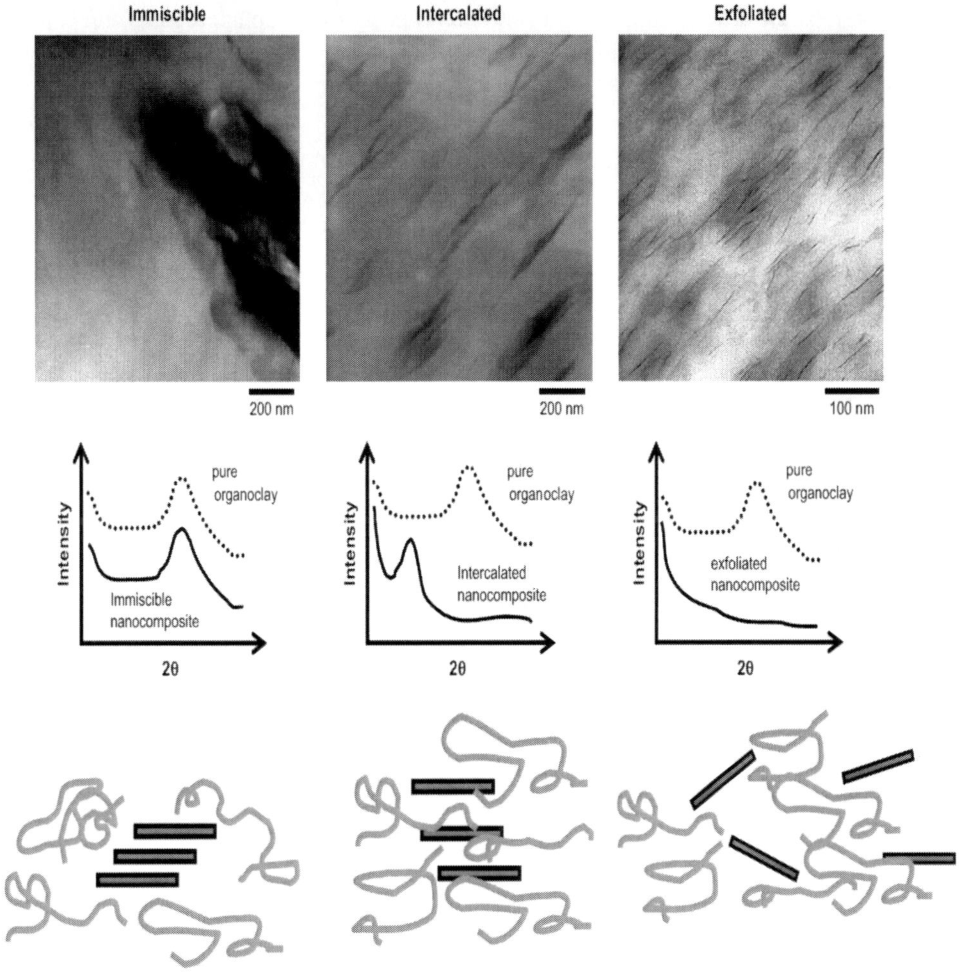

Figure 3. Illustration of different states of dispersion of organoclays with corresponding WAXS and TEM results.

X-ray scans of polymer nanocomposites often show a peak reminiscent of the organoclay peak but shifted to lower 2q or larger d-spacing. The fact that there is a peak indicates that the platelets are not exfoliated. The peak shift indicates that the gallery has expanded, and it is usually assumed that polymer chains have entered or have been intercalated in the gallery. Placing polymer chains in such a confined space would involve a significant entropy penalty that presumably must be driven by an energetic attraction between the polymer and the organoclay [76–79]. It is possible that the gallery expansion may in some cases be caused by intercalation of oligomers or low molecular weight polymer chains. The early literature seemed to suggest that ''intercalation'' would be useful and perhaps a precursor to exfoliation. Subsequent research has suggested alternative ideas about how the exfoliation process may occur in melt processing and how the details of the mixing equipment and conditions alter the state of the dispersion that is achieved [54, 82, 84 and 140]. These ideas are summarized in the cartoon shown in Figure 4 [84]. As made commercially, the particles of an organoclay powder are about 8 mm in size and consist of aggregates of tactoids, or stacks of platelets; the stresses imposed during melt mixing break up aggregates and can shear the stack into smaller ones as suggested in Figure 4. However, there evidently is a limit to how finely the clay can be dispersed just by mechanical forces. If the polymers and organoclay have an ''affinity'' for one another, the contact between polymers and organoclay can be increased by peeling the platelets from these stacks one by one until, given enough time in the mixing device, all the platelets are individually dispersed as suggested in Figure 4. This notion is supported by many TEM images at various locations in the extruder and is more plausible than imagining the polymer chains diffusing into the galleries, i.e., intercalation, and eventually pushing them further and further apart until an exfoliated state is reached. The nature of the extruder and the screw configuration are important to achieve good organoclay dispersion [83]. Longer residence times in the extruder favor better dispersion [83].

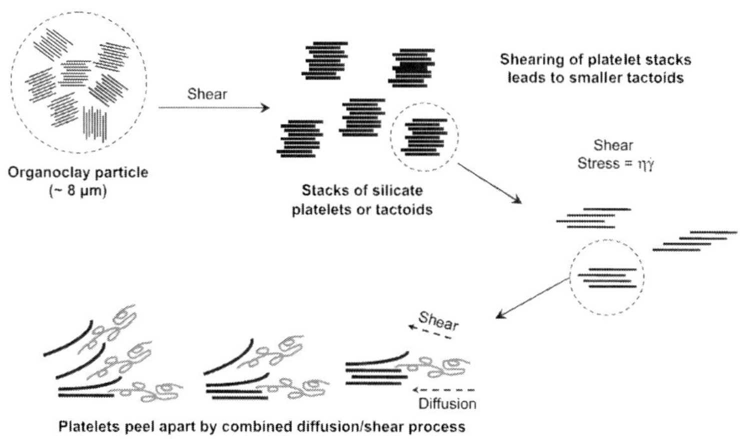

Figure 4. Mechanism of organoclay dispersion during melt processing.

In some cases, having a higher melt viscosity is helpful in achieving dispersion apparently because of the higher stresses that can be imposed on the clay particles [84, 126]. However, this effect is not universally observed. The location of where the organoclay is introduced into the extruder has also been shown to be important [120]. However, no matter how well these process considerations are optimized, it is clear that complete or almost complete exfoliation cannot be achieved unless there is a good thermodynamic affinity between the organoclay and the polymer matrix. This affinity can be affected to a very significant extent by optimizing the structure of the surfactant used to form the organoclay [85, 88, 99, 100, 109, 113, 119, 141] and possibly certain features of the clay itself like its CEC [115], as this affects the density of surfactant molecules over the silicate surface.

A key factor in the polymer–organoclay interaction is the affinity that polymer segments have for the silicate surface [84, 85, 94, 113, 141, 142]. Nylon 6 appears to have good affinity for the silicate surface, perhaps by hydrogen bonding, and as a result very high levels of exfoliation can be achieved in this matrix provided the processing conditions and melt rheology are properly selected [83, 84, 120]. Surfactants with a single long alkyl tail give the best exfoliation [141]. As more long chain alkyls are added to the surfactant, the extent of exfoliation is decreased [141]. It has been proposed that at least one alkyl tail is needed to reduce the platelet– platelet cohesion while adding more than one tends to block Figure 2. Aspect ratio distribution of native sodium montmorillonite platelets [61]. Reproduced with permission of the American Chemical Society. Access of the polyamide chains from the silicate surface, diminishes these favorable interactions while increasing the very unfavorable alkyl–polyamide interaction. On the other hand, non-polar polyolefin segments have no attraction to the polar silicate surface, and in this case, increasing the number of alkyls on the surfactant improves dispersion of the organoclay in the polyolefin matrix since a larger number of alkyls decrease the possible frequency of the unfavorable polyolefin–silicate interaction and increases the frequency of more favorable polyolefin–alkyl contacts [96, 100, 105]. Even under the best of circumstances, exfoliation of organoclays in neat polyolefins like polypropylene, PP, or polyethylene, PE, is not very good and is far less than that observed in polyamides, polyurethanes, and some other polar polymers [84,96,101]. It has been found that a small amount of a polyolefin that has been lightly grafted with maleic anhydride, w1% MA by weight is typical, can act as a very effective ‘‘compatibilizer’’ for dispersing the organoclay in the parent polyolefin [84, 101, 103, 106, 117, 118, 143–148]. This does not lead to the high level of exfoliation that can be achieved in polyamides. However, this approach has allowed such nanocomposites to move forward in commercial applications, particularly in automotive parts [54, 99, 106]. In the case of olefin copolymers with polar monomers like vinyl acetate and methacrylic acid (and corresponding ionomers), the degree of exfoliation that can be achieved progressively improves as the polar monomer content increases [113, 119]. In all cases, the best exfoliation is achieved when the structure of the surfactant and the process parameters are optimized.

1.3.3. Characterization of Nanocomposite Morphology

An important issue is to relate the performance of nanocomposites to their morphological structure. Experimental evaluation of performance is much easier than characterization of their morphology. Wide angle X-ray scattering (WAXS) is frequently used because such analyses are relatively simple to do. However, such analyses can be misleading and are not quantitative [149– 151]. As indicated in Figure 3, the organoclay has a characteristic peak indicative of the platelet separation or d-spacing; other peaks may be seen resulting from multiple reflections as predicted by Bragg's Law.

The presence of the same peak in the nanocomposite is irrefutable evidence that the nanocomposite contains organoclay tactoids as suggested in Figure 3. However, the absence of such a peak is not conclusive evidence for a highly exfoliated structure as has been repeatedly pointed out in the literature [151]. Many factors must be considered to interpret WAXS scans. If the sensitivity or counting time of the scan is low, then an existing peak may not be seen. When the tactoids are internally disordered or not well aligned to one another, the peak intensity will be low and may appear to be completely absent. These issues can be illustrated by analyses of polyolefin nanocomposites, which are never fully exfoliated, that have been injection molded. X-ray scans of the molded surface reveal a peak indicating the presence of tactoids. However, after milling away the surface of these specimens, subsequent scans of the milled surface in the core of the bar may not reveal a peak because the tactoids are more randomly oriented in the interior than near the as-molded surface [103, 118]. However, if a more sensitive scan is made, the peak can usually be seen. In some cases, the

WAXS scan may reveal a shift in the peak location relative to that of the neat organoclay. The peak may shift to lower angles, or larger d-spacing, and is generally taken as evidence of ''intercalation'' of polymers (or perhaps other species) into the galleries [48–53, 73 and 79]. However, an opposite shift may also occur which is usually attributed to loss of unbound surfactant from the gallery or to surfactant degradation [89, 107]. All of these processes may take place simultaneously rendering uncertainty in the interpretation.

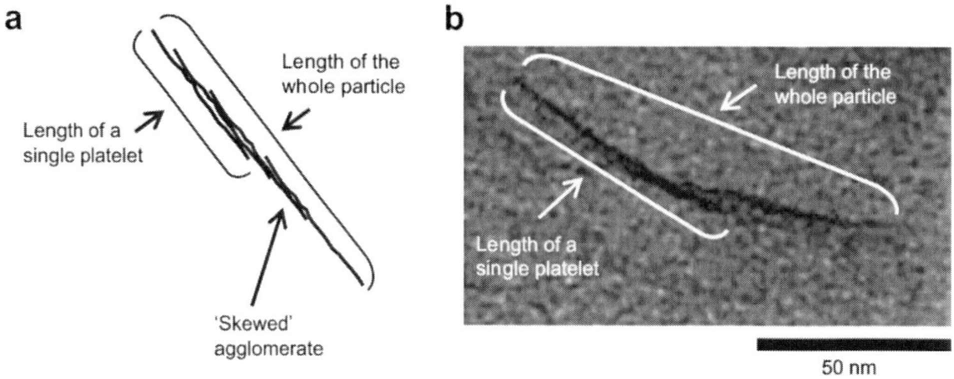

Figure 5. Examples of skewed platelets such that particles appear longer than platelets of MMT.

In any case, intercalation per se does not seem to be a contributor to develop useful nanocomposite performance. Small angle X-ray scattering, SAXS, can be more informative and somewhat quantitative as explained by numerous authors [17, 152– 156]. However, this technique has not been widely used, except in a few laboratories, probably because most laboratories do not have SAXS facilities or experience in interpreting the results. Other techniques like solid-state NMR and neutron scattering have also been used on a limited basis to explore clay dispersion [95, 157– 62]. A far more direct way of visualizing nanocomposite morphology is via transmission electron microscopy, TEM. However, this approach requires considerable skill and patience but can be quantitative. The use of TEM is often criticized because it reveals the morphology in such a small region. However, this can be overcome by taking images at different magnifications and from different locations and orientations until a representative picture of the morphology is established. The major obstacle in obtaining good TEM images is not in the operation of the microscope but in microtoming sections that are thin and uniform enough to reveal the morphology. Fortunately, the elemental composition of the clay compared to that of the polymer matrix is such that no staining is required. When exfoliation is almost complete as in the case of nylon 6 one can see the w1 nm thick clay platelets as dark lines when the microtome cut is perpendicular to the platelets. Image analysis can be used to quantify the distribution of platelet lengths, but meaningful statistics require analyzing several hundred particles [58, 84, 93 and 119]. However, it must be remembered that the dimensions observed reflect a random cut through an irregular platelet and only rarely will the maximum dimension be seen [163, 164]. Thus, the aspect ratio distribution seen in this way will lead to smaller values than true dimensions like those given by Figure 2.

Even for the best nylon 6 nanocomposites, exfoliation is generally never complete and one can see particles consisting of two, three or more platelets [58]. In some cases, these platelets may be skewed relative to one another as suggested in Figure 5 [93]. Thus, some particles may appear to be longer than the platelets really are. These kinds of issues should be kept in mind when interpreting quantitative analyses of particle aspect ratios and in comparison of observed performance with that predicted by composite theory [58]. Nanocomposites made from polyolefins, styrenics, and other polymers that lead to lower degrees of exfoliation reveal particles much thicker than single clay platelets as expected [101, 110, 111, 117, 119]. However, the clay particles are also much longer than the individual clay platelets indicated in Figure 2. As the polymer–organoclay affinity is increased by adding a compatibilizer, e.g., PP-g-MA or PE-g-MA, or increasing the content of a polar comonomer, e.g., vinyl acetate, the clay particles not only become thinner (fewer platelets in the stack) but also become shorter [103, 106, 117–120]. However, generally the particle thickness decreases more rapidly than the length such that the aspect ratio increases; this generally improves performance.

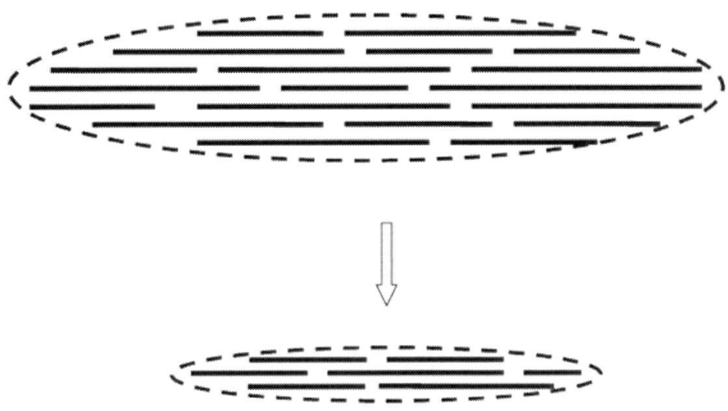

Figure 6. A more realistic picture of clay tactocids and how they become shorter as the level of dispersion increases.

The fact that the particles become shorter does not mean that clay platelets are breaking or being attributed during processing, although, this may occur under some extreme conditions [120]. Instead, considerable evidence indicates that the vision of tactoids as usually drawn, see Figures 3 and 4, where the platelets are all of the same length and in registry with one another is not correct. Figure 6 shows a more realistic vision of a tactoid where the particle length can be much longer than individual platelets and how these particles evolve as dispersion improves [106, 117].

Complications arise when calculating an average aspect ratio of particles when there is a distribution of both length and thickness. First, one can calculate a number average, a weight average, or other weightings of the distribution [58, 113, 119]. Second, one can average the aspect ratios or average separately the lengths and thickness and calculate an aspect ratio from these averages [113, 119]. There is no theoretical guidance on which is the better predictor of performance or for use in composite modeling [113, 117, 119].

To take full advantage of the reinforcement or tortuosity clay platelets or particles can provide to mechanical and thermal or barrier properties of nanocomposites, they must be oriented in the appropriate direction and not curled or curved. The alignment of particles is affected by the type of processing used to form the test specimen, e.g., extrusion, injection molding, etc. This is a separate issue from the degree of dispersion or exfoliation which is usually determined in the mixing process. Techniques like compression molding usually do not lead to good alignment or straightening of the high aspect ratio particles, and measurements made on such specimens often underestimate the potential performance. TEM can be used to assess and even quantify particle orientation and curvature and this information can, in principle, be factored into appropriate models to ascertain their effect on performance [86, 110, 131, 134, 138].

1.3.4. Nanocomposite Mechanical Properties: Reinforcement

A common reason for adding fillers to polymers is to increase the modulus or stiffness via reinforcement mechanisms described by theories for composites [58, 165–185]. Properly dispersed and aligned clay platelets have proven to be very effective for increasing stiffness. This is illustrated in Figure 7 by comparing the increase in the tensile modulus, E, of injection molded composites based on nylon 6, relative to the modulus of the neat polyamide matrix, Em, when the filler is an organoclay versus glass fibers [58]. In this example, increasing the modulus by a factor of two relative to that of neat nylon 6 requires approximately three times more mass of glass fibers than that of montmorillonite, MMT or platelets. Thus, the nanocomposite has a weight advantage over the conventional glass fiber composite. Furthermore, if the platelets are aligned in the plane of the sample, the same reinforcement should be seen in all directions within the plane, whereas fibers reinforce only along a single axis in the direction of their alignment [165].

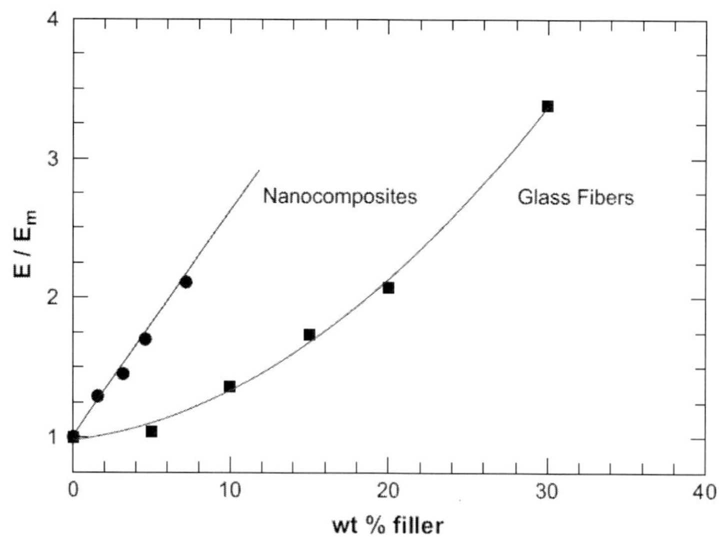

Figure 7. Comparison of modulus reinforcement (relative to matrix polymer) increases for nanocomposites based on MMT versus glass fiber (aspect ratio ~20) for a nylon 6 matrix.

In addition, the surface finish of the nanocomposite is much better than that of the glass fiber composite due to the nanometer size of the clay platelets versus the 10–15 m diameter of the glass fibers. A central question is whether the greater efficiency of the clay has anything to do with its nanometric dimensions, i.e., a ''nano-effect''. Answering this question requires considering many issues which we will review later in this section; however, the short answer is that we can explain essentially all of the experimental trends using composite theory without invoking any ''nano-effects'' [58]. Figure 8 shows an analogous comparison of nanocomposites based on the thermoplastic polyolefin or TPO matrix, polypropylene plus an ethylene-based elastomer, with conventional talc-filled TPO [103]. The latter is widely used in automotive applications. However, in some cases, they are being replaced with TPO nanocomposites.

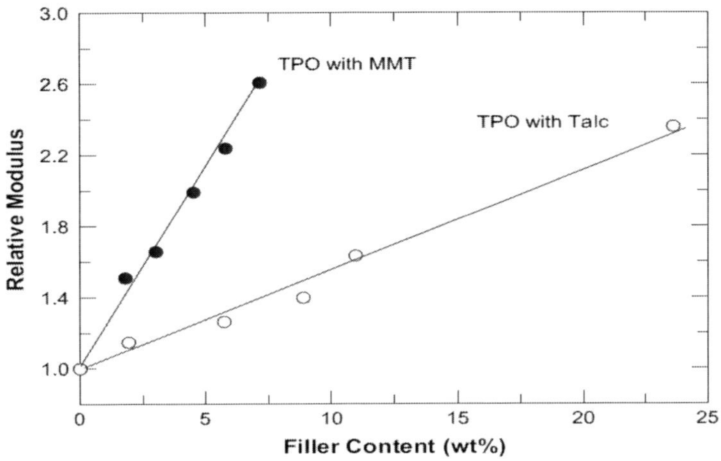

Figure 8. Comparison of modulus reinforcement for nanocomposites based on MMT versus talc for a TPO matrix

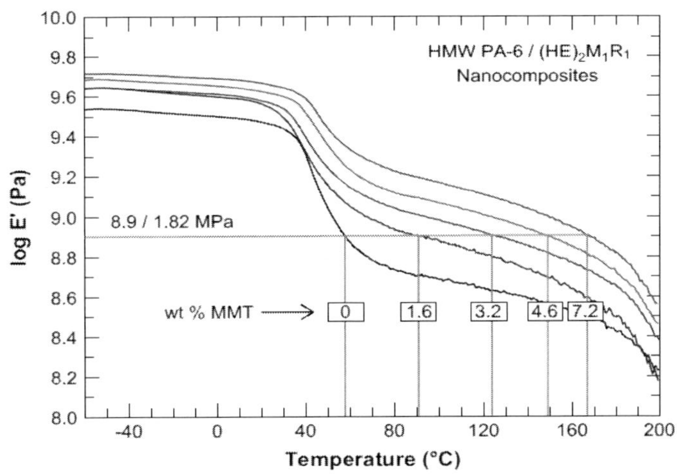

Figure 9. Experimental storage modulus data versus temperature for nylon 6 nanocomposites. The horizontal line is used to estimate the heat distortion temperature (HDT) at an applied stress of 1.82 MPa or 264psi.

In this case, doubling the modulus of the TPO requires more than four times more talc than MMT. This presents a weight, and consequently fuel savings along with the improved surface finish [103, 106]. The polyamide nanocomposites in Figure 7 are very highly exfoliated whereas the exfoliation of the clay in the TPO of Figure 8 is not nearly so perfect [103, 106, 117, 118]. However, it should be recognized that the talc particles do not have as high an aspect ratio as the glass fibers used in these comparisons [58, 186].

Another factor at play here is the lower modulus of TPO than nylon 6. The lower the matrix modulus, the greater is the relative increase in reinforcement caused by adding a filler [94]. Figure 9 shows dynamic mechanical moduli of the nylon 6 nanocomposites from Figure 7 versus temperature. The intersection of these curves with the horizontal line shown is a good approximation to the heat distortion temperature, HDT, of these materials [58].

This temperature, used as a benchmark for many applications, can be increased by approximately 100 _C by the addition of about 7% by weight of MMT. This effect has been explained by simple reinforcement, as predicted by composite theory, without invoking any special ''nano-effects''. The effect of MMTon the glass transition of these materials is very slight if any at all [58]. Indeed, glass fibers cause an analogous increase in HDT. The addition of fillers, including clay, can also increase strength as well as modulus [84]. However, the opposite may also occur [99]. A main issue is the level of adhesion of the filler to the matrix. For glass fiber composites, chemical bonding at the interface using silane chemistry is used to achieve high strength composites [186]. On the other hand, the modulus of glass fiber composites is not significantly affected by the level of interfacial adhesion [186]. Unfortunately, at this time, there is no effective way to measure the level of adhesion of clay particles with polymer matrices. In addition, there are no effective methods at this time to create chemical bonds between clay particles and polymer matrices analogous to those used for glass fibers. Generally, the addition of organoclays to ductile polymers increases the yield strength. However, for brittle matrices failure strength is typically decreased [84, 99–101].

The addition of fillers generally decreases the ductility of polymers, e.g., elongation at break. For glass fibers, talc, etc. this is well known and expected. Similar trends are also seen for nanocomposites [84, 100], but this seems to have been unexpected and disappointing to some working in this field. Impact strength is an energy measurement, i.e., it is force acting through a distance. A reduced elongation at break often means a reduced energy to break but there are exceptions to this [84, 100, 116]. The addition of clay may increase the stress levels via reinforcement more than the reduction in deformation as recently demonstrated for some nanocomposites [116]. Generally speaking, the reduction in ductility or energy to break is more severe when the polymer matrix is below its glass transition, whereas the effects of adding clay may not be so dramatic when the matrix is above its glass transition temperature [82, 84, 100 and 119]. This involves a shift in fracture mechanisms that is beyond the scope of this book. Melt rheological properties of polymers can be dramatically altered in the low shear rate or frequency region such that these fluids appear to have a yield stress [84, 117, 118, 187–189]. The effects in the high shear rate region are usually much less dramatic [84]. A great deal has been written about these effects and their causes which will not be reviewed here. Interestingly, the addition of clay seems to be an effective way to increase ''melt strength'' which can be useful in some polymer processing operations like film blowing or blow molding [54, 96]. To answer the question of whether the large increase in modulus caused by clay platelets or particles relative to conventional fillers, like that illustrated in Figs. 7 and 8, is due to some ''nanoeffect'', one must first determine whether effects of this magnitude can be predicted by composite theories. By a "nano-effect", we mean some change in the local properties of the matrix caused by the extremely high surface area filler and the small distances between nanofiller particles even at low mass loadings. It is well known that clay particles are effective nucleating agents which greatly change the crystalline morphology and crystal type for polymers like nylon 6 or PP [87]. Potential "confinement" effects are also discussed in the context of nanocomposites.

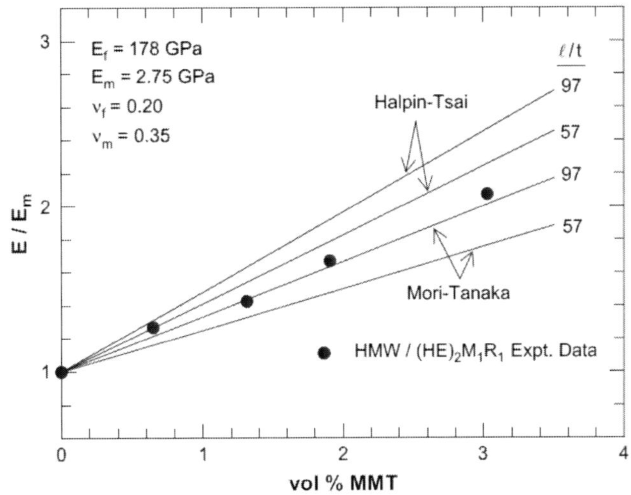

Figure 10. Experimental and theoretical stiffness data for nylon 6 nanocomposites. Model predictions are based on unidirectional reinforcement of pure MMT having a filter modulus of 178 GPa and aspect ratio of 57(experimentally determined number average value) and 97, corresponding to complete explosion.

A basic premise of composite theories is that the matrix and filler have the same properties as when the other component is not there. These theories only predict the effects of simple reinforcement and do not allow for any "nano-effects" of the type mentioned. Clearly, reinforcement does occur and the issue is whether that alone can explain the observations or not. Composite theories consider only the aspect ratio, orientation and volume fraction of filler in the matrix. The absolute filler particle size does not enter into the calculations. We have already mentioned the difficulties of experimentally determining the aspect ratio (sectioning issues, averaging of distributions, etc.). Furthermore, determining what values to assign to the properties of the clay platelets (like its modulus) is not trivial. Finally, the various composite theories differ somewhat in their predictions due to the assumptions and simplifications used in their mathematical formulation. These and other issues are worth remembering as we proceed with their analysis using data for nylon 6 nanocomposites. We wish to compare composite calculations with experimental data for the modulus of nylon 6 nanocomposites where the degree of exfoliation is very high but not perfect. An image analysis of many TEM photomicrographs was used to construct a platelet length distribution which looks very similar to that in Figure 2 [58]. The average platelet length was found to be 91 nm. While the majority of the clay particles was single platelets (thickness w0.94 nm), there were some doublets, some triplets, and a few quadruplets. It was estimated, by a rather involved analysis, that the average platelet thickness was 1.61 nm [58]. Thus, an upper bound on the aspect ratio for perfect exfoliation (using number averages of the distribution) would be about 91/0.94 ¼ 97 while a more realistic estimate might be 91/1.61 ¼57. We also need the in-plane modulus of a montmorillonite platelet. Information from a variety of sources suggests that a reasonable value is 178 GPa [58, 190]. However, some molecular dynamics calculations suggest significantly larger values [191].

A density for MMT of 2.83 g/cm3 was used to convert weight fractions to volume fractions. The properties of the matrix (modulus, Poisson ratio, density, etc.) were experimentally measured values [58]. The equations of Halpin–Tsai [168] and Mori–Tanaka [167] are frequently used for composite calculations. The former predicts higher levels of reinforcement for the cases of interest here than the latter as seen in Figure 10.

Interestingly, the predictions via these two theories and the two estimates of the aspect ratio give results that bracket the experimental data. Therefore, simple reinforcement considerations adequately explain the observations given all the issues involved in making these calculations. Any ''nano-effect'' is relatively minor, if at all. More needs to be said about the comparisons that are shown in Figures 7 and 8. The aspect ratio of the glass fibers in the nylon 6 matrix is about 20, whereas the aspect ratio of MMT platelets is 3–5 times larger than this. However, calculations using composite theory reveal that the larger aspect ratio of the clay versus glass fibers is not enough to explain all of the large differences in modulus reinforcement shown in Figure 7. A significant part of the difference in modulus enhancement stems from the much higher modulus of MMT than glass fibers, i.e., 178 versus 72.4 GPa [58]. The comparison of MMT versus talc in a TPO matrix shown in Figure 8 is more difficult to explain, but similar factors are at play [103, 117]. 3.5. Nanocomposite thermal properties: dimensional stability. The high thermal expansion coefficients of neat plastics causes dimensional changes during molding and as the ambient temperature changes that are either undesirable or in some cases unacceptable for certain applications. The latter is a particular concern for automotive parts where plastics must be integrated with metals which have much lower coefficients of thermal expansion, CTE. Fillers are frequently added to plastics to reduce the CTE. For low aspect ratio filler particles, the reduction in CTE follows, more or less, a simple additive rule and is not very large. In these cases, the linear CTE changes are similar in all three coordinate directions. However, when high aspect ratio fillers, like fibers or platelets, are added and well oriented, the effects can be much larger. In these cases, the CTE in the three coordinate directions may be very different.

The fibers or platelets typically have a higher modulus and a lower CTE than the matrix polymer. As the temperature of the composite changes, the matrix tries to extend or contract in its usual way. However, the fibers or platelets resist this change by creating opposing stresses in the two phases. When the filler to matrix modulus is large, the restraint to dimensional change can be quite significant within the direction of alignment. Platelets can provide their restraint in two directions, when appropriately oriented, while fibers can only do so in one direction. Because of their shape differences, fibers can cause a greater reduction in the direction of their orientation than platelets can [166, 170]. The CTE in the direction normal to the fibers or the platelet plane can actually increase when such fillers are added. Theories based on the mechanisms described above are available for quantitatively predicting CTE behavior [166, 170].

Montmorillonite platelets are particularly effective for reducing CTE of plastics as shown in Figure 11 for well-exfoliated nylon 6 nanocomposites [86].

These data were measured in the flow direction of the injection molded bars. When the semicrystalline nylon 6 matrix is above its glass transition temperature, the CTE reduction is greater than when below the Tg. The neat nylon 6 has a higher CTE above Tg than below. However, because of its lower modulus above Tg, the MMT platelets are more effective for reducing CTE. Note that the two curves in Figure 11 seem to cross at about 7 wt% MMT. For these specimens, the CTE in the transverse direction is also reduced by adding MMT but not quite as efficiently as in the flow direction since platelet orientation is not as great in the former as in the latter direction. The CTE in the normal direction actually increases as MMT is added. These trends are quantitatively predicted by the theories mentioned earlier [86]. CTE behavior is also a major consideration for the TPO materials used in automotive applications [103, 106, 117]. As seen in Figure 12, MMT is much more efficient at reducing CTE than talc in these materials [106]. Composite theories capture these trends reasonably well [117].

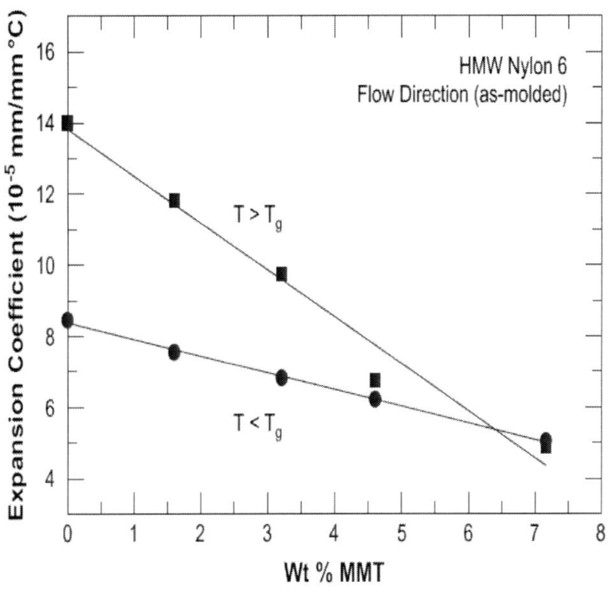

Figure 11. Linear thermal expansion coefficients of as-molded nylon 6 nanocomposites determined in the flow direction for T >Tg.

1.4. Variations and Applications of Polymer-Based Nanocomposites: Properties Other Than Reinforcement

Polymer composites comprising nanoparticles (including nanofibers where the fiber diameter is in the nanodimension range) are often investigated where reinforcement of the polymer matrix is achieved. While the reinforcement aspects are a major part of the nanocomposite investigations reported in the literature, many other variants and property enhancements are under active study and in some cases, commercialization. The advantages of nanoscale particle incorporation can lead to a myriad of application possibilities where the analogous larger scale particle incorporation would not yield the sufficient property profile for utilization. These areas include barrier properties, membrane separation, UV screens, flammability resistance, polymer blend compatibilization, electrical conductivity, impact modification, and biomedical applications. Examples of nanoparticle, nanoplatelet and nanofiber incorporation into polymer matrices are listed in Table 2 along with potential utility where properties other than mechanical property reinforcement are relevant.

Table 2. Examples of nanoscale filler incorporated in polymer composites for property enhancement other than reinforcement

Nanofiller	Property enhancement(s)	Application/utility
Exfoliated clay	Flame resistance, barrier, compatibilizer for polymer blends	
SWCNT; MWCNT	Electrical conductivity, charge transport,	Electrical/electronics/ optoelectronics
Nanosilver	Antimicrobial	
ZnO	UV adsorption	UV screens
Silica	Viscosity modification	Paint, adhesives
CdSe, CdTe	Charge transport	Photovoltaic cells
Graphene	Electrical conductivity, barrier, charge transport	Electrical/electronic
POSS	Improved stability, flammability resistance	Sensors, LEDs

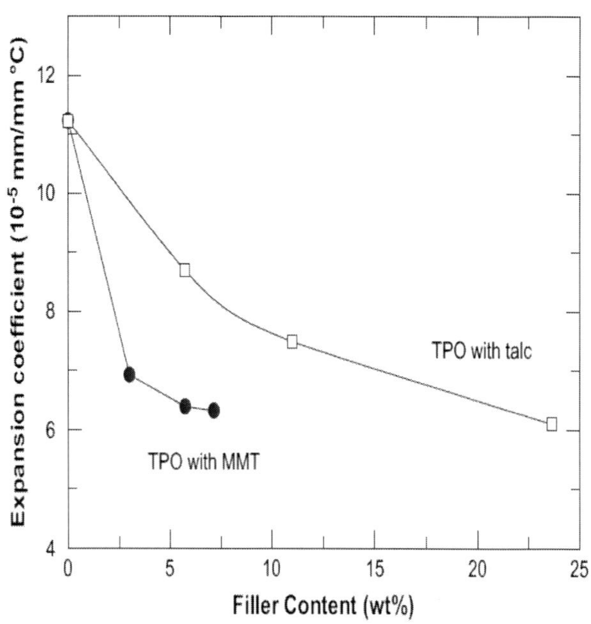

Figure 12. Comparison of linear coefficients of thermal expansion (flow directions) as a function of filter content of TPO composites formed from MMT and Talc.

1.4.1. Barrier and Membrane Separation Properties

The barrier properties of polymers can be significantly altered by inclusion of inorganic platelets with sufficient aspect ratio to alter the diffusion path of penetrant molecules as illustrated in Figure 13.

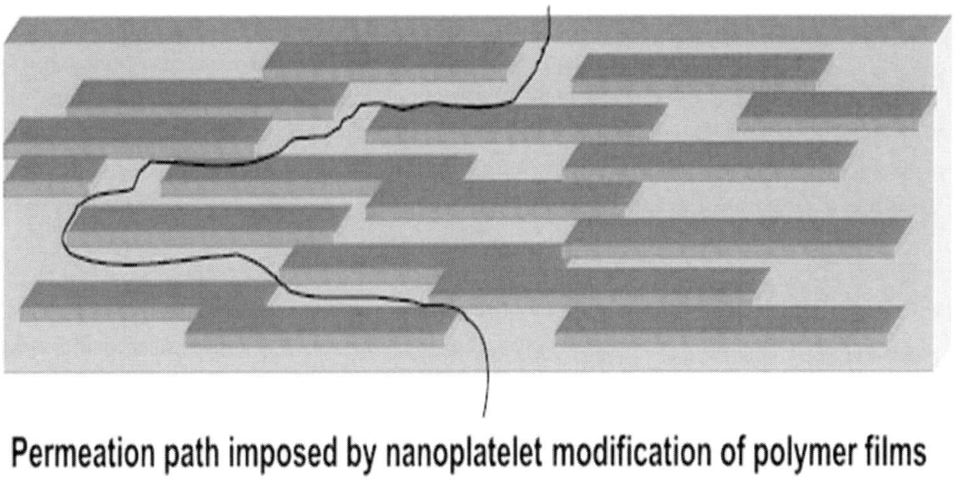

Figure 13. Barrier to permeation imposed by nanoparticles imbedded in a polymeric matrix.

Various continuum models have been proposed to predict the permeability of platelet filled composites as listed in Table 3. These models are generally based on random, parallel platelets perpendicular to the permeation direction (random in only two directions). The model by Bharadwaj introduces an orientation factor [196].

Table 3. Models for predicting barrier properties of platelet filled nanocomposites

Model	Filler type	Particle geometry	Formulas	Reference
Nielsen	Ribbon[a]	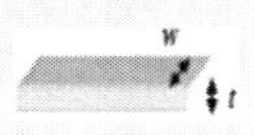	$(P_0/P)(1-\phi)=1+\alpha\phi/2$	[192]
Cussler (Regular array)	Ribbon[a]	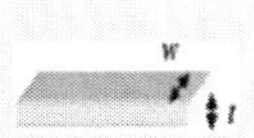	$(P_0/P)(1-\phi)=1+(\alpha\phi)^2/4$	[193]
(Random array)	Ribbon[a]	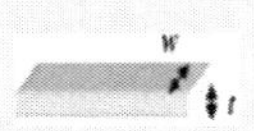	$(P_0/P)(1-\phi)=(1+\alpha\phi/3)^2$	[193]
Gusev and Lusti	Disk[b]		$(P_0/P)(1-\phi)=\exp[(\alpha\phi/3.47)^{0.71}]$	[194]
Fredrickson and Bicerano	Disk[b]	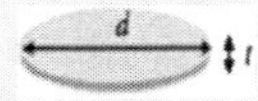	$(P_0/P)(1-\phi)=4(1+x+0.1245x^2)/(2+x))^2$ where $x=\alpha\phi/2\ln(\alpha/2)$	[195]
Bharadwaj	Disk[b]	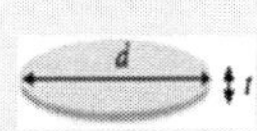	$(P_0/P)(1-\phi)=1+0.667\alpha\phi(S+(1/2))$ where S = orientation factor (from −1/2 to 1)	[196]

At the high aspect ratio which can be achieved (such as with exfoliated clay) in nanocomposites, significant decreases in permeability are predicted and observed in practice. Four of these models have been applied to polyisobutylene/vermiculite nanocomposites with aspect ratios predicted in the range of expectations [70]. The variability in the models was, however, shown to be substantial. In many cases, the nanocomposites investigated can be approximated by the continuum models, thus the "nano-effect" is not observed. This should not be surprising as the dimensions of permeating gas molecules are still much lower than the nanodimension modification. Differences would be expected in those cases where the Tg of the matrix polymer is changed. However, for practical applications the nanoscale dimensions are still quite important as transparency can be maintained along with surface smoothness for thin films. This is critical for food packaging applications.

Exfoliated clay modified poly-ethylene terephthalate (PET) is one of the more prevalent nanocomposites investigated by both academic and industrial laboratories for barrier applications [197–199]. In situ polymerized PET-exfoliated clay composites were found to show a 2-fold reduction in permeability with only 1 wt% clay versus the control PET [197]. PET-exfoliated clay composites also prepared via in situ polymerization using a clay-supported catalyst exhibited a 10 to 15-fold reduction in O2 permeability with 1–5 wt% clay [198]. The moisture vapor transmission, however, did not show any significant change. Exfoliated clay modified chlorobutyl rubber showed decreased diffusion for several organic chemicals suggesting utility for chemical protective gloves/clothing [200]. Exfoliated clay added to polyamide 6/polyolefin (polyethylene or polypropylene) blends yielded an improved barrier to styrene permeation for melt blown films [201]. It was noted that the polymer blend nanocomposite was a better barrier than the control polyamide nanocomposite. While most of the papers investigating barrier properties incorporate low levels of exfoliated clay, a novel approach employed producing a self-supporting clay fabric film followed by infiltration with an epoxy resin/amine hardener mixture and polymerization [202]. The resultant semitransparent nanocomposite film contained up to 77% volume fraction clay with an oxygen permeability 2–3 orders of magnitude lower than the control epoxy. The concept of mixed matrix membranes involving molecular sieve inclusions in a polymer film to enhance the permselectivity properties for membrane separation was developed by Koros et al. [203] to address the limits imposed by upper bound limits typically observed with polymer membranes [204]. These inclusions (carbon molecular sieves, zeolite structures) need to be at nanolevel dimensions as the dense layer thickness of commercial membranes is in the range of 100 nm. This approach has shown promise in exceeding the noted upper bound in various studies [205, 206]. The addition of silica nanoparticles to poorly packing polymer membranes (specifically poly(4-methyl-2-pentyne)) has been shown to yield even poorer packing and higher free volume [44]. This leads to increased permeability for larger organic molecules and a selectivity reversal for mixtures of these molecules with smaller molecules (e.g., n-butane/methane). This indicates that the separation process has changed from molecular sieving expected at low free volume and small void diameters to surface diffusion as the free volume and void diameters increase. The addition of nanoparticle TiO_2 to poly(trimethyl silylpropyne) (also a high free volume polymer with poor chain packing) showed a decrease in gas permeability up to 7 vol% TiO_2 with increasing permeability and higher free volume observed above 7 vol% loading [207].

1.4.2. Flammability Resistance

Increased flammability resistance has been noted as an important property enhancement involving nanoplatelet/nanofiber modification of polymeric matrices. While the specific reasons for this are under continuing investigation, a qualitative explanation observed in many studies involves the formation of a stable carbon/ nanoplatelet or nanofiber surface. This surface exhibits analogous characteristics to intumescent coatings whereby the resultant "char" provides protection to the interior of the specimen by preventing continual surface regeneration of available fuel to continue the combustion process. The primary advantage noted with nanofiller incorporation is the reduction in the maximum heat release rate (determined by cone calorimetry) [137, 208]. While significant reductions can be observed in the maximum heat release rates, the total heat release remains constant with nanofiller addition.

The relevance of reducing the maximum heat release rate is to minimize the flame propagation to adjacent areas in the range of the ignited material (dimension range of meter(s)). The flammability improvements for nanofiller addition are less advantageous when the more common empirical regulatory (pass/fail) flammability tests are conducted (UL94, ASTM flammability tests) [137, 208, 209].

In specific cases, the nanoparticle addition can result in a reduced flammability rating due to the melt viscosity increase preventing dripping as a mechanism of flame extinguishment (e.g., change UL94 rating from V-2 to HB) [209]. The primary advantage of nanofiller addition for these tests generally involves the reduction in the flame retardant additives that need to be incorporated to pass the specific test [137, 209, 210]. This has been observed in

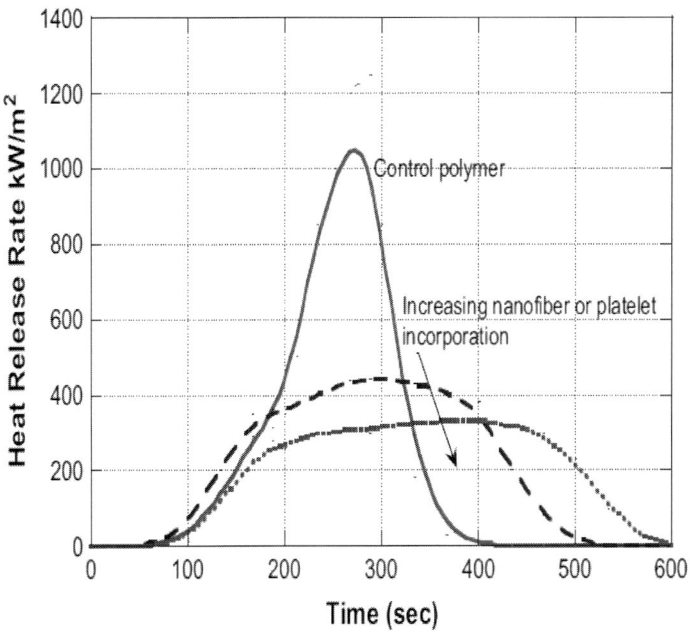

Figure 14. Generalized behavior for nanofiber or platelet modified polymers in the cone calorimetry heat release rate test.

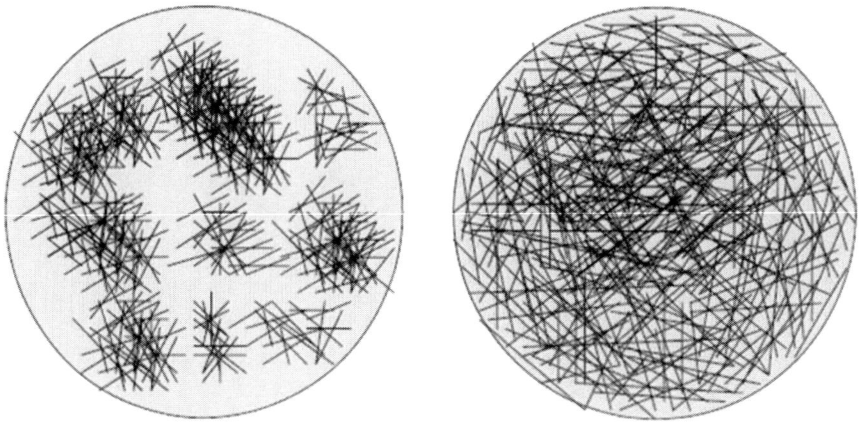

Figure 15. Surface structure of nanocomposites during /after forced combustion: (a) incomplete surface coverage due to low nanofiber level, agglomeration during combustion and or low aspect ratio; (b) desired surface dispersion during /after forced combustion.

various nanoparticle modified composites including exfoliated clay with halogen-based flame retardants/Sb2O3 [211] and EVA (ethylene– vinyl acetate copolymer) nanocomposites with magnesium hydroxide nanoparticles and microcapsulated red phosphorus [212]. The majority of the flame retardant studies on nanofiller incorporation in polymers involve exfoliated clay. Studies involving polyamide 6 [213, 214] and polypropylene [215] yielded similar observations with reduced peak heat release rates but no change in the total heat release with exfoliated clay addition. This is illustrated in Figure 14 for generalized data for nanofiller modified polymers. While the curve position and shape will vary for different polymer matrix materials and nanofiller incorporation, the generalized behavior of decreased peak heat release rate with no basic change in the overall heat release (area under the curve) is very typical.

The surface characteristics during and after forced combustion show that incomplete surface coverage will lead to poorer flammability resistance and can be related to low nanofiller level, low aspect ratio, poor dispersion and/or agglomeration during combustion. This generalized behavior, illustrated in Figure 15, is true for exfoliated clay as well as carbon nanotubes as discussed below.

Studies involving carbon nanotubes have also shown a decrease in the peak heat release rate with no change in the total heat release [216, 217] with effectiveness equal to or better than exfoliated clay. The level of dispersion of the carbon nanotubes in the polymer matrix was shown to be an important variable [216]. Upon combustion, the surface layer was enriched with a protective nanotube network providing a thermal and structural barrier to the combustion process. Continuity of the network (as illustrated in Figure 15) was important to achieve optimum performance as very low levels of nanotube incorporation or poor dispersion did not allow a continuous surface network during the combustion process. It is noted that the incorporation of nanoclay and carbon nanotubes often results in slightly earlier ignition than the unmodified polymer presumably due to the increased thermal conductivity. However, at the later stages of combustion, the reinforcement of the char layer provides a stable thermal barrier preventing regeneration of polymers at the surface available for rapid combustion.

1.4.3. Polymer Blend Compatibilization

A primary mechanism in compatibilization of phase separated polymer blends involves lowering the interfacial tension between the phases and preventing coalescence of the particles during melt processing. This can be achieved by addition of graft or block copolymers with constituents equal to or compatible with the blend components. It has been observed in many cases that the addition of nanoparticles (particularly exfoliated clay) can also prevent the coalescence retaining improved dispersion after shear mixing. Specific examples involving exfoliated clay compatibilization include polycarbonate/poly (methyl methacrylate) [218], poly (2,6-dimethyl-1,4-phenylene oxide)/polyamide 6 [219], polyamide 6/ethylene–propylene rubber [220], polystyrene/poly- (methyl methacrylate) [221] and poly(vinylidene fluoride)/ polyamide 6 blends [222]. An example of exfoliated clay compatibilization of a poly (methyl methacrylate)/polystyrene (70/30 by weight) blend is illustrated in Figure 16 [221]. After shearing and annealing above the blend component Tgs, the blend containing exfoliated clay shows the ability to resist coalescence. Nanoscale SiO2 particle compatibilization was noted for polystyrene/polypropylene blends where a significant reduction in the polystyrene phase dimensions was observed [223]. The compatibilization was hypothesized to be due to increased viscosity retarding coalescence.

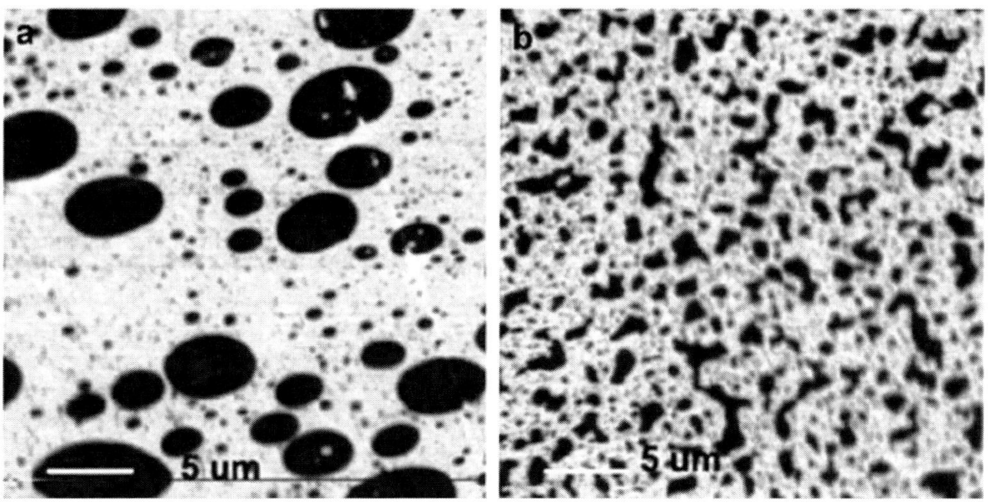

Figure 16. Scanning transmission X-ray microscopic images (30μm x 30μm) of PS/PMMA blends annealed at 190 °C for 14h, (a) PS/PMMA(30/70), (b) PS/PMMA/Closite 6A(27/63/10).

Several additional hypotheses have been noted in the literature. One explanation is that the nanoparticles concentrate at the interface preventing coalescence by a barrier-type mechanism.

Another hypothesis notes that both polymers bound by physical [224] or chemical interactions [221] on the nanoparticle will concentrate at the interface similar to a block or graft copolymer comprising both components of the blend. In the case of physical interactions, it was noted that for a polypropylene/polystyrene blend both polymers were intercalated into the clay gallery with the polymer chain extending outside the particle [224]. The nanoparticle size is important for both explanations and offers an advantageous property not achievable at higher length scales.

1.4.4. Biomedical Applications

The applicability of polymer nanotechnology and nanocomposites to emerging biomedical/biotechnological applications is a rapidly emerging area of development which this discussion can only briefly cover. One area of intense research involves electrospinning for producing bioresorbable nanofiber scaffolds for tissue engineering applications. This might be construed as a nanocomposite since the resultant scaffold allows for cell growth yielding a unique composite system.

Another area also involving nanofibers is the utilization of electrically conducting nanofibers based on conjugated polymers for the regeneration of nerve growth in a biological living system. (Figure 15) Surface structure of nanocomposites during/after forced combustion: (a) incomplete surface coverage due to low nanofiber level, agglomeration during combustion and/or low aspect ratio; (b) desired surface dispersion during/after forced combustion. Nanoparticle silver, silver oxide and silver salts have been incorporated into polymer matrices to provide antimicrobial/biocidal activity [225–227]. Nanoscale silver was shown to be a much better antimicrobial additive to nylon 6 than microscale silver particles [227] due to the much higher rate of silver ion release. Nanosilver (5–50 nm) at levels of 0.1–1.0 wt% in poly (methyl methacrylate) bone cement exhibited high antibacterial activity for joint arthroplasty utility [228] without the cytotoxicity of silver salts. Polymer nanocomposites based on hydroxyapatite ($Ca10(PO4)6(OH)2$) have been investigated for bone repair and implantation [229]. Hydroxyapatite, a major constituent of hard tissue, exhibits undesirable mechanical properties if directly employed. As a result, polymer-based matrix composites are desired. Biodegradation of the matrix is also desired to allow for the infiltration of new bone growth at the repair site. Often natural polymers (polysaccharides, polypeptides, collagen and chitosan) or synthetic biodegradable polymers are employed as the matrix in these studies. Collagen derived gelatin [230] and poly-2-hydroxyethylmethacrylate/poly (3-caprolactone) [231] nanocomposites based on hydroxyapatite are examples of systems studied for bone repair systems.

Electrospinning of biodegradable polymer solutions is a popular method for producing nanofiber scaffolds for tissue engineering applications. Poly (L-lactic acid)/exfoliated montmorillonite clay/salt solutions were electrospun followed by salt leaching/gas foaming [232]. The resultant scaffold structure contained both nano- and micro-sized pores offering a combination of cell growth and blood vessel invasion micro-dimensions along with nanodimensions for nutrient and metabolic waste transport. Polymer matrix nanocomposites have been proposed for drug delivery/release applications. The addition of nanoparticles can provide an impediment to drug release allowing slower and more controlled release, and reduced swelling [233] and improved mechanical integrity [234] of hydrogel-based nanocomposites. Iron oxide nanoparticles have been investigated for various applications including drug delivery, magnetic resonance imaging contrast enhancement, immunoassay and cellular therapy. These investigations often employ magnetite ($Fe3O4$) dispersed in a polymeric microsphere or microcapsule involving biodegradable and/or natural polymers [235, 236]. Poly (L-lysine) microspheres containing magnetic nanoparticles prepared by coacervation were prepared and characterized for potential use in targeted drug delivery applications [237]. Iron and cobalt nanoparticles encapsulated in polydimethylsiloxane have been noted for treating retinal detachment disorders [238, 239]. A novel imprint lithography method based on a crosslinked perfluoroether templating mold has been employed to fabricate uniform micro- and nanoparticles from organic polymers including biodegradable polymers such as poly (lactic acid) [240]. Inclusion of inorganic nanoparticles such as discussed above has the potential for a diverse range of biomedical applications using this method.

1.4.5. Fuel Cell Applications

Fuel cell applications involve polymers in the proton exchange membrane, binder for the electrodes and matrix for bipolar plates. The electrodes typically comprise carbon black particles (0.5– 1.0 mm) with Pt catalyst particles of 2–5 nm and a polymeric binder (usually Nafion_). Platinum nanoparticles deposited onto single-walled carbon nanotubes with Nafion_ as a binder were shown to give improved performance over the conventional carbon blackbased electrodes [241]. Nanoparticle incorporation in the proton exchange membrane has been noted in numerous publications in order to improve mechanical properties as well as to enhance proton conductivity. In direct methanol fuel cells, nanoparticles have been incorporated to reduce methanol crossover [242]. Heteropolyacids (HPA) (e.g., H3PW12O40; H3PMo12O40) have been added to proton exchange membranes to yield improved proton conductivity at higher temperatures while retaining good mechanical properties [243–246]. The particle size of the HPA inclusions was generally in the nanorange. Silica nanoparticle inclusion in proton exchange membranes gave lower methanol crossover in several studies [247, 248]. Zirconium phosphate [249], zirconium hydrogen phosphate [250] and TiO2 [251] nanoparticle incorporation in proton exchange membranes exhibited promise in direct methanol fuel cells. Nanoclay modified Nafion_ (laponite [252] and montmorillonite [253]) membranes have been reported to offer improvements over the unmodified controls. Sulfonated carbon nanofibers incorporated into a sulfonated EPDM proton exchange membrane yielded an order of magnitude improvement in the proton conductivity achieving performance comparable to the state-of-the-art Nafion_ 117 [254].

4.6. Electrical/electronics, optoelectronics, and sensors. Nanotechnology is deeply embedded in the design of advanced devices for electronic and optoelectronic applications. The dimensional scale for electronic devices has now entered the nanorange. The utility of polymer-based nanocomposites in these areas is very diverse involving many potential applications as well as types of nanocomposites. One specific nanocomposite type receiving considerable interest involves conjugated polymers and carbon nanotubes. A recent review of this area has found a litany of potential applications including photovoltaic (PV) cells and photodiodes, supercapacitors, sensors, printable conductors, light emitting diodes (LEDs) and field effect transistors [255]. This paper can only briefly discuss the nanocomposite technology applied to this broad field.

The electrical conductivity of carbon nanotubes in insulating polymers has also been a topic of considerable interest. The potential applications include electromagnetic interference shielding, transparent conductive coatings, electrostatic dissipation, supercapacitors, electromechanical actuators and various electrode applications [256, 257]. The percolation threshold for electrical conductivity of epoxy composites containing multi-wall carbon nanotubes was found to be 0.0025 wt% MWCNT [258] considerably lower than nanoscale dispersed carbon black particles. The threshold conductivity of single-walled carbon nanotubes in epoxy composites was noted to be a function of the SWCNT type with values as low as 0.00005 vol fraction [259].Water dispersed carbon black particles (42 nm) added to acrylic emulsions yielded electrical conductivity percolation levels as low as 1.5 vol% in dried films [260]. In this system, the carbon black particles concentrate at the interface between the emulsion particles during drying yielding a percolation network. The modulus of the emulsion system chosen as the matrix was noted to be an important variable with higher modulus (higher Tg) yielding lower threshold percolation values. The basic device structure for organic/polymeric PV and LED devices is shown in Figure 17. The major difference between the two devices is the light emitting layer and the light harvesting layer. While all the potential layers are noted, actual devices often do not contain all the layers. Polymer-based nanocomposites are most relevant for the anode, hole injection layer, light emitting layer and light harvesting layer. In the case where flexible transparent substrates are desired, nanocomposite or nanolayered combinations of polymer and barrier nanomaterials can be employed to provide barrier properties to oxygen and water permeation. Silicon-based photovoltaic (PV) devices offer high efficiency, excellent stability and proven commercial utility. Organic/polymerbased PV devices offer the potential for lower cost and more flexible manufacture but need significant improvements in both efficiency and long term stability. Conjugated polymers are often employed in the organic-based device light harvesting layer but have limitations in charge transport. Combinations of conjugated polymer with inorganic semiconductors have been proposed as a resolution to this deficiency. Cadmium–selenium, CdSe, nanorod incorporation in poly (3-hexylthiophene) yielded power conversion efficiency of 1.7% [261].

Silicon solar cells, however, typically deliver power conversion efficiencies of 10% or greater. Layer-by-layer assembly of functionalized poly (phenylene vinylene) and CdSe nanoparticle composites yielded uniform thin films with a power conversion efficiency of 0.71% [262]. A functionalized polystyrene containing an electroactive carbazole pendant group and an amine salt pendant group capable of electrostatic interaction with CdTe was described for potential PV applications [263]. Poly (3-hexylthiophene)– ZnO nanofiber composites exhibited a power conversion efficiency of 0.53% which is significantly better than the analogous bilayer structure of the noted components [264]. Spherical ZnO nanoparticle (20–40 nm) incorporation into poly (3-hexylthiophene) was ineffective at low levels of incorporation (20–30 wt %) but large improvements were observed at 50–60 wt% addition yielding a power conversion efficiency increase of 35 times to a value of 0.42% [265]. SWCNT incorporation into poly (3-octylthiophene) increased the short circuit current by two orders of magnitude and improved the important fill factor [266]. Ink jet printing is a potential method for producing low cost, high volume PV and LED devices. The ink jet printing of the nanocomposite of poly (vinyl alcohol) and CdTe for these devices was demonstrated by Tekin et al. [267].

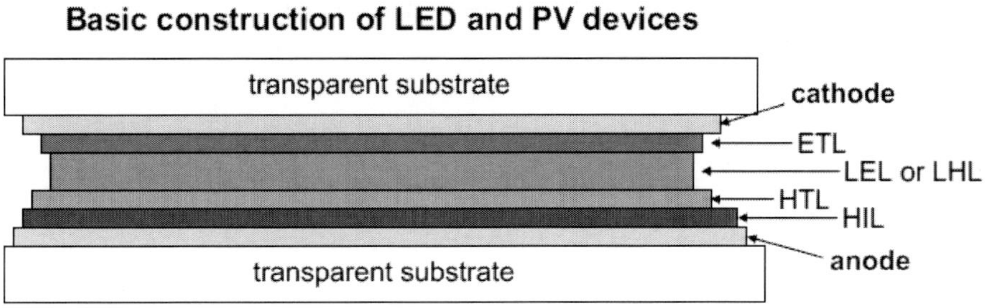

Figure 17. Generalized diagram of LED and PV device construction.

The primary material presently employed for the hole injection layer (HIL) is PEDOT:PSS (poly(3,4-ethylene dioxythiophene):poly(styrene sulfonic acid)) but has limitations in life-time performance. The addition of silver nanoparticles (3–6 nm) to PEDOT:PSS resulted in maximum device luminance at 20 wt% [268]. The addition of nickel nanoparticles (30 nm) into the PEDOT: PSS HIL was also shown to improve device performance attributed to the improved hole current [269]. Gold nanoparticle (5–10 nm) incorporation into poly (9,90-dioctylfluorene) (LEL layer) gave higher external quantum efficiencies and improved oxygen stability compared to the unmodified polyfluorene [270, 271]. Poly (9,90- dioctylfluorene)-layered silicate nanocomposites were also demonstrated to improve device stability as well as offering higher quantum efficiencies over the unmodified polyfluorene [272]. Poly (phenylene) with polyhedral oligomeric silsesquioxane (POSS) side groups inhibited interchain interaction leading to improved fluorescence quantum yields in solution and photoluminescent thermal stability in bulk [273]. Exfoliated graphite sheets (graphene) have been only recently investigated in nanocomposites [274] at least partly due to the lack of methods to achieve high levels of exfoliation [275, 276]. It was noted that the electrical conductivity for polystyrene–grapheme nanocomposites had a percolation threshold of 0.1 vol% rivaling carbon nanotubes [274]. The electrical conductivity at 1 vol% was 0.1 S/cm. The emerging interest of graphene in electronics applications parallels that which occurred with carbon nanotube discovery. Graphene, in sheet form, may offer promise in replacing silicon as Moore's law limits are achieved [277].

Conjugated polymers with various nanoscale filler inclusions have been investigated for sensor applications including gas sensors, biosensors and chemical sensors. The nanofillers employed include metal oxide nanowires, carbon nanotubes, nanoscale gold, silver, nickel, copper, platinum and palladium particles [278]. With carbon nanotubes, the electrical resistance was found to be significantly changed by exposure to specific gases such as NO_2 and NH3 [279]. A nanocomposite of SWCNT/polypyrrole yielded gas sensor sensitivity similar to SWCNT alone [280]. The sensing capability of these nanocomposites can be based on conductivity changes due to gas or chemical interactions with either the nanofiller or the conjugated polymer, pH changes, electrochromic or electro-optical property changes, catalytic activity, chemiluminescent property or biological recognition. Examples of sensors for dopamine detection include a poly (anilineboronic acid)/carbon nanotube composite [281] and a polyaniline/gold composite hollow sphere system [282]. A review of conjugated polymer nanocomposites employed as sensors has been recently published [278].

1.4.6. Commercial Applications of Polymer-Based Nanocomposites

A question often posed is ''with all the interest and associated large R&D expenditures in nanotechnology (including polymer nanocomposites) why isn't there more commercial impact?''

Often major discoveries take several decades to have a large commercial impact (polyethylene, carbon fiber composites as examples) as the cost/performance variables upon discovery are outside the realm of commodity utility and additional advances are required to achieve an economically competitive position. The commodities of today were the specialties of the past and even the automobile was a specialized article of commerce for several decades. This is the case with the technologies being developed with polymer-based nanocomposites. While many of the applications being commercialized today will remain specialties, there are areas where the specialty polymer nanocomposites of today will be the commodities of the future. Examples of commercial polymer based nanocomposites are listed in Table 4. The most publicized application for polymer nanocomposites was an automotive application by Toyota for a timing belt cover. The utilization of exfoliated clay reinforcement of TPO (thermoplastic polyolefin) was later utilized by General Motors for an exterior step assist for another automotive application. There are many more applications involving exfoliated clay reinforcement. However, many of them have not been publicized such as those noted above.

Table 4. Examples of nanocomposites commercial utility

Polymer matrix	Nanoparticle	Property improvement	Application	Company and/or product trade name
Polyamide 6	Exfoliated clay	Stiffness	Timing belt cover: automotive	Toyota/Ube
TPO (thermoplastic polyolefin)	Exfoliated clay	Stiffness/strength	Exterior step assist	General Motors
Epoxy	Carbon nanotubes	Strength/stiffness	Tennis rackets	Babolat
Epoxy	Carbon nanotubes	Strength/stiffness	Hockey sticks	Montreal: Nitro Hybtonite®
Polyisobutylene	Exfoliated clay	Permeability barrier	Tennis balls, tires, soccer balls	InMat LLC
SBR, natural rubber, polybutadiene	Carbon black (20–100 nm: primary particles)	Strength, wear and abrasion	Tires	Various
Various	MWCNT	Electrical conductivity	Electrostatic dissipation	Hyperion
Unknown	Silver	Antimicrobial	Wound care/bandage	Curad®
Nylon MXD6, PP	Exfoliated clay	Barrier	Beverage containers, film	Imperm™: Nanocor
SBR rubber	Not disclosed	Improved tire performance in winter	Winter tires	Pirelli
Natural rubber	Silver	Antimicrobial	Latex gloves	
Various	Silica	Viscosity control, thixotropic agent	Various	
Polyamides nylon 6, 66, 12	Exfoliated clay	Barrier	Auto fuel systems	Ube

One of the initial uses for exfoliated clay in barrier applications involved a 20 mm coating on the interior of a tennis ball to prevent depressurization. The product was developed by InMat LLC and introduced in 2001. Sports equipment was one of the initial areas where carbon fiber composites were commercialized. This is also true for carbon nanotubes where the carbon nanotubes (at low levels) reinforce the epoxy matrix of the carbon fiber composite in specialty tennis rackets and hockey sticks. In these applications, performance overrides the economic disadvantages of the expensive carbon nanotube inclusion.

1.4.7. Comments on the Future of Polymer Matrix Based Nanocomposites

An area where nanocomposites could achieve a dramatic commercial prominence is in advanced composites. Carbon fiber reinforced composites have a limit on the achievable properties (particularly in cross-ply composites) due to the low modulus and strength of the matrix phase. Modification of the matrix phase with carbon nanotubes at the lower scale of dimensions and carbon nanofibers at a higher dimensional scale would allow for significant increases in the modulus and strength contributions of the matrix to the overall composite properties. While this would offer some improvement in unidirectional composites, it could be dramatic in the case of cross-ply composites which are the major type of composite structure utilized in advanced composite applications. This concept is under current consideration and could allow a step change in the advanced composite field. The basic concept has already been commercially employed in specialty sports equipment (tennis rackets and hockey sticks). These improvements are keys to future aircraft and wind energy turbine applications. This approach is analogous to naturally occurring composite structures where the hierarchical construction method employs several dimensional scales beginning at the nanolevel. The same concept is also relevant to the more commodity reinforced composites where exfoliated clay could be added to the matrix for unsaturated polyester–fiberglass composites or other fiberglass reinforced matrix polymers.

As the secrets of nature's methodology to optimize material properties by nanolevel construction are unlocked (biomimetics), translation of these findings to polymer nanocomposites should allow for further advances. Nanostructured surfaces have been noted to yield superhydrophobic character (lotus leaf) and exceptional adhesion (gecko foot). The confluence of the biological and polymer material science disciplines often involves the design of nanoscale polymeric blends and composite systems to mimic the biological systems. Carbon nanotube or exfoliated graphite (graphene) offers substantial opportunities in the electrical/electronics/optoelectronics areas as well as potential in specific emerging technologies.

One specific area would be replacement of ITO as a transparent conductor for lower cost and flexible devices. Carbon nanotube sheets have been proposed [283] and the potential for carbon nanotube conjugated polymer composites would be of interest if sufficient electrical conductivity can be obtained (>103 S/cm). The potential of low cost graphene production is yet to be realized and large scale utility will be awaiting the synthetic breakthrough. An additional area not discussed in detail in this review involves the importance of morphology control which includes both dispersion and alignment. These issues have been discussed in many of the papers already cited in this review as well as in recent reviews [156, 284]. Obtaining the optimum properties for nanocomposites will usually require excellent dispersion of the nanoparticles. The tendency for nanoparticles (including platelets and fibers of nanoscale dimensions) to coalesce into macro size agglomerates can seriously impact the achievable properties.

In specific cases, excellent isotropic dispersion may not be the desired morphology but rather a hierarchical morphology that offers unique properties such as those observed with percolation pathways to maximize electrical conductivity or patterned morphology to achieve novel optical or electronic properties [284].

1.5. Optical Properties of Polymer Nanocomposites

The characteristic properties of metal nanoparticles and nanocomposites have been the subject of study for a long time because of their unique optical, thermal, mechanical, electronic and electrical properties. Plasmon resonance allows for the enhancement and manipulation of local electromagnetic fields at the nanoparticle surface due to surface effects and as a result of these singular optical features, noble metal nanoparticles stimulate great interest for implementation in photonics, biotechnology and space applications. Due to the interesting optical properties of nanocomposites which are different from those of individual metals, these materials have drawn much attention. Combining metal and polymer together enhances the optical properties of nanometals and also alters the mechanical behavior of the polymer.

The former is of prime importance due to its possible application in polarizers, light stable color filters, solar cells and optical sensors. If the properties of the polymer can also be tuned in the correct direction with enhanced optical properties of metals, these materials could be the best candidates for space applications. The size dependent structural, optical, thermal and electronic properties can be tailored to be suitable for device applications by varying the particle size. Small changes of the configuration of the composite concerning the metal fraction as well as the size and shape of the nanoparticles can lead to dramatic changes in the electrical and optical properties. In the last few decades, the study of optical properties of nanocomposites has been stimulated by promising applications. There have been significant changes in thermal as well as optical properties of the nanoparticles.

However, as attractive as such nanocomposites may be, the process of blending or dispersing nanoparticles in a polymer matrix has proven to be problematic. Once well dispersed in a matrix, the intrinsic properties of a nanostructured material are determined by its size, shape, composition, crystallinity and structure. The polymer system of the noble metals of copper, gold and silver, shows very intense colors, which are absent for the bulk materials as well as for the individual atoms. Their origin is attributed to the surface plasmon resonance (SPR) and the collective oscillation of free conduction electrons induced by an interacting electromagnetic field. SPR was first quantitatively described by Mie by solving Maxwell's equation with the appropriate boundary conditions for spherical particles. The total extinction cross-section composed of absorption and scattering is given as a summation over all electric and magnetic multipole oscillations. Mie's (1908) theory is successful in reproducing the extinction spectra from gold nanoparticles for size > 20 nm. However, as size decreases, the additional effects take place and the plasmon peak shifts towards lower wavelength, which can be explained by introducing a size dependent dielectric constant in the Mie theory.

1.6. Viscoelastic Properties of Polymer Nanocomposites

The addition of fillers is one of the avenues for altering the viscoelastic properties of polymers while maintaining the desired characteristics of a given polymer matrix, such as chemical inertness, temperature stability, electrical properties, and self-healing properties. In contrast to conventional particle–polymer composites, the addition of nanofillers with large specific interfacial areas between solid surfaces and a polymer matrix offers additional capability for the control of nanocomposite properties through the modification of the viscoelastic properties of the interfacial polymer. Specifically, experimental studies have found that the viscoelastic properties of nanocomposite are strongly influenced not only by the nanoparticle surface fraction but also by the nanoparticle surface chemistry (modification).

Coarse-grained and atomistic molecular dynamics (MD) and Monte Carlo simulations have provided valuable insight into factors responsible for the alteration of polymer matrix properties with the addition of interfaces. The main conclusion of those studies is that the strength of the nanoparticle–polymer interaction, the nanoparticle specific surface area, the surface structure, and the proximity to the glass transition are the most important factors controlling the properties of the interfacial polymer. For example, the addition of nanoparticles with attractive interactions leads to decreased polymer dynamics and increased viscosity and time dependent shear modulus of the polymer matrix, whereas the addition of nanoparticles with repulsive, or excluded-volume, interactions leads to increased polymer dynamics and decreased viscosity and time-dependent shear modulus. These effects scale linearly with the specific surface area of nanoparticles and increase as the temperature decreases. Another important factor controlling interfacial-polymer dynamics is the structure of the surface. A polymer next to structureless, or flat, surfaces with attractive surface–polymer interactions exhibits very different dynamics than a polymer near structured surfaces. In the latter case, polymer motion near the surface is strongly correlated to the surface structure.

Although these studies have provided detailed recipes for altering a polymer matrix to modify nanocomposite properties, no simple methodology exists to accurately predict the viscoelastic properties of random and aggregated polymer nanocomposites containing tens and hundreds of nanoparticles. Indeed, theoretical and empirical expressions work well for regular composites, in which the majority of the polymer has bulk-polymer properties, but they fail to predict the viscoelastic properties of nanocomposites, in which a large fraction of the polymer is situated at the nanoparticle–polymer interface and which, therefore, have properties different from those of the bulk polymer. Even when the third component (interfacial polymer) is introduced to the theoretical estimates, it is not clear how to estimate the size of the interfacial layer and its time-dependent properties and how they depend on the filler concentration. MD simulations are well suited for predicting the viscosity and time-dependent shear modulus of polymer nanocomposites with unentangled or weakly entangled polymers, but they are limited to systems with only a few nanoparticles at best. They provide the overall mechanical properties of composites, but it is less clear how to obtain the position- and time-dependent shear modulus from MD simulations for systems with complicated geometries, which is needed as an input for the interfacial-polymer properties in theoretical models or in larger scale simulations with finite elements or similar techniques. In this short report, we discuss a simple methodology for upscaling position-dependent polymer properties from MD simulations into continuum level modeling with the subsequent validation against MD simulations. The key supposition tested in this work is that the influence of a solid material embedded in a polymer matrix on the interfacial-polymer shear modulus is similar to its influence on the polymer-bead mean-square displacements (MSDs). This approximation allows us to infer the position dependent shear modulus of the interfacial polymer from an analysis of the MSDs as a function of separation from the surface. The latter can be straightforwardly performed in MD simulations.

The position-dependent interfacial-polymer shear modulus from such an approximation and the properties of the bulk like polymer and solid material are then used in the material-point method (MPM) simulations of nanocomposites to obtain the average mechanical properties of the composites. Finally, a comparison of the average properties from MD and MPM simulations is made to draw conclusions about the validity of the approximations made in the calculations of the position-dependent shear modulus used in the MPM simulations. The methodology is tested on six systems consisting of a nanocylinder embedded in a polymer matrix with three polymer volume fractions for two strengths of the polymer–surface interactions. We have chosen MPM to perform simulations on the upper scale, with the properties of all components taken from previously performed MD simulations (lower scale). In brief, MPM, which is a particle-based method, provides a new and demonstrated method for solving equations for solid mechanics. It has significant advantages over finite element analysis (FEA) when complicated structures such as nanocomposite membranes, foams, and fabric-reinforced composites are discretized. These structures are difficult to analyze by FEA but are very easily implemented in MPM. MPM has further advantages when position-sensitive material properties, such as the transport properties of the polymer matrix, are implemented as a function of the distance from the polymer–particle interface. In MPM, each material point can be assigned different material properties, including different properties to particles within the same background grid element. Handling position dependent material properties in FEA, is much more difficult and further complicates the meshing process. In summary, MD input into full scale MPM simulations provides a good approach to the multiscale modeling of complicated geometries with interaction-dependent mechanical and transport properties (Figure 18).

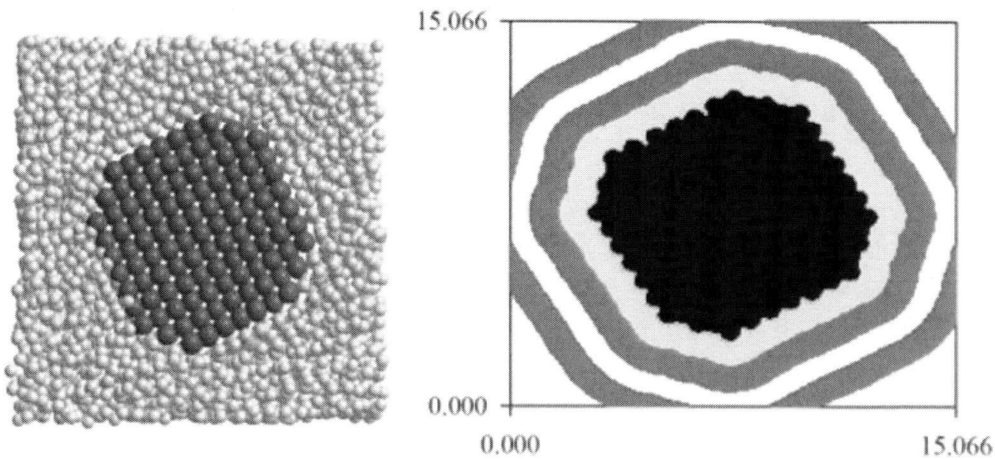

Figure 18. Snapshots of MD simulations of a cylinder embedded in a polymer matrix and an equivalent system used in MPM simulations with the filler VF value of 30%. The material points within 0.41s of the cylinder surface bead positions and inside the cylinder have been assigned the FCC crystal properties. Four interfacial polymer layers and bulk polymer are shown in different colors.

REFERENCES

[1] Mark JE, Jiang CY, Tang MY. *Macromolecules* 1984;17:2613–6.
[2] Wilkes GL, Orler B, Huang H. *Polym. Prep.* 1985;26:300–1.
[3] Wen J, Wilkes GL. *Chem. Mater.* 1996;8:1667–81.
[4] Kojima Y, Usuki A, Kawasumi M, Okada A, Kurauchi T, Kamigaito O. *J. Polym. Sci. Part A. Polym. Chem.* 1993;31:983–6.
[5] Okada A, Fukushim Y, Kawasumi M, Inagaki S, Usuki A, Sugiyama S, et al. US Patent 4,739,007; 1988.
[6] Kawasumi M. *J. Polym. Sci. Part A. Polym. Chem.* 2004;42:819–24.
[7] Loo YL, Register RA, Ryan AJ. *Phys. Rev. Lett.* 2000;84:4120–3.
[8] Loo YL, Register RA, Ryan AJ. *Macromolecules* 2002;35:2365–74.
[9] Santana OO, Muller AJ. *Polym. Bull* 1994;32:471–7.
[10] Woo E, Huh J, Jeong YG, Shin K. *Phys Rev Lett* 2007;98:136103.
[11] Shin K, Woo E, Jeong YG, Kim C, Huh J, Kim KW. *Macromolecules* 2007;40: 6617–23.
[12] Wu H, Wang W, Yang H, Su Z. *Macromolecules* 2007;40:4244–9.
[13] Di Maio E, Iannace S, Sorrentio L, Nicolais L. *Polymer* 2004;45:8893–900.
[14] Zhang QX, Yu ZZ, Yang M, Ma J, Mai YW. *J. Polym. Sci. Part B. Polym. Phys.* 2003; 41:2861–9.
[15] Mu B, Wang Q, Wang H, Jian L. *J. Macromol. Sci. Part B Phys* 2007;46:1093– 104.
[16] Nam JY, Ray SS, Okamoto M. *Macromolecules* 2003;36:7126–31.
[17] Lincoln DM, Vaia RA, Krishnamoorti R. *Macromolecules* 2004;37:4554–61.
[18] Li L, Li CY, Ni C, Rong L, Hsiao B. *Polymer* 2007;48:3452–60.
[19] Phang IY, Pramoda KP, Liu T, He C. *Polym. Int 2004*;53:1282–9.
[20] Wu D, Zhou C, Fan X, Mao D, Bian Z. *Polym Polym Compos* 2005;13:61–71.
[21] Bilotti E, Fischer HR, Peijs T. *J. Appl. Polym. Sci.* 2008;107:1116–23.
[22] Xu D, Wang Z. *Polymer* 2008;49:330–8.
[23] Homminga D, Goderis B, Dolbnya I, Reynaers H, Groeninckx G. *Polymer* 2005;46:11359–65.
[24] Chen EC, Wu TM. *J. Polym. Sci. Part B. Polym. Phys.* 2008;46:158–69.
[25] Kim B, Lee SH, Lee D, Ha B, Park J, Char K. *Ind Eng Chem Res* 2004;43:6082–9.
[26] Harrats C, Groeninckx G. *Macromol Rapid Commun* 2008;29:14–26.
[27] Gue' rin G, Prud'homme RE. *J. Polym. Sci. Part B. Polym. Phys.* 2007;45:10–7.
[28] Rittigstein P, Torkelson JM. *J. Polym. Sci. Part B. Polym. Phys.* 2006;44: 2935–43.
[29] Pluta M, Jeszka JK, Boiteux G. *Eur. Polym. J.* 2007;43:2819–35.
[30] Lee KJ, Lee DK, Kim YW, Choe WS, Kim JH. *J. Polym. Sci. Part B. Polym. Phys.* 2007;45:2232–8.
[31] Xu H, Yang B, Wang J, Guang S, Li C. *J. Polym. Sci. Part A. Polym. Chem.* 2007;45: 5308–17.
[32] Ramasundaram SP, Kim KJ. *Macromol. Symp.* 2007;249–250:295–302.
[33] Huang JC, He CB, Xiao Y, Mya KY, Dai J, Siow YP. *Polymer* 2003;44:4491–9.
[34] Pham JQ, Mitchell CA, Bahr JL, Tour JM, Krishanamorrti R, Green PF. *J. Polym. Sci. Part B. Polym. Phys.* 2003;41:3339–45.
[35] Bohning M, Goering H, Hao N, Mach R, Oleszak F, Schonhals A. *Rev. Adv. Mater. Sci.* 2003;5:155–9.

[36] Xu W, Ge M, Pan WP. *J. Therm. Anal. Calorim* 2004;78:91–9
[37] Fragiadakis D, Pissis P. *J. Non-Cryst Solids* 2007;353:4344–52.
[38] Shi X, Gan Z. *Eur. Polym. J.* 2007;43:4852–8.
[39] Huskic' M, Zˇ igon M. *Eur. Polym. J.* 2007;43:4891–7.
[40] Yuen SM, Ma CCM, Lin YY, Kuan HC. *Compos. Sci. Technol.* 2007;67:2564–73.
[41] Uthirakumar P, Nahm KS, Hahn YB, Lee YS. *Eur. Polym. J.* 2004;40:2437–44.
[42] Sun Y, Luo Y, Jia D. *J. Appl. Polym. Sci.* 2008;107:2786–92.
[43] Chang JH, Mun MK, Kim JC. *J. Appl. Polym. Sci.* 2007;106:1248–55.
[44] Merkel TC, Freeman BD, Spontak RJ, He Z, Pinnau I, Meakin P, et al. *Science* 2002;296:519–22.
[45] Kim JH, Koros WJ, Paul DR. *Polymer* 2006;47:3094–103 and 3104–3111.
[46] Huang Y, Paul DR. *J. Polym. Sci. Part B. Polym. Phys.* 2007;45:1390–8.
[47] Van Olphen H. An introduction to clay colloid chemistry. New York: Interscience; 1963.
[48] LeBaron PC, Wang Z, Pinnavaia TJ. *Appl. Clay Sci.* 1999;15:11–29.
[49] Pinnavaia TJ, Beall GW, editors. Polymer–clay nanocomposites. New York:John Wiley & Sons; 2000.
[50] Yariv S, Cross H, editors. Organo-clay complexes and interactions. New York:Marcel Dekker; 2002.
[51] Ray SS, Okamoto M. *Prog. Polym. Sci.* 2003;28:1539–641.
[52] Mai Y, Yu Z, editors. Polymer nanocomposites. Cambridge: Woodhead; 2006.
[53] Hussain F, Hojjati M. *J. Compos Mater.* 2006;40(17):1511–65.
[54] Hunter DL, Kamena KW, Paul DR. *MRS Bull* 2007;32:2806.
[55] Xie W, Gao Z, Liu K, PanW-P, Vaia R, Hunter D, et al. *Thermochim Acta* 2001; 367–368:339–50.
[56] Xie W, Gao Z, Pan W-P, Hunter D, Singh A, Vaia R. *Chem. Mater.* 2001;13: 2979–90.
[57] Xie W, Xie R, Pan W-P, Hunter D, Koene B, Tan L-S, et al. *Chem. Mater.* 2002; 14:4837–45.
[58] Fornes TD, Paul DR. *Polymer* 2003;44:4993.
[59] Paul DR, Zeng QH, Yu AB, Lu GQ. *J. Colloid Interface Sci* 2005;292:462.
[60] Heinz H, Vaia RA, Krishnamoorti R, Farmer BL. *Chem. Mater.* 2007;19:59–68.
[61] Ploehn HJ, Liu C. *Ind. Eng. Chem. Res.* 2006;45:7025–34.
[62] Fukushima Y, Inagaki S. *J. Inclusion Phenom* 1987;5:473–82.
[63] Kojima Y, Usuki A, Kawasumi M, Okada A, Kurauchi T, Kamigaito O. *J. Polym. Sci. Part A. Polym. Chem.* 1993;31:1755–8.
[64] Biasci L, Aglietto M, Ruggeri G, Ciardelli F. *Polymer* 1994;35(15):3296–309.
[65] Huang X, Lewis S, Brittain WJ. *Macromolecules* 2000;33:2000–4.
[66] Zhou Q, Fan X, Xia C, Mays J, Advincula R. *Chem. Mater.* 2001;13:2465–7.
[67] Albrecht M, Ehrler S, Muhlebach A. *Macromol Rapid Commun* 2003;24:382–7.
[68] Kiersnowski A, Piglowski J. *Eur. Polym. J.* 2004;40:1199–207.
[69] Goldberg HA, Feeney CA, Karim DP, Farrell M. Rubber World 2002;226:1–17. see also p. 20 and 37.
[70] Takahashi S, Goldberg HA, Feeney CA, Karim DP, Farrell M, O'Leary K, et al. *Polymer* 2006;47:3083–93.
[71] Vaia RA, Ishii H, Giannelis EP. *Chem. Mater.* 1993;5:1694.

[72] Vaia RA, Teukolsky RK, Giannelis EP. *Chem. Mater.* 1994;6:1017–22.
[73] Vaia RA, Jandt KD, Kramer EJ, Giannelis EP. *Macromolecules* 1995;28:8080–5.
[74] Giannelis EJ. *Adv. Mater.* 1996;8(1):29.
[75] Vaia RA, Jandt KD, Kramer EJ, Giannelis EP. *Chem. Mater.* 1996;8:2628–35.
[76] Vaia RA, Giannelis EP. *Macromolecules* 1997;30:7990–9.
[77] Vaia RA, Giannelis EP. *Macromolecules* 1997;30:8000–9.
[78] Lee JY, Baljon ARC, Lorin RF. *J. Chem. Phys.* 1990;111(21):9754–60.
[79] Manias E, Chen H, Krishnamoorti R, Genzer J, Kramer EJ, Giannelis EP. *Macromolecules* 2000;33:7955–66.
[80] Anastasiadis SH, Karatasos K, Vlachos G. *Phys. Rev. Lett* 2000;84(5):915–8.
[81] Vaia RA, Giannelis EP. *Polymer* 2001;42:1281–5.
[82] Cho JW, Paul DR. *Polymer* 2001;42:1083.
[83] Dennis HR, Hunter DL, Chang D, Kim S, White JL, Cho JW, et al. *Polymer* 2001;42:9513.
[84] Fornes TD, Yoon PJ, Keskkula H, Paul DR. *Polymer* 2001;42:9929.
[85] Fornes TD, Yoon PJ, Hunter DL, Keskkula H, Paul DR. *Polymer* 2002;43:5915.
[86] Yoon PJ, Fornes TD, Paul DR. *Polymer* 2002;43:6727.
[87] Fornes TD, Paul DR. *Polymer* 2003;44:3945.
[88] Yoon PJ, Fornes TD, Paul DR. *Polymer* 2003;44:5323.
[89] Yoon PJ, Hunter DL, Paul DR. *Polymer* 2003;44:5341.
[90] Fornes TD, Yoon PJ, Paul DR. *Polymer* 2003;44:7545.
[91] Fornes TD, Hunter DL, Paul DR. *Polymer* 2004;45:2321.
[92] Shah RK, Paul DR. Polymer 2004;45:2991.
[93] Chavarria F, Paul DR. *Polymer* 2004;45:8501.
[94] Fornes TD, Paul DR. *Macromolecules* 2004;37:7698.
[95] Bourbigot S, Vanderhart D, Gilman J, Stretz HA, Paul DR. *Polymer* 2004;45: 7627.
[96] Hotta S, Paul DR. *Polymer* 2004;45:7639.
[97] Ibanes C, David L, Seguela R, Rochas C, Robert G. *J. Polym. Sci. Part B. Polym. Phys.* 2004;42:2633–48.
[98] Vlasveld DPN, Groenewold J, Bersee HEN, Mendes E, Picken SJ. *Polymer* 2005; 46:6102–13.
[99] Stretz HA, Paul DR, Li R, Keskkula H, Cassidy PE. *Polymer* 2005;46:2621.
[100] Shah RK, Hunter DL, Paul DR. *Polymer* 2005;46:2646.
[101] Stretz HA, Paul DR, Cassidy PE. *Polymer* 2005;46:3818.
[102] Zeng QH, Yu AB, Lu GQ, Paul DR. *J. Nanosci. Nanotechnol.* 2005;46:3818.
[103] Lee H-S, Fasulo PD, Rodgers WR, Paul DR. *Polymer* 2005;46:11673.
[104] Ahn YC, Paul DR. Polymer 2006;47:2830.
[105] Shah RK, Paul DR. *Macromolecules* 2006;39:3327.
[106] Lee H-S, Fasulo PD, Rodgers WR, Paul DR. *Polymer* 2006;47:3528.
[107] Shah RK, Paul DR. *Polymer* 2006;47:4074.
[108] Shah RK, Krishnaswarmy RK, Takahashi S, Paul DR. *Polymer* 2006;47: 6187.
[109] Chavarria F, Paul DR. *Polymer* 2006;47:7760.
[110] Stretz HA, Paul DR. *Polymer* 2006;47:8123.
[111] Stretz HA, Paul DR. *Polymer* 2006;47:8527.
[112] Shah RK, Cui L, Williams KL, Bauman B, Paul DR. *J. Appl. Polym. Sci.* 2006;102:2980.

[113] Shah RK, Kim DH, Paul DR. *Polymer* 2007;48:1047.
[114] Cui L, Paul DR. *Polymer* 2007;48:1632.
[115] Chavarria K, Nairn K, White P, Hill AJ, Hunter DL, Paul DR. *J. Appl. Polym. Sci.* 2007;105:2910.
[116] Yoo Y, Shah RK, Paul DR. *Polymer* 2007;48:4867.
[117] Kim DH, Fasulo PD, Rodgers WR, Paul DR. *Polymer* 2007;48:5308.
[118] Kim DH, Fasulo PD, Rodgers WR, Paul DR. *Polymer* 2007;48:5960.
[119] Cui L, Ma X, Paul DR. *Polymer* 2007;48:6325.
[120] Chavarria K, Shah RK, Hunter DL, Paul DR. *Polym Eng Sci* 2007;47:1847.
[121] Liu L, Qi Z, Zhu X. *J. Appl. Polym. Sci.* 1999;71:1133–8.
[122] Heinemann J, Reichert P, Thomann R, Mu lhaupt R. *Macromol Rapid Commun* 1999;20:423–30.
[123] Garce' s JM, Moll DJ, Bicerano J, Fibiger R, McLeod DG. *Adv. Mater.* 2000;12(23): 1835–9.
[124] Varlot K, Reynaud E, Kloppfer MH, Vigier G, Varlet J. *J. Polym. Sci. Part B. Polym. Phys.* 2001;39:1360–70.
[125] Reichert P, Hoffmann B, Bock T, Thomann R, Mulhaupt R, Friedrich C. *Macromol. Rapid. Commun.* 2001;22(7):519–23.
[126] Kim SW, Jo WH, Lee MS, Ko MB, Jho JY. *Polymer* 2002;34(3):103–11.
[127] Ray SS, Yamada K, Ogami A, Okamoto M, Ueda K. *Macromol. Rapid. Commun.* 2002;23:943–7.
[128] Ray SS, Yamada K, Okamoto M, Fujimoto Y, Ogami A, Ueda K. *Polymer* 2003; 44:6633–46.
[129] Ray SS, Okamoto K, Okamoto M. *Macromolecules* 2003;36:2355–67.
[130] Yalcin B, Valladares D, Cakmak M. *Polymer* 2003;44:6913–25.
[131] Yalcin B, Cakmak M. *Polymer* 2004;45:2691–710.
[132] Konishi Y, Cakmak M. *Polymer* 2005;46:4811–26.
[133] Hsieh AJ, Moy P, Beyer FL, Madison P, Napadensky E. *Polym. Eng. Sci.* 2004; 44(5):825–37.
[134] Weon J-I, Sue H-J. *Polymer* 2005;46:6325–34.
[135] Vermogen A, Masenelli-Varlot K, Se'gue' la R, Duchet-Rumeau J, Boucard S, Prele P. *Macromolecules* 2005;38:9661–9.
[136] Kim Y, White JL. *J. Polym. Sci.* 2005;96:1888–96.
[137] Morgan AB. *Polym Adv Technol* 2006;17:206–17.
[138] Masenelli-Varlot K, Vigier G, Vermogen A, Gauthier C, Cavaille' JY. *J. Polym. Sci. Part B. Polym. Phys.* 2007;45:1243–51.
[139] Picard E, Vermogen A, Ge' rard J-F, Espuche E. *J. Membr. Sci.* 2007;292:133–44.
[140] Krishnamoorti R. *MRS Bull* 2007;32:341.
[141] Fornes TD, Hunter DL, Paul DR. *Macromolecules* 2004;37:1793.
[142] Tanaka G, Goettler LA. *Polymer* 2002;43:541–53.
[143] Kawasumi M, Hasegawa N, Kato M, Usuki A, Okada A. *Macromolecules* 1997; 30:6333–8.
[144] Hasegawa N, Kawasumi M, Kato M, Usuki A, Okada A. *J. Appl. Polym. Sci.* 1998; 67:87–92.
[145] Kato M, Okamoto H, Hasegawa N, Tsukigase A, Usuki A. *Polym Eng Sci* 2003; 43(6):1312.

[146] Hasegawa N, Usuki A. *J. Appl. Polym. Sci.* 2004;93:464–70.
[147] Toth R, Coslanich A, Ferrone M, Fermeglia M, Pricl S, Miertus S, et al. *Polymer* 2004;45:8075–83.
[148] Sza' zdi L, Puka'nszky B, Foldes E, Puka'nszky B. *Polymer* 2005;46:8001–10.
[149] Davis CH, Mathias LJ, Gilman JW, Schiraldi DA, Shields JR, Trulove P, et al. *J. Polym. Sci. Part B. Polym. Phys.* 2002;40:2661–6.
[150] Vaia RA, Liu W. *J. Polym. Sci. Part B. Polym. Phys.* 2002;40:1590–600.
[151] Morgan AB, Gilman JW. *J. Appl. Polym. Sci.* 2003;87:1329–38.
[152] Lincoln DM, Vaia RA, Wang Z-G, Hsiao BS, Krishnamoorti R. *Polymer* 2001; 42:9975–85.
[153] Lincoln DM, Vaia RA, Wang Z-G, Hsiao BS. *Polymer* 2001;42:1621–31.
[154] Vaia RA, Liu W, Koerner H. *J. Polym. Sci. Part B. Polym. Phys.* 2003;41:3214–36.
[155] Justice RS, Schaefer DW, Vaia RA, Tomlin DW, Bunning TJ. *Polymer* 2005;46: 4465–73.
[156] Schaefer DW, Justice RS. *Macromolecules* 2007;40(24):8501–17.
[157] VanderHart DL, Asano A, Gilman JW. *Chem. Mater.* 2001;13:3796–809.
[158] VanderHart DL, Asano A, Gilman JW. *Chem. Mater.* 2001;13:3781–95.
[159] VanderHart DL, Asano A. *Macromolecules* 2001;34:3819–22.
[160] Davis RD, Gilman JW, VanderHart DL. *Polym. Degrad Stab.* 2003;79: 111–21.
[161] Ho DL, Briber RM, Glinka CJ. *Chem. Mater.* 2001;13:1923–31.
[162] Schmidt G, Nakatani AI, Butler PD, Han CC. *Macromolecules* 2002;35: 4725–32.
[163] Oshinski AJ, Keskkula H, Paul DR. *Polymer* 1996;37(22):4891–907.
[164] Corte' L, Leibler L. *Polymer* 2005;46:6360–8.
[165] Lee KY, Paul DR. *Polymer* 2005;46:9064.
[166] Lee KY, Kim KH, Jeoung SK, Ju SI, Shim JH, Kim NH, et al. *Polymer* 2007;48: 4174.
[167] Mori T, Tanaka K. *Acta Metall* 1973;21:571.
[168] Halpin JC, Kardos JL. *Polym Eng Sci* 1976;16(5):344.
[169] Chow TS. *J. Polym. Sci.* 1978;16:959–65.
[170] Chow TS. *J. Polym. Sci.* 1978;16:967–70.
[171] Chow TS. *J. Mater. Sci.* 1980;15:1873–88.
[172] Hine PJ, Lusti HR, Gusev AA. *Compos. Sci. Technol.* 2002;62:1445–53.
[173] Lusti HR, Hine PJ, Gusev AA. *Compos. Sci. Technol.* 2002;62:1927–34.
[174] Van Es M, Xiqiao F, van Turnhout J, van der Giessen E. In: Al-Malaika S, Golovoy A, Wilkie CA, editors. Specialty polymer additives: principles and applications. Oxford: Blackwell Science; 2001. p. 391–414.
[175] Brune DA, Bicerano J. *Polymer* 2002;43:369–87.
[176] Luo J-J, Daniel IM. *Compos. Sci. Technol.* 2003;63:1607–16.
[177] Zhu L, Narh KA. *J. Polym. Sci. Part B. Polym. Phys.* 2004;42:2391–406.
[178] Tsai J, Sun CT. *Compos Mater* 2004;38(7):567–79.
[179] Sharaf MA, Mark JE. *Polymer* 2004;45:3943–52.
[180] Wang J, Pyrz R. *Compos. Sci. Technol.* 2004;64:925–34.
[181] Wang J, Pyrz R. *Compos. Sci. Technol.* 2004;64:935–44.
[182] Shepard PD, Golemba FJ, Maine FW. Adv Chem Ser 1973;134:41–51.
[183] Sheng N, Boyce MC, Parks DM, Rutledge GC, Abes JI, Cohen RE. *Polymer* 2004; 45:487–506.
[184] Hbaieb K, Wang QX, Chia YHJ, Cotterell B. *Polymer* 2007;48:901–9.

[185] Sen S, Thomin JD, Kumar SK, Keblinski P. *Macromolecules* 2007;40:4059–67.
[186] Laura DM, Keskkula H, Barlow JW, Paul DR. *Polymer* 2002;43:4673–87.
[187] Krishnamoorti R, Giannelis EP. *Macromolecules* 1997;30:4097–102.
[188] Medellin-Rodriguez FJ, Burger C, Hsiao BS, Chu B, Vaia R, Phillips S. *Polymer* 2001;42:9015–23.
[189] Wagener R, Reisinger TJG. *Polymer* 2003;44:7513–8.
[190] Chen B, Evans JRG. *Scr. Mater.* 2006;54:1581–5.
[191] Manevitch OL, Rutledge GC. *J. Phys. Chem. B* 2004;108:1428–35.
[192] Nielsen LE. *J. Macromol. Sci.* (Chem) 1967;A1:929–42.
[193] Lape NK, Nuxoll EE, Cussler EL. *J. Membr. Sci.* 2004;236:29–37.
[194] Gusev AA, Lusti HR. *Adv. Mater.* 2001;13:1641–3.
[195] Fredrickson GH, Bicerano J. *J Chem Phys* 1999;110:2181–8.
[196] Bharadwaj K. *Macromolecules* 2001;34:9189–92.
[197] Kim SH, Kim SC. *J. Appl. Polym. Sci.* 2007;103:1262–71.
[198] Choi WJ, Kim HJ, Yoon KH, Kwon OH, Hwang CI. *J. Appl. Polym. Sci.* 2006;100: 4875–9.
[199] Matayabas Jr JC, Turner SR. In: Pinnavaia TJ, Beall GW, editors. Polymer–clay nanocomposites. John Wiley & Sons Ltd.; 2000. p. 207–25; See also: Matayabas Jr JC, Turner SR, Sublett BJ, Connell GW, Gilmer JW, Barbee RB. US Patent 6,084,019, assigned to Eastman Chemical Corp.; July 4, 2000.
[200] Sridhar V, Tripathy DK. *J. Appl. Polym. Sci.* 2006;101:3630–7.
[201] Brule' B, Flat JJ. *Macromol. Symp.* 2006;233:210–6.
[202] Triantafyllidis KS, LeBaron PC, Park I, Pinnavaia TJ. *Chem. Mater.* 2006;18: 4393–8.
[203] Zimmerman CM, Singh A, Koros WJ. *J. Membr. Sci.* 1997;137:145–54.
[204] Robeson LM. *J. Membr. Sci.* 1991;62:165–85.
[205] Husain S, Koros WJ. *J. Membr. Sci.* 2007;288:195–207.
[206] Chung TS, Jiang LY, Li Y, Kulprathipanja S. *Prog. Polym. Sci.* 2007;32:483–507.
[207] Matteucci S, Kusuma VA, Sanders D, Swinnea S, Freeman BD. *J. Membr. Sci.* 2008;307:196–217.
[208] Bourbigot S, Duquesne S, Jama C. Macromol Symp 2006;233:180–90.
[209] Schartel B, Bartholmai M, Knoll U. *Polym. Adv. Technol.* 2006;17:772–7.
[210] Nazare S, Kandola BK, Horrocks AR. *Polym. Adv. Technol.* 2006;17:294–303.
[211] Zanetti M, Camino G, Canavese D, Morgan AB, Lamelas FJ, Wilkie CA. *Chem. Mater.* 2002;14:189–93.
[212] Lv JP, Liu WH. *J. Appl. Polym. Sci.* 2007;105:333–40.
[213] Dasari A, Yu ZZ, Mai YW, Liu S. *Nanotechnology* 2007;18:445602 (1–10).
[214] Kashiwagi T, Harris Jr RH, Zhang X, Briber RM, Cipriano BH, Raghavan SR, et al. *Polymer* 2004;45:881–91.
[215] Qin H, Zhang S, Zhao C, Hu G, Yang M. *Polymer* 2005;46:8386–95.
[216] Kashiwagi T, Du F, Winey KI, Groth KM, Shields JR, Bellayer SP, et al. *Polymer* 2005;46:471–81.
[217] Kashiwagi T, Grulke E, Hilding J, Harris R, Awad W, Douglas J. *Macromol. Rapid Commun.* 2002;23:761–5.
[218] Ray SS, Bousmina M. *Macromol. Rapid Commun.* 2005;26:450–5.
[219] Li Y, Shimizu H. *Polymer* 2004;45:7381–8.
[220] Khatua BB, Lee DJ, Kim HY, Kim JK. *Macromolecules* 2004;37:2454–9.

[221] Si M, Araki T, Ade H, Kilcoyne ALD, Fisher R, Sokolov JC, et al. *Macromolecules* 2006;39:4793–801.
[222] Vo LT, Giannelis EP. *Macromolecules* 2007;40:8271–6.
[223] Zhang Q, Yang H, Fu Q. *Polymer* 2004;45:1913–22.
[224] Wang Y, Zhang Q, Fu Q. *Macromol. Rapid Commun.* 2003;24:231–5.
[225] Hung HS, Hsu SH. *Nanotechnology* 2007;18:475101 (9 pp).
[226] Chen CZ, Cooper SL. *Adv. Mater.* 2000;12:843–6.
[227] Damm C, Munstedt H, Rosch A. *Mater. Chem. Phys.* 2008;108:61–6.
[228] Alt V, Bechert T, Steinru cke P, Wagener M, Seidel P, Dingeldein E, et al. *Biomaterials* 2004;25:4383–91.
[229] Hule RA, Pochan DJ. *MRS Bull* 2007;32:354–8.
[230] Kim HW, Kim HE, Salih V. *Biomaterials* 2005;26:5221–30.
[231] Huang J, Lin YW, Fu XW, Best SM, Brooks RA, Rushton N, et al. *J. Mater. Sci. Mater. Med.* 2007;18:2151–7.
[232] Lee YH, Lee JH, An IG, Kim C, Lee DS, Lee YK, et al. *Biomaterials* 2005;26: 3165–72.
[233] Zhang Q, Zha L, Ma J, Liang B. *Macromol. Rapid Commun.* 2007;28:116–20.
[234] Haraguchi K, Li HJ. *Macromolecules* 2006;39:1898–905.
[235] Gupta AK, Gupta M. *Biomaterials* 2005;26:3995–4021.
[236] Berry CC. *J. Mater. Chem.* 2005;15:543–7.
[237] Toprak MS, McKenna BJ, Waite JH, Stucky GD. *Chem. Mater.* 2007;19: 4263–9.
[238] Vadala ML, Rutnakornpituk M, Zalich MA, Pierre St TG, Riffle JS. *Polymer* 2004;45:7449–61.
[239] Rutnakornpituk M, Baranauskas VV, Riffle JS, Connolly J, Pierre St TG, Dailey JP. *Eur. Cells Mater.* 2002;3:102–5.
[240] Rolland JP, Maynor BW, Euliss LE, Exner AE, Denison GM, Desimone JM. *J. Am. Chem. Soc.* 2005;127:10096–100.
[241] Kongkanand A, Kuwabata S, Girishkumar G, Kamat P. *Langmuir* 2006;22: 2392–6.
[242] DeLuca NW, Elabd YA. *J. Polym. Sci. Part B. Polym. Phys.* 2006;44:2201–25.
[243] Kim YS,Wang F, Hickner M, Zadwodzinski TA, McGrath JE. *J. Membr. Sci.* 2003; 212:263–82.
[244] Zaidi SMJ, Mikhailenko SD, Robertson GP, Guiver MD, Kaliaguine S. *J. Membr. Sci.* 2000;173:17–34.
[245] Wang Z, Ni H, Zhao C, Li X, Fu T, Na H. *J. Polym. Sci. Part B. Polym. Phys.* 2006;44: 1967–78.
[246] Li X, Xu D, Zhang G, Wang Z, Zhao C, Na H. *J. Appl. Polym. Sci.* 2007;103:4020–6.
[247] Su YH, Liu YL, Sun YM, Lai JY, Guiver MD, Gao Y. *J. Power Sources* 2006;155: 111–7.
[248] Jiang R, Kunz HR, Fenton JM. *J. Membr. Sci.* 2006;272:116–24.
[249] Yang C, Srinivasan S, Arico AS, Creti P, Baglio V, Antonucci V. Electrochem Solid-*State Lett* 2001;4:A31–4.
[250] Hill ML, Kim YS, Einsla BR, McGrath JE. *J. Membr. Sci.* 2006;283:102–8.
[251] Prashantha K, Park SG. *J. Appl. Polym. Sci.* 2005;98:1875–8.
[252] Be' bin P, Caravanier M, Galiano H. *J. Membr. Sci.* 2006;278:35–42.
[253] Thomassin JM, Pagnoulle C, Bizzari D, Caldarella G, Germain A, Je' ro^me R. *Solid State Ionics* 2006;177:1137–44.

[254] Barroso-Bujans F, Verdejo R, Arroyo M, Lopez-Gonzalez MM, Riande E, Lopez-Manchado MA. *Macromol Rapid Commun* 2008;29:234–8.
[255] Baibarac M, Go' mez-Romero P. *J Nanosci Nanotechnol* 2006;6:1–14.
[256] Baughman RH, Zakhidov AA, De Heer WA. *Science* 2002;297:787–92.
[257] Moniruzzaman M, Winey KI. *Macromolecules* 2006;39:5194–205.
[258] Sandler JKW, Kirk JE, Kinloch IA, Shaffer MSP, Windle AH. *Polymer* 2003;44: 5893–9.
[259] Bryning MB, Islam MF, Kikkawa JM, Yodh AG. *Adv. Mater.* 2005;17: 1186–91.
[260] Kim YS, Wright JB, Grunlan JC. *Polymer* 2008;49:570–8.
[261] Huynh WU, Dittmer JJ, Alivisatos AP. *Science* 2002;295:2425–7.
[262] Liang Z, Dzienis KL, Xu J, Wang Q. *Adv Funct Mater* 2006;16:542–8.
[263] Qi XY, Pu KY, Fang C,Wen GA, Zhang H, Boey FYC, et al. *Macromol Chem Phys* 2007;208:2007–17.
[264] Olson DC, Shaheen SS, Collins RT, Ginley DS. *J. Phys. Chem.* C 2007;111:16670–8.
[265] Kwong CY, Choy WCH, Djuris˘ic' AB, Chui PC, Cheng KW, Chan WK. *Nanotechnology* 2004;15:1156–61.
[266] Kymakis E, Amaratunga GAJ. *Appl. Phys. Lett.* 2002;80:112–5.
[267] Tekin E, Smith PJ, Hoeppener S, van den Berg AMJ, Susha AS, Rogach AL, et al. *Adv. Funct. Mater.* 2007;17:23–8.
[268] Park JW, Ullah MH, Park SS, Ha CS. *J. Mater. Sci. Mater Electron* 2007;18:S393–7.
[269] Oey CC, Djuris˘ic' AB, Kwong CY, Cheung CH, Chan WK, Nunzi JM, et al. *Thin Solid Films* 2005;492:253–8.
[270] Park JH, Lim YT, Park OO, Kim JK, Yu JW, Kim YC. *Chem. Mater.* 2004;16: 688–92.
[271] Park JH, Lim YT, Park OO, Kim YC. *Macromol. Rapid Commun.* 2003;24: 331–4.
[272] Park JH, Lim YT, Park OO, Kim JK, Yu JW, Kim YC. *Adv. Funct. Mater* 2004;14: 377–82.
[273] Miyake J, Chujo Y. *Macromol. Rapid Commun.* 2008;29:86–92.
[274] Stankovich S, Dikin DA, Dommett GHB, Kohlhaas KM, Zimney EJ, Stach EA, et al. *Nature* 2006;442:282–6.
[275] Novoselov KS, Geim AK, Morozov SV, Jiang D, Zhang Y, Dubonos SV, et al. *Science* 2004;306:666–9.
[276] Gilje S, Han S, Wang M, Wang KL, Kaner RB. *Nano Lett.* 2007;7:3394–8.
[277] Van Noorden R. *Nature* 2006;442:228–9.
[278] Hatchett DW, Josowicz M. *Chem Rev* 2008. on web 01/03/2008.
[279] Kong J, Franklin NR, Zhou C, Chapline MG, Peng S, Cho K, et al. *Science* 2000; 287:622–5.
[280] An KH, Jeong SY, Hwang HR, Lee YH. *Adv. Mater. 2004*;16:1005–9.
[281] Ali SR, Ma Y, Parajuli RR, Balogun Y, Lai WYC, He H. *Anal. Chem.* 2007;79: 2583–7.
[282] Feng X, Mao C, Yang G, Hou W, Zhu JJ. *Langmuir* 2006;22:4384–9.
[283] Zhang M, Fang S, Zakhidov AA, Lee SB, Aliev AE, Williams CD, et al. *Science* 2005;309:1215–9.
[284] Vaia RA, Maguire JF. *Chem. Mater.* 2007;18:2736–51.

Chapter 2

TYPES OF NANOPARTICLES

2.1. INTRODUCTION

Nanoparticles are defined as particulate dispersions or solid particles with a size in the range of 10-100nm. Depending upon the method of preparation, nanoparticles, nanospheres or nanocapsules can be obtained. Nanocapsules are systems in which the drug is confined to a cavity surrounded by a unique polymer membrane, while nanospheres are matrix systems in which the drug is physically and uniformly dispersed. In recent years, biodegradable polymeric nanoparticles, particularly those coated with hydrophilic polymer such as poly (ethylene glycol) (PEG) known as long-circulating particles, have been used as potential drug delivery devices because of their ability to circulate for a prolonged period of time targeting a particular organ, as carriers of DNA in gene therapy, and their ability to deliver proteins, peptides and genes. The major goals in designing nanoparticles as a delivery system are to control particle size, surface properties and release of pharmacologically active agents in order to achieve the site-specific action of the drug at the therapeutically optimal rate and dose regimen.

Though liposomes have been used as potential carriers with unique advantages including protecting drugs from degradation, targeting to the site of action and reduction of toxicity or side effects, their applications are limited due to inherent problems such as low encapsulation efficiency, rapid leakage of water-soluble drugs in the presence of blood components and poor storage stability. On the other hand, polymeric nanoparticles offer some specific advantages over liposomes. For instance, they help to increase the stability of drugs/proteins and possess useful controlled release properties. The advantages of using nanoparticles as a drug delivery system include the following:

1. Particle size and surface characteristics of nanoparticles can be easily manipulated to achieve both passive and active drug targeting after parenteral administration.
2. They control and sustain release of the drug during the transportation and at the site of localization, altering organ distribution of the drug and subsequent clearance of the drug so as to achieve increases in drug therapeutic efficacy and reduction in side effects.
3. Controlled release and particle degradation characteristics can be readily modulated by the choice of matrix constituents. Drug loading is relatively high and drugs can be

incorporated into the systems without any chemical reaction; this is an important factor for preserving the drug activity.

4. Site-specific targeting can be achieved by attaching targeting ligands to the surface of particles or the use of magnetic guidance.
5. The system can be used for various routes of administration including oral, nasal, parenteral, intra-ocular etc.

In spite of these advantages, nanoparticles do have limitations. For example, their small size and large surface area can lead to particle aggregation, making physical handling of nanoparticles difficult in liquid and dry forms. In addition, small particle size and large surface areas readily result in limited drug loading and burst release. These practical problems have to be overcome before nanoparticles can be used clinically or made commercially the available details of the latest development of nanoparticulate drug delivery systems, surface modification issues, drug loading strategies, release control and potential applications of nanoparticles [1, 2].

2.2. Preparation of Nanoparticles

Nanoparticles can be prepared from a variety of materials such as proteins, polysaccharides and synthetic polymers. The selection of matrix materials is dependent on many factors including: (a) size of nanoparticles required; (b) inherent properties of the drug, e.g., aqueous solubility and stability; (c) surface characteristics such as charge and permeability; (d) degree of biodegradability, biocompatibility and toxicity; (e) drug release profile desired; and (f) antigenicity of the final product.

Nanoparticles have been prepared most frequently by three methods: (1) dispersion of preformed polymers; (2) polymerization of monomers; and (3) ionic gelation or coacervation of hydrophilic polymers. However, other methods such as supercritical fluid technology and particle replication in non-wetting templates (PRINT) have also been described in the literature for production of nanoparticles. The latter was claimed to have absolute control of particle size, shape and composition, which could set an example for the future mass production of nanoparticles in industry [3-6].

Dispersion of preformed polymers: Dispersion of preformed polymers is a common technique used to prepare biodegradable nanoparticles from poly (lactic acid) (PLA); poly (D, L-glycolide), PLG; poly (D, L-lactide-co-glycolide) (PLGA) and poly (cyanoacrylate) (PCA). This technique can be used in various ways as described below.

Solvent evaporation method: In this method, the polymer is dissolved in an organic solvent such as dichloromethane, chloroform or ethyl acetate which is also used as the solvent for dissolving the hydrophobic drug. The mixture of polymer and drug solution is then emulsified in an aqueous solution containing a surfactant or emulsifying agent to form oil in water (o/w) emulsion. After the formation of stable emulsion, the organic solvent is evaporated either by reducing the pressure or by continuous stirring. Particle size was found to be influenced by the type and concentrations of stabilizer, homogenizer speed and polymer concentration. In order to produce small particle size, a high-speed homogenization or ultra sonication may be employed [7-11].

Spontaneous emulsification or solvent diffusion method: This is a modified version of the solvent evaporation method. In this method, the water miscible solvent along with a small amount of the water immiscible organic solvent is used as an oil phase. Due to the spontaneous diffusion of solvents, an interfacial turbulence is created between the two phases leading to the formation of small particles. As the concentration of water miscible solvent increases, a decrease in the size of particle can be achieved. Both solvent evaporation and solvent diffusion methods can be used for hydrophobic or hydrophilic drugs. In the case of a hydrophilic drug, a multiple w/o/w emulsion needs to be formed with the drug dissolved in the internal aqueous phase [12-15].

2.3. Polymerization Method

In this method, monomers are polymerized to form nanoparticles in an aqueous solution. The drug is incorporated either by being dissolved in the polymerization medium or by adsorption onto the nanoparticles after polymerization is completed. The nanoparticle suspension is then purified to remove various stabilizers and surfactants employed for polymerization by ultracentrifugation and re-suspending the particles in an isotonic surfactant-free medium. This technique has been reported for making polybutylcyanoacrylate or poly (alkyl cyanoacrylate) nanoparticles. Nanocapsule formation and their particle size depends on the concentration of the surfactants and stabilizers used [16-18].

2.3.1. Coacervation or Ionic Gelation Method

Much research has been focused on the preparation of nanoparticles using biodegradable hydrophilic polymers such as chitosan, gelatin and sodium alginate. Calvo and co-workers [19, 20] developed a method for preparing hydrophilic chitosan nanoparticles by ionic gelation. The method involves a mixture of two aqueous phases, of which one is the polymer chitosan, a di-block co-polymer ethylene oxide or propylene oxide (PEO-PPO) and the other is a polyanion sodium tripolyphosphate. In this method, positively charged amino group of chitosan interacts with negative charged tripolyphosphate to form coacervates with a size in the range of nanometer. Coacervates are formed as a result of electrostatic interaction between two aqueous phases, whereas, ionic gelation involves the material undergoing transition from liquid to gel due to ionic interaction conditions at room temperature.

2.3.2. Production of Nanoparticles Using Supercritical Fluid Technology

Conventional methods such as solvent extraction-evaporation, solvent diffusion and organic phase separation methods require the use of organic solvents which are hazardous to the environment as well as to physiological systems. Therefore, the supercritical fluid technology has been investigated as an alternative to prepare biodegradable micro- and nanoparticles because supercritical fluids are environmentally safe [21].

A supercritical fluid can be generally defined as a solvent at a temperature above its critical temperature, at which the fluid remains a single phase regardless of pressure. Supercritical CO_2 (SC CO_2) is the most widely used supercritical fluid because of its mild critical conditions (Tc = 31.1 °C, Pc = 73.8 bars), nontoxicity, non-flammability, and low price. The most common processing techniques involving supercritical fluids are supercritical anti-solvent (SAS) and rapid expansion of critical solution (RESS). The process of SAS employs a liquid solvent, which is completely miscible with the supercritical fluid (SC CO_2), to dissolve the solute to be micronized; at the process conditions, because the solute is insoluble in the supercritical fluid, the extract of the liquid solvent by supercritical fluid leads to the instantaneous precipitation of the solute, resulting in the formation of nanoparticles. Thote and Gupta (2005) [22] reported the use of a modified SAS method for formation of hydrophilic drug dexamethasone phosphate drug nanoparticles for microencapsulation purpose.

RESS differs from the SAS process in that its solute is dissolved in a supercritical fluid (such as supercritical methanol) and then the solution is rapidly expanded through a small nozzle into a region of lower pressure. Thus, the solvent power of supercritical fluids dramatically decreases and the solute eventually precipitates. This technique is clean because the precipitate is basically solvent free. RESS and its modified process have been used for the product of polymeric nanoparticles. The supercritical fluid technology technique, although environmentally friendly and suitable for mass production, requires specially designed equipment and is more expensive [23, 24].

2.3.3. Effect of Characteristics of Nanoparticles on Drug Delivery

Particle size and size distribution are the most important characteristics of nanoparticle systems. They determine the *in vivo* distribution, biological fate, toxicity and the targeting ability of nanoparticle systems. In addition, they can also influence the drug loading, drug release and stability of nanoparticles. Many studies have demonstrated that nanoparticles of sub-micron size have a number of advantages over microparticles as a drug delivery system. Nanoparticles generally have a relatively higher intracellular uptake compared to microparticles and are available to a wider range of biological targets due to their small size and relative mobility. Desai et al. found that 100 nm nanoparticles had a 2.5 fold greater uptake than 1 μm microparticles, and 6 fold greater uptake than 10 μm microparticles in a Caco-2 cell line [25-27] In a subsequent study, the nanoparticles penetrated throughout the submucosal layers in a rat in situ intestinal loop model, while microparticles were predominantly localized in the epithelial lining. It was also reported that nanoparticles can cross the blood-brain barrier following the opening of tight junctions by hyper osmotic mannitol, which may provide sustained delivery of therapeutic agents for difficult-to-treat diseases like brain tumors. Tween 80 coated nanoparticles have been shown to cross the blood-brain barrier [28]. In some cell lines, only submicron nanoparticles can be taken up efficiently but not the larger size microparticles.

Drug release is affected by particle size. Smaller particles have a larger surface area. Therefore, most of the associated drug would be at or near the particle surface, leading to fast drug release. Larger particles have large cores which allow more drug to be encapsulated and slowly diffuse out [29, 30]. Smaller particles also have greater risk of aggregation of particles during storage and transportation of nanoparticle dispersion. It is always a challenge to formulate nanoparticles with the smallest size possible but maximum stability.

Polymer degradation can also be affected by the particle size. For instance, the rate of PLGA polymer degradation was found to increase with increasing particle size *in vitro*. It was thought that in smaller particles, degradation products of PLGA formed can diffuse out of the particles easily while in large particles. Degradation products are more likely to remain within the polymer matrix for a longer period to cause autocatalytic degradation of the polymer material. Therefore, it was hypothesized that larger particles will contribute to faster polymer degradation as well as the drug release [31]. However, Panyam et al. prepared PLGA particles with different size ranges and found that the polymer degradation rates *in vitro* were not substantially different for different size particles [32]. Currently, the fastest and most routine method of determining particle size is by photon-correlation spectroscopy or dynamic light scattering. Photon-correlation spectroscopy requires the viscosity of the medium to be known and determines the diameter of the particle by Brownian motion and light scattering properties. The results obtained by photon-correlation spectroscopy are usually verified by scanning or transmission electron microscopy (SEM or TEM).

2.3.4. Surface Properties of Nanoparticles

When nanoparticles are administered intravenously, they are easily recognized by the body immune systems, and are then cleared by phagocytes from the circulation. Apart from the size of nanoparticles, their surface hydrophobicity determines the amount of adsorbed blood components, mainly proteins (opsonins) [33]. This in turn influences the *in vivo* fate of nanoparticles. The binding of these opsonins onto the surface of nanoparticles called opsonization acts as a bridge between nanoparticles and phagocytes. The association of a drug to conventional carriers leads to modification of the drug biodistribution profile, as it is mainly delivered to the mononuclear phagocytes system (MPS) such as the liver, spleen, lungs and bone marrow. Once in the blood stream, surface non-modified nanoparticles (conventional nanoparticles) are rapidly opsonized and massively cleared by the macrophages of MPS rich organs.

Generally, it is IgG, compliment C3 components that are used for recognition of foreign substances, especially foreign macromolecules [34]. In order to increase the likelihood of the success in drug targeting by nanoparticles, it is necessary to minimize the opsonization and to prolong the circulation of nanoparticles *in vivo*. This can be achieved by (a) surface coating of nanoparticles with hydrophilic polymers/surfactants; (b) formulation of nanoparticles with biodegradable copolymers with hydrophilic segments such as polyethylene glycol (PEG), polyethylene oxide, polyoxamer, poloxamine and polysorbate 80 (Tween 80) [35].

Studies show that PEG conformation at the nanoparticle surface is of utmost importance for the opsonin repelling function of the PEG layer. PEG surfaces in brush-like and intermediate configurations reduce phagocytosis and complement activation whereas PEG surfaces in mushroom-like configuration were potent complement activators and favored phagocytosis. The zeta potential of a nanoparticle is commonly used to characterize the surface charge property of nanoparticles [38]. It reflects the electrical potential of particles and is influenced by the composition of the particle and the medium in which it is dispersed. Nanoparticles with a zeta potential above (+/-) 30 mV have been shown to be stable in suspension, as the surface charge prevents aggregation of the particles. The zeta potential can also be used to determine whether a charged active material is encapsulated within the center of the nanocapsule or adsorbed onto the surface [36-38].

2.3.5. Drug Loading

Ideally, a successful nanoparticulate system should have a high drug-loading capacity thereby reducing the quantity of matrix materials for administration. Drug loading can be done by two methods:

- Incorporating at the time of nanoparticle production (incorporation method)
- Absorbing the drug after formation of nanoparticles by incubating the carrier with a concentrated drug solution (adsorption /absorption technique). Drug loading and entrapment efficiency very much depend on the solid-state drug solubility in matrix material or polymer (solid dissolution or dispersion), which is related to the polymer composition, the molecular weight, the drug polymer interaction and the presence of end functional groups (ester or carboxyl) . The PEG moiety has no or little effect on drug loading. The macromolecule or protein shows greatest loading efficiency when it is loaded at or near its isoelectric point when it has minimum solubility and maximum adsorption. For small molecules, studies show the use of ionic interaction between the drug and matrix materials can be a very effective way to increase the drug loading [39-42].

Drug Release

To develop a successful nanoparticulate system, both drug release and polymer biodegradation are important consideration factors. In general, the drug release rate depends on: (1) solubility of the drug; (2) desorption of the surface bound/ adsorbed drug; (3) drug diffusion through the nanoparticle matrix; (4) nanoparticle matrix erosion/degradation; and (5) a combination of the erosion/diffusion process. Thus solubility, diffusion and biodegradation of the matrix materials govern the release process. In the case of nanospheres, where the drug is uniformly distributed, the release occurs by diffusion or erosion of the matrix under sink conditions [43]. If the diffusion of the drug is faster than matrix erosion, the mechanism of release is largely controlled by a diffusion process. The rapid initial release or 'burst' is mainly attributed to a weakly bound or adsorbed drug to the large surface of nanoparticles. It is evident that the method of incorporation has an effect on the release profile. If the drug is loaded by the incorporation method, the system has a relatively small burst effect and better sustained release characteristics. If the nanoparticle is coated by a polymer, the release is then controlled by diffusion of the drug from the core across the polymeric membrane. The membrane coating acts as a barrier to release, therefore, the solubility and diffusivity of the drug in the polymer membrane becomes a determining factor in drug release. Furthermore, the release rate can also be affected by the ionic interaction between the drug and the addition of auxiliary ingredients. When the drug is involved in interaction with auxiliary ingredients to form a less water soluble complex, then the drug release can be very slow with almost no burst release effect. However, if the addition of auxiliary ingredients e.g., the addition of ethylene oxide-propylene oxide block copolymer

(PEO-PPO) to chitosan, reduces the interaction of the model drug bovine serum albumin (BSA) with the matrix material (chitosan) due to competitive electrostatic interaction of PEO-PPO with chitosan, then an increase in drug release could be observed[44].

Various methods which can be used to study the *in vitro* release of the drug are: (1) side-by-side diffusion cells with artificial or biological membranes; (2) dialysis bag diffusion technique; (3) reverse dialysis bag technique; (4) agitation followed by ultracentrifugation/ centrifugation and (5) ultra-filtration or centrifugal ultra-filtration techniques. Usually the release study is carried out by controlled agitation followed by centrifugation. Due to the time-consuming nature and technical difficulties encountered in the separation of nanoparticles from release media, the dialysis technique is generally preferred [45, 46].

2.4. Applications of Nanoparticulate Delivery Systems

Tumor Targeting Using Nanoparticulate Delivery Systems

The rationale of using nanoparticles for tumor targeting is based on the following factors: 1) the nanoparticles will be able to deliver a concentrated dose of drug in the vicinity of the tumor targets via the enhanced permeability and retention effect or active targeting by ligands on the surface of the nanoparticles and 2) the nanoparticles will reduce the drug exposure of healthy tissues by limiting the drug distribution to the target organ. Verdun et al. [47] demonstrated in mice treated with doxorubicin incorporated into poly (isohexylcyanoacrylate) nanospheres that higher concentrations of doxorubicin manifested in the liver, spleen and lungs than in mice treated with free doxorubicin. Studies show that the polymeric composition of nanoparticles such as type, hydrophobicity and biodegradation profile of the polymer along with the associated drug's molecular weight, its localization in the nanospheres and mode of incorporation technique and adsorption or incorporation, have a major influence on the drug distribution pattern *in vivo* [48]. The exact underlying mechanism is not fully understood but the biodistribution of nanoparticles is rapid, within ½ hour to 3 hours, and likely involves MPS and the endocytosis/phagocytosis process.

Recently Bibby et al. [49] reported on the biodistribution and pharmacokinetics (PK) of a cyclic RGD doxorubicin- nanoparticle formulation in tumor bearing mice. Their biodistribution studies revealed decreasing drug concentrations over time in the heart, lung, kidney and plasma and accumulating drug concentrations in the liver, spleen and tumor. The majority injected dose appeared in the liver (56%) and only 1.6% in the tumor at 48 hrs post injection, confirming that nanoparticles have a greater tendency to be captured by liver. This indicates that the greatest challenge of using nanoparticles for tumor targeting is to avoid particle uptake by the mononuclear phagocytic system (MPS) in liver and spleen. Such propensity of MPS for endocytosis/phagocytosis of nanoparticles provides an opportunity to effectively deliver therapeutic agents to these cells. This biodistribution can be of benefit for the chemotherapeutic treatment of MPS- rich organs/tissues localized tumors like hepatocarcinoma, hepatic metastasis arising from digestive tract or gynecological cancers, brochopulmonary tumors, primitive tumors and metastasis, small cell tumors, myeloma and leukemia. It has been proven that using doxorubicin loaded conventional nanoparticles was effective against the hepatic metastasis model in mice [50]. It was found that there was a greater reduction in the degree of metastasis than when the free drug was used. The underlying mechanism responsible for the increased therapeutic efficacy of the formulation was the transfer of doxorubicin from healthy tissue, acting as a drug reservoir to the malignant tissues. Histological examination showed a considerable accumulation of nanoparticles in the lysosomal vesicles of Kupffer cells, whereas nanoparticles could not be clearly identified in tumoral cells [51-53]. Thus Kupffer cells, after a massive uptake of nanoparticles by phagocytosis, were able to induce the release of doxorubicin, leading to a gradient of drug concentration that was favorable for a prolonged diffusion of the free and still active drug towards the neighboring metastatic cells. When conventional nanoparticles are used as carriers in chemotherapy, some cytotoxicity against the Kupffer cells can be expected, which would result in deficiency of Kupffer cells and naturally lead to reduced liver uptake and decreased therapeutic effect with intervals of less than 2 weeks administration. Moreover, conventional nanoparticles can also target bone marrow (MPS tissue), which is an important but unfavorable site of action for most anticancer drugs because chemotherapy with such carriers may increase the myelosuppresive effect [54]. Therefore, the ability of conventional nanoparticles to enhance anticancer drugs efficacy is limited to targeting tumors at the level of MPS-rich organs. Also, directing anticancer drug-loaded nanoparticles to other tumoral sites is not feasible if a rapid clearance of nanoparticles occurs shortly after intravenous administration [55].

2.5. Long Circulating Nanoparticles

To be successful as a drug delivery system, nanoparticles must be able to target tumors which are localized outside MPS-rich organs. In the past decade, a great deal of work has been devoted to developing so-called “stealth” particles or PEGylated nanoparticles, which are invisible to macrophages or phagocytes. A major breakthrough in the field came when the use of hydrophilic polymers (such as polyethylene glycol, poloxamines, poloxamers, and polysaccharides) to efficiently coat conventional nanoparticle surface produced an opposing effect to the uptake by the MPS.

These coatings provide a dynamic “cloud” of hydrophilic and neutral chains at the particle surface which repel plasma proteins. As a result, those coated nanoparticles become invisible to MPS and therefore remained in circulation for a longer period of time. Hydrophilic polymers can be introduced at the surface either by adsorption of surfactants or by the use of block or branched copolymers for the production of nanoparticles. Studies show that nanoparticles containing a coat of PEG not only have a prolonged half-life in the blood compartment but also are able to selectively extravasate in pathological sites such as tumors or inflamed regions with a leaky vasculature.

As a result, such long-circulating nanoparticles have increased the potential to directly target tumors located outside of MPS-rich regions [56]. The size of the colloidal carriers, as well as their surface characteristics, is critical to the biological fate of nanoparticles. A size less than 100 nm and a hydrophilic surface are essential in achieving the reduction of opsonisation reactions and subsequent clearance by macrophages. Coating conventional nanoparticles with surfactants or PEG to obtain a long-circulating carrier has now been used as a standard strategy for drug targeting *in vivo*.

Extensive efforts have been devoted to achieving "active targeting" of nanoparticles in order to deliver drugs to the right targets, based on molecular recognition processes such as ligand-receptor or antigen-antibody interaction. Considering the fact that folate receptors are over expressed on the surface of some human malignant cells and the cell adhesion molecules such as selectins and integrins are involved in metastatic events, nanoparticles bearing specific ligands such as folate may be used to target ovarian carcinoma while specific peptides or carbohydrates may be used to target integrins and selectins. Oyewumi et al. [57] demonstrated that the benefits of folate ligand coating were to facilitate tumor cell internalization and retention of Gd-nanoparticles in the tumor tissue. Targeting with small ligands appears more likely to succeed since they are easier to handle and manufacture. Furthermore, it could be advantageous when the active targeting ligands are used in combination with the long-circulating nanoparticles to maximize the likelihood of the success in active targeting of nanoparticles [58].

2.6. Reversion of Multidrug Resistance in Tumor Cells

Anticancer drugs, even if they are located in the tumor interstitium, can turn out to be of limited efficacy against numerous solid tumor types, because cancer cells are able to develop mechanisms of resistance.

These mechanisms allow tumors to evade chemotherapy. Multidrug resistance (MDR) is one of the most serious problems in chemotherapy. MDR occurs mainly due to the over expression of the plasma membrane p-glycoprotein (Pgp), which is capable of extruding various positively charged xenobiotics, including some anticancer drugs, out of cells. In order to restore the tumoral cells' sensitivity to anticancer drugs by circumventing Pgp-mediated MDR, several strategies including the use of colloidal carriers have been applied. The rationale behind the association of drugs with colloidal carriers, such as nanoparticles, against drug resistance derives from the fact that Pgp probably recognizes the drug to be effluxed out of the tumoral cells only when this drug is present in the plasma membrane, and not when it is located in the cytoplasm or lysosomes after endocytosis [59, 60].

2.7. Nanoparticles for Oral Delivery of Peptides and Proteins

Significant advances in biotechnology and biochemistry have led to the discovery of a large number of bioactive molecules and vaccines based on peptides and proteins. Development of suitable carriers remains a challenge due to the fact that the bioavailability of these molecules is limited by the epithelial barriers of the gastrointestinal tract and their susceptibility to gastrointestinal degradation by digestive enzymes. Polymeric nanoparticles allow encapsulation of bioactive molecules and protect them against enzymatic and hydrolytic degradation. For instance, it has been found that insulin-loaded nanoparticles have preserved insulin activity and produced blood glucose reduction in diabetic rats for up to 14 days following the oral administration [61].

The surface area of human mucosa extends to 200 times that of skin. The gastrointestinal tract provides a variety of physiological and morphological barriers against protein or peptide delivery, e.g., (a) proteolytic enzymes in the gut lumen like pepsin, trypsin and chymotrypsin; (b) proteolytic enzymes at the brush border membrane (endopeptidases); (c) bacterial gut flora; and (d) the mucus layer and epithelial cell lining itself. The histological architecture of the mucosa is designed to efficiently prevent uptake of particulate matter from the environment. One important strategy to overcome the gastrointestinal barrier is to deliver the drug in a colloidal carrier system, such as nanoparticles, which is capable of enhancing the interaction mechanisms of the drug delivery system and the epithelia cells in the GI tract [62-65].

2.8. Targeting of Nanoparticles to Epithelial Cells in the GI Tract Using Ligands

Targeting strategies to improve the interaction of nanoparticles with adsorptive enterocytes and M-cells of Peyer's patches in the GI tract can be classified into those utilizing specific binding to ligands or receptors and those based on nonspecific adsorptive mechanisms. The surface of enterocytes and M cells display cell-specific carbohydrates, which may serve as binding sites to colloidal drug carriers containing appropriate ligands. Certain glycoproteins and lectins bind selectively to this type of surface structure by specific receptor-mediated mechanisms. Different lectins, such as bean lectin and tomato lectin, have been studied to enhance oral peptide adsorption [66]. Vitamin B-12 absorption from the gut under physiological conditions occurs via receptor-mediated endocytosis. The ability to increase the oral bioavailability of various peptides (e.g., granulocyte colony stimulating factor, erythropoietin) and particles by covalent coupling to vitamin B-12 has been studied. For this intrinsic process, mucoprotein is required, which is prepared by the mucus membrane in the stomach and binds specifically to cobalamin. The mucoprotein completely reaches the ileum where resorption is mediated by specific receptors [67].

Absorption Enhancement Using Non-Specific Interactions

In general, the gastrointestinal absorption of macromolecules and particulate materials involves either a paracellular route or endocytotic pathway. The paracellular route of absorption of nanoparticles utilizes less than 1% of the mucosal surface area. Using polymers such as chitosan, starch or poly (acrylate) can increase the paracellular permeability of macromolecules. The endocytotic pathway for absorption of nanoparticles is either by receptor-mediated endocytosis, that is, active targeting, or adsorptive endocytosis which does not need any ligands. This process is initiated by an unspecific physical adsorption of material to the cell surface by electrostatic forces such as hydrogen bonding or hydrophobic interactions. Adsorptive endocytosis depends primarily on the size and surface properties of the material. If the surface charge of the nanoparticles is positive or uncharged, it will provide an affinity to adsorptive enterocytes though hydrophobic, whereas if it is negatively charged and hydrophilic, it shows greater affinity to adsorptive enterocytes and Mcells. This shows that a combination of size, surface charge and hydrophilicity play a major role in affinity. This is demonstrated with polystyrene nanoparticles and when it is carboxylated [68, 60].

2.9. Nanoparticles for Gene Delivery

Polynucleotide vaccines work by delivering genes encoding relevant antigens to host cells where they are expressed, producing the antigenic protein within the vicinity of professional antigen presenting cells to initiate immune response. Such vaccines produce both humoral and cell-mediated immunity because intracellular production of protein, as opposed to extracellular deposition, stimulates both arms of the immune system. The key ingredient of polynucleotide vaccines, DNA, can be produced cheaply and has much better storage and handling properties than the ingredients of the majority of protein-based vaccines [70-73]. Polynucleotide vaccines are set to supersede many conventional vaccines particularly for immunotherapy. However, there are several issues related to the delivery of polynucleotides which limit their application. These issues include the efficient delivery of the polynucleotide to the target cell population and its localization to the nucleus of these cells, and ensuring that the integrity of the polynucleotide is maintained during delivery to the target site [74].

Nanoparticles loaded with plasmid DNA could also serve as an efficient sustained release gene delivery system due to their rapid escape from the degradative endo-lysosomal compartment to the cytoplasmic compartment. Hedley et al. [75] reported that following their intracellular uptake and endolysosomal escape, nanoparticles could release DNA at a sustained rate resulting in sustained gene expression. This gene delivery strategy could be applied to facilitate bone healing by using PLGA nanoparticles containing therapeutic genes such as bone morphogenic protein.

2.10. Nanoparticles for Drug Delivery into the Brain

The blood-brain barrier (BBB) is the most important factor limiting the development of new drugs for the central nervous system. The BBB is characterized by relatively impermeable endothelial cells with tight junctions, enzymatic activity and active efflux transport systems. It effectively prevents the passage of water-soluble molecules from the blood circulation into the CNS, and can also reduce the brain concentration of lipid-soluble molecules by the function of enzymes or efflux pumps [76].

Consequently, the BBB only permits selective transport of molecules that are essential for brain function. Strategies for nanoparticle targeting to the brain rely on the presence of and nanoparticle interaction with specific receptor-mediated transport systems in the BBB. For example polysorbate LDL, transferrin receptor binding antibody (such as OX26), lactoferrin, cellpenetrating peptides and melanotransferrin have been shown to be capable of the delivery of a self nontransportable drug into the brain via the chimeric construct that can undergo receptor-mediated transcytosis [77]. It has been reported that poly (butylcyanoacrylate) nanoparticles were able to deliver hexapeptide dalargin, doxorubicin and other agents into the brain which is significant because of the great difficulty for drugs to cross the BBB. Despite some reported success with polysorbate 80 coated NPs, this system does have many shortcomings including desorption of polysorbate coating, rapid NP degradation and toxicity caused by the presence of high concentrations of polysorbate[78]. OX26 MAbs (anti-transferrin receptor MAbs), the most studied BBB targeting antibody, have been used to enhance the BBB penetration of lipsosomes. However, recently, Ji et al. [79] demonstrated that brain uptake of lactoferrin, an iron-binding glycoprotein belonging to the transferrin (Tf) family, is twice that of OX26 and transferrrin *in vivo*. It is possible that we will soon see these BBB specific molecules used for targeting nanoparticles to the brain [80-92].

References

[1] Langer R. Biomaterials in drug delivery and tissue engineering: one laboratory's experience. *Acc Chem. Res* 2000; 33: 94-101.

[2] Bhadra D, Bhadra S, Jain P, Jain NK. Pegnology: a review of PEG-ylated systems. *Pharmazie* 2002; 57: 5-29.

[3] Kommareddy S, Tiwari SB, Amiji MM. Long-circulating polymeric nanovectors for tumor-selective gene delivery. *Technol Cancer Res Treat* 2005; 4: 615- 25.

[4] Lee M, Kim SW. Polyethylene glycol-conjugated copolymers for plasmid DNA delivery. *Pharm Res* 2005; 22: 1-10.

[5] Vila A, Sanchez A, Tobio M, Calvo P, Alonso MJ. Design of biodegradable particles for protein delivery. *J. Control Release* 2002; 78: 15-24.

[6] Mu L, Feng SS. A novel controlled release formulation for the anticancer drug paclitaxel (Taxol(R)): PLGA nanoparticles containing vitamin E TPGS. *J Control Release* 2003; 86: 33-48.

[7] Kreuter J. Nanoparticles. In *Colloidal drug delivery systems*, J, K., Ed. Marcel Dekker: New York, 1994; pp 219-342.

[8] Reverchon E, Adami R. Nanomaterials and supercritical fluids. *The Journal of Supercritical Fluids* 2006; 37: 1-22.

[9] Rolland JP, Maynor BW, Euliss LE, Exner AE, Denison GM, DeSimone JM. Direct fabrication and harvesting of monodisperse, shape-specific nanobiomaterials. *J. Am. Chem. Soc.* 2005; 127: 10096-10100

[10] Kompella UB, Bandi N, Ayalasomayajula SP. Poly (lactic acid) nanoparticles for sustained release of budesonide. *Drug Deliv. Technol.* 2001; 1: 1-7.

[11] Ravi MN, Bakowsky U, Lehr CM. Preparation and characterization of cationic PLGA nanospheres as DNA carriers. B*iomaterials* 2004; 25: 1771-1777.

[12] Li YP, Pei YY, Zhou ZH, Zhang XY, Gu ZH, Ding J, Zhou JJ, Gao, XJ, PEGylated polycyanoacrylate nanoparticles as tumor necrosis factor-[alpha] carriers. *J Control Release* 2001; 71: 287-296.

[13] Kwon, HY, Lee JY, Choi SW, Jang Y, Kim JH. Preparation of PLGA nanoparticles containing estrogen by emulsification-diffusion method. Colloids Surf. A: Physicochem. *Eng. Aspects* 2001; 182: 123-130.

[14] Zambaux M, Bonneaux F, Gref R, Maincent P, Dellacherie E, Alonso M, Labrude P, Vigneron C. Influence of experimental parameters on the characteristics of poly(lactic

acid) nanoparticles prepared by double emulsion method. *J. Control. Release* 1998; 50: 31-40.

[15] Niwa T, Takeuchi H, Hino T, Kunou N, Kawashima Y. Preparation of biodegradable nanoparticles of water-soluble and insoluble drugs with D,Llactide/ glycolide copolymer by a novel spontaneous emulsification solvent diffusion method, and the drug release behavior. *J. Control. Release* 1993; 25: 89-98.

[16] Zhang Q, Shen Z, Nagai T. Prolonged hypoglycemic effect of insulin-loaded polybutylcyanoacrylate nanoparticles after pulmonary administration to normal rats. *Int. J. Pharm.* 2001; 218: 75-80.

[17] Boudad H, Legrand P, Lebas G, Cheron M, Duchene D, Ponchel G. Combined hydroxypropyl-[beta]- cyclodextrin and poly(alkylcyanoacrylate) nanoparticles intended for oral administration of saquinavir. *Int. J. Pharm.* 2001; 218: 113-124.

[18] Puglisi G, Fresta M, Giammona G, Ventura CA. Influence of the preparation conditions on poly(ethylcyanoacrylate) nanocapsule formation. *Int. J. Pharm.* 1995; 125: 283-287.

[19] Calvo P, Remunan-Lopez C, Vila-Jato JL, Alonso MJ. Novel hydrophilic chitosan-polyethylene oxide nanoprticles as protein carriers. *J. Appl. Polymer Sci.* 1997; 63: 125-132.

[20] Calvo P, Remunan-Lopez C, Vila-Jato JL, Alonso MJ. Chitosan and chitosan/ethylene oxide-propylene oxide block copolymer nanoparticles as novel carriers for proteins and vaccines. *Pharm Res.* 1997; 14: 1431-1436.

[21] Jung J, Perrut M. Particle design using supercritical fluids: Literature and patent survey. *J. Supercritical Fluids* 2001; 20: 179-219.

[22] Thote AJ, Gupta RB. Formation of nanoparticles of a hydrophilic drug using supercritical carbon dioxide and microencapsulation for sustained release. Nanomedicine: *Nanotech. Biology Medicine* 2005; 1: 85-90.

[23] Sun Y, Mezian M, Pathak P, Qu L. Polymeric nanoparticles from rapid expansion of supercritical fluid solution. *Chemistry* 2005; 11: 1366-73.

[24] Panyam J, Labhasetwar V. Biodegradable nanoparticles for drug and gene delivery to cells and tissue. *Adv. Drug Deliv Rev.* 2003; 55: 329-47.

[25] Desai MP, Labhasetwar V, Walter E, Levy RJ, Amidon G L, The mechanism of uptake of biodegradable microparticles in Caco-2 cells is size dependent. *Pharm Res* 1997; 14: 1568-73.

[26] Desai MP, Labhasetwar V, Amidon GL, Levy RJ. Gastrointestinal uptake of biodegradable microparticles: effect of particle size. *Pharm Res.* 1996; 13: 1838-45.

[27] Kroll RA, Pagel MA, Muldoon LL, Roman-Goldstein S, Fiamengo SA, Neuwelt EA. Improving drug delivery to intracerebral tumor and surrounding brain in a rodent model: a comparison of osmotic versus bradykinin modification of the blood-brain and/or blood-tumor barriers. *Neurosurgery* 1998; 43: 879-86; discussion 886-9.

[28] Kreuter J, Ramge P, Petrov V, Hamm S, Gelperina SE, Engelhardt B, Alyautdin R, von Briesen H, Begley DJ. Direct evidence that polysorbate-80-coated poly(butylcyanoacrylate) nanoparticles deliver drugs to the CNS via specific mechanisms requiring prior binding of drug to the nanoparticles. *Pharm Res* 2003; 20: 409-16.

[29] Zauner W, Farrow NA, Haines AM. In vitro uptake of polystyrene microspheres: effect of particle size, cell line and cell density. *J. Control Release* 2001; 71: 39-51.

[30] Redhead HM, Davis SS, Illum L. Drug delivery in poly(lactide-co-glycolide) nanoparticles surface modified with poloxamer 407 and poloxamine 908: in vitro characterisation and in vivo evaluation. *J. Control Release* 2001; 70: 353-363.

[31] Dunne M, Corrigan OI, Ramtoola Z. Influence of particle size and dissolution conditions on the degradation properties of polylactide-co-glycolide particles. *Biomaterials* 2000; 21: 1659-1668.

[32] Panyam J, Dali MM, Sahoo S K, Ma W, Chakravarthi SS, Amidon GL, Levy RJ, Labhasetwar V. Polymer degradation and in vitro release of a model protein from poly(,-lactide-co-glycolide) nano- and microparticles. *J. Control Release* 2003; 92: 173-187.

[33] Swarbrick J, Boylan J. *Encyclopedia of pharmaceutical technology*. 2nd ed.; Marcel Dekker: New York, 2002.

[34] Muller RH, Wallis KH. Surface modification of i.v. injectable biodegradable nanoparticles with poloxamer polymers and poloxamine 908. *Int. J. Pharm.* 1993; 89: 25-31.

[35] Brigger I, Dubernet C, Couvreur P. Nanoparticles in cancer therapy and diagnosis. *Adv. Drug Deliv. Rev.* 2002; 54: 631-651.

[36] Grislain L, Couvreur P, Lenaerts V, Roland M, Deprez- Decampeneere D, Speiser P. Pharmacokinetics and distribution of a biodegradable drug-carrier. *Int. J. Pharm.* 1983; 15: 335-345.

[37] Olivier JC. Drug transport to brain with targeted nanoparticles. *NeuroRx* 2005; 2: 108-119.

[38] Couvreur P, Barratt G, Fattal E, Legrand P, Vauthier C. Nanocapsule technology: a review. *Crit. Rev. Ther Drug Carrier Syst* 2002; 19: 99-134.

[39] Govender T, Stolnik S, Garnett MC, Illum L, Davis SS. PLGA nanoparticles prepared by nanoprecipitation: drug loading and release studies of a water soluble drug. *J. Control. Rel.* 1999; 57: 171-185.

[40] Govender T, Riley T, Ehtezazi T, Garnett MC, Stolnik S, Illum L, Davis SS. Defining the drug incorporation properties of PLA-PEG nanoparticles. *Int. J. Pharm* 2000; 199: 95-110.

[41] Panyam J, Williams D, Dash A, Leslie-Pelecky D, Labhasetwar V. Solid-state solubility influences encapsulation and release of hydrophobic drugs from PLGA/PLA nanoparticles. *J. Pharm. Sci.* 2004; 93: 1804-14.

[42] Peracchia M, Gref R, Minamitake Y, Domb A, Lotan N, Langer R. PEG-coated nanospheres from amphiphilic diblock and multiblock copolymers: investigation of their drug encapsulation and release characteristics. *J. Control Release* 1997; 46: 223–231.

[43] Chen Y, McCulloch, RK, Gray BN. Synthesis of albumin-dextran sulfate microspheres possessing favourable loading and release characteristics for the anti-cancer drug doxorubicin. *J. Control Release* 1994; 31: 49-54.

[44] Chen Y, Mohanraj VJ, Parkin JE. Chitosan-dextran sulfate nanoparticles for delivery of an antiangiogenesis peptide. *Letters in Peptide Science* 2003; 10: 621-627.

[45] Magenheim B, Levy MY, Benita S. A new in vitro technique for the evaluation of drug release profilefrom colloidal carriers - ultrafiltration technique at low pressure. *Int. J. Pharm.* 1993; 94: 115-123.

[46] Fresta M, Puglisi G, Giammona G, Cavallaro G, Micali N, Furneri PM. Pefloxacin mesilate- and ofloxacinloaded polyethylcyanoacrylate nanoparticles; characterization of the colloidal drug carrier formulation. *J. Pharm. Sci.* 1995; 84: 895-902.

[47] Verdun C, Brasseur F, Vranckx H, Couvreur P, Roland M. Tissue distribution of doxorubicin associated with polyhexylcyanoacrylate nanoparticles. *Cancer Chemother. Pharmacol* 1990; 26: 13-18.

[48] Couvreur P, Kante B, Lenaerts V, Scailteur V, Roland M, Speiser P. Tissue distribution of antitumor drugs associated with polyalkylcyanoacrylate nanoparticles. *J. Pharm. Sci.* 1980; 69: 199-202.

[49] Bibby DC, Talmadge JE, Dalal MK, Kurz SG, Chytil KM, Barry SE, Shand DG, Steiert M. Pharmacokinetics and biodistribution of RGD-targeted doxorubicinloaded nanoparticles in tumor-bearing mice. i. 2005; 293: 281-290.

[50] Chiannilkulchai N, Ammoury N, Caillou B, Devissaguet JP, Couvreur P. Hepatic tissue distribution of doxorubicin-loaded nanoparticles after i.v. administration in reticulosarcoma M 5076 metastasis-bearing mice. *Cancer Chemother Pharmacol* 1990; 26: 122-6.

[51] Moghimi SM, Hunter AC, Murray JC. Long-circulating and target-specific nanoparticles: theory to practice. *Pharmacol Rev* 2001; 53: 283-318.

[52] Storm G, Belliot S, Daemen T, Lasic D. Surface modification of nanoparticles to oppose uptake by the mononuclear phagocyte system. *Adv. Drug Deliv. Rev.* 1995; 17: 31-48.

[53] Torchilin V, Trubetskoy V. Which polymer can make nanoparticulate drug carriers long circulating? *Adv. Drug Deliv. Rev.* 1995; 16: 141-155.

[54] Jeon SI, Lee JH, Andrade JD, De Gennes PG. Protein-- surface interactions in the presence of polyethylene oxide : I. Simplified theory. *J. Colloid Interface Sci.* 1991; 142: 149-158.

[55] Jeon SI, Andrade JD. Protein--surface interactions in the presence of polyethylene oxide: II. Effect of protein size. *J. Colloid Interface Sci.* 1991; 142: 159-166.

[56] Stella B, Arpicco S, Peracchia M, Desmaele D, Hoebeke J, Renoir M, d'Angelo J, Cattel L, Couvreur P. Design of folic acid-conjugated nanoparticles for drug targeting. *J. Pharm. Sci.* 2000; 89: 1452-1464.

[57] Oyewumi MO, Yokel RA, Jay M, Coakley T, Mumper RJ. Comparison of cell uptake, biodistribution and tumor retention of folate-coated and PEG-coated gadolinium nanoparticles in tumor-bearing mice. *J. Control Release* 2004; 95: 613-626.

[58] Krishna R, Mayer L. Multidrug resistance (MDR) in cancer-mechanisms, reversal using modulators of MDR and the role of MDR modulators in influencing the pharmacokinetics of anticancer drugs. *Eur. J. Cancer Sci* 2000; 11: 265-283.

[59] Larsen AK, Escargueil AE, Skladanowski A. Resistance mechanisms associated with altered intracellular distribution of anticancer agents. *Pharmacol Ther* 2000; 85: 217-29.

[60] Bennis S, Chapey C, Couvreur P, Robert J. Enhanced cytotoxicity of doxorubicin encapsulated in polyisohexylcyanoacrylate nanospheres against multidrug-resistant tumour cells in culture. *Eur. J. Cancer* 1994; 30A: 89-93.

[61] Damge C, Michel C, Aprahamian M, Couvreur P, Devissaguet JP. Nanocapsules as carriers for oral peptide delivery. *J. Control. Release* 1990; 13: 233-239.

[62] Brandtzaeg P, Berstad A, Farstad I, Haraldsen G, Helgeland L, Jahnsen F, Johansen F, Natvig I, Nilsen E, Rugtveit J. Mucosal immunity – a major adaptive defense mechanism. *Behring Inst. Mitt* 1997; 98: 1-23.

[63] Lee V, Yamamoto A. Penetration and enzymatic barriers to peptide and protein absorption. *Adv. Drug Deliv. Rev.* 1990; 4: 171-207.

[64] Haltner E, Easson J, Lehr C. Lectins and bacterial invasion factors for controlling endo- and transcytosis of bioadhesive drug carrier systems. *Eur. J. Pharm. Biopharm* 1997; 44: 3-13.

[65] Hussain N, Jani PU, Florence AT. Enhanced oral uptake of tomato lectin-conjugated nanoparticles in the rat. *Pharm Res* 1997; 14: 613-8.

[66] Russell-Jones GJ, Arthur L, Walker H. Vitamin B12- mediated transport of nanoparticles across Caco-2 cells. *Int. J. Pharm.* 1999; 179: 247-255.

[67] Russell-Jones GJ. The potential use of receptormediated endocytosis for oral drug delivery. *Adv. Drug Deliv. Rev.* 2001; 46: 59-73.

[68] Schipper N, Olsson S, Hoogstrate J, de Boer A, Varum K, Artursson P. Chitosans as absorption enhancers for poorly absorbable drugs. 3: Influence of mucus on absorption enhancement. *Eur. J. Pharm Sci* 1999; 8: 335-43.

[69] Lehr C, Bowstra J, Tukker J, Junginer H. Intestinal transit of bioadhesive microspheres in an in situ loop in the rat. *J. Control Release* 1990; 13: 51-62.

[70] Bjork E, Isakkson U, Edman P, Artursson P. Starch microspheres induce pulsatile delivery of drugs and peptides across the epithelial barrier by reversible separation of the tight junctions. *J. Drug Target* 1995; 6: 501-507.

[71] Florence AT, Hussain, N. Transcytosis of nanoparticle and dendrimer delivery systems: evolving vistas. *Adv. Drug Deliv Rev.* 2001; 50 Suppl 1: S69-89.

[72] Jani P, Halbert GW, Langridge J, Florence AT. The uptake and translocation of latex nanospheres and microspheres after oral administration to rats. *J. Pharm Pharmacol* 1989; 41: 809-12.

[73] Gurunathan S, Wu C, Freidag B. DNA vaccines: a key for inducing long-term cellular immunity. *Curr. Opin. Immunol*, 2000; 12: 442-447.

[74] Panyam J, Zhou WZ, Prabha S, Sahoo SK, Labhasetwar V. Rapid endo-lysosomal escape of poly(DL-lactide-co-glycolide) nanoparticles: implications for drug and gene delivery. *Faseb J.* 2002; 16: 1217-26.

[75] Hedley M, Curley J, Urban R. Microspheres containing plasmid-encoded antigens elicit cytotoxic T-cell responses. *Nat Med* 1998; 4: 365-368.

[76] Chen Y, Dalwadi G, Benson H. Drug delivery across the blood-brain barrier. *Current Drug Delivery* 2004; 1: 361-376.

[77] Kreuter J. Influence of the surface properties on nanoparticle-mediated transport of drugs to the brain. *J. Nanosci Nanotechnol.* 2004; 4: 484-8.

[78] Pardridge WM. Drug and gene targeting to the brain with molecular Trojan horses. *Nat. Rev Drug Discov.* 2002; 1: 131-9.

[79] Ji B, Maeda J, Higuchi M, Inoue K, Akita H, Harashima H, Suhara T. Pharmacokinetics and brain uptake of lactoferrin in rats. *Life Sciences* 2006; 78: 851-855.

[80] Scherrmann JM, Temsamani J, The use of Pep: Trans vectors for the delivery of drugs into the central nervous system. *International Congress Series* 2005; 1277: 199-211.

[81] Gabathuler R, Arthur G, Kennard M, Chen Q, Tsai S, Yang J, Schoorl W, Vitalis TZ, Jefferies WA. Development of a potential protein vector (NeuroTrans) to deliver drugs across the blood brain barrier. *International Congress Series* 2005; 1277: 171-184.

[82] Pardridge WM. Drug and gene targeting to the brain via blood-brain barrier receptor-mediated transport systems. *International Congress Series* 2005; 1277: 49-62.

Chapter 3

NANOPARTICLE APPLICATIONS

INTRODUCTION

Nanotechnology offers unique approaches to probe and control a variety of biological and medical processes that occur at nanometer length scales, and is expected to have a revolutionary impact on biology [1] and medicine [2]. Among the approaches for exploiting nanotechnology in medicine, nanoparticles offer some unique advantages as sensing, image enhancement, and delivery agents [3, 4]. Several varieties of nanoparticles with biomedical relevance are available including, polymeric nanoparticles, metal nanoparticles, liposomes, micelles, quantum dots, dendrimers, and nanoassemblies. To further the application of nanoparticles in disease diagnosis and therapy, it is important that the systems are biocompatible and capable of being functionalized for recognition of specific target sites in the body after systemic administration.

3.1. Biomedical Applications of Gold Nanoparticles Functionalized Using Hetero-Bifunctional Poly (ethylene glycol) Spacer

Nanoparticles based on gold chemistry have recently attracted significant research and practical attention. They are versatile agents with a variety of biomedical applications including use in highly sensitive diagnostic assays [5], thermal ablation and radiotherapy enhancement [6, 7], as well as drug and gene delivery [8]. For instance, antibody-modified gold nanoparticles when used for detection of prostate specific antigen, had an almost a million-fold higher sensitivity than a conventional ELISA-based assay [9]. Near-infrared radiation absorbing gold-silica nanoshells have been prepared and evaluated for thermal ablation of tumors after systemic administration [10].

For biomedical applications, surface functionalization of gold nanoparticles is essential in order to target them to specific disease areas and allow them to selectively interact with cells or biomolecules. Surface conjugation of antibodies and other targeting moieties is usually achieved by adsorption of the ligand to the gold surface. Surface adsorption, however, can denature the proteins or, in some cases, limit the interactions of the ligand with the target on the cell surface due to steric hinderance. Additionally, for systemic applications, long-circulating nanoparticles are desired for passive targeting to tumors and inflammatory sites.

Poly (ethylene glycol) (PEG)-modification of nanoparticles affords a long circulating property by evading macrophage-mediated uptake and removal from the systemic circulation [11]. Surface modification of gold nanoparticles through the PEG spacer would, therefore, allow the modified nanoparticles to remain in the systemic circulation for the prolonged period and provide flexibility to the attached ligand for efficient interaction with its target. Hetero-bifunctional PEG having a thiol group on one terminus and conjugated with coumarin carbamate on the other was efficiently synthesized and purified. Gold nanoparticles were prepared by reduction of gold hypochlorite ($HAuCl_4$) with freshly-prepared trisodium citrate. As the gold nanoparticles are formed, the color of the dispersion changes from pale yellow to bright red.

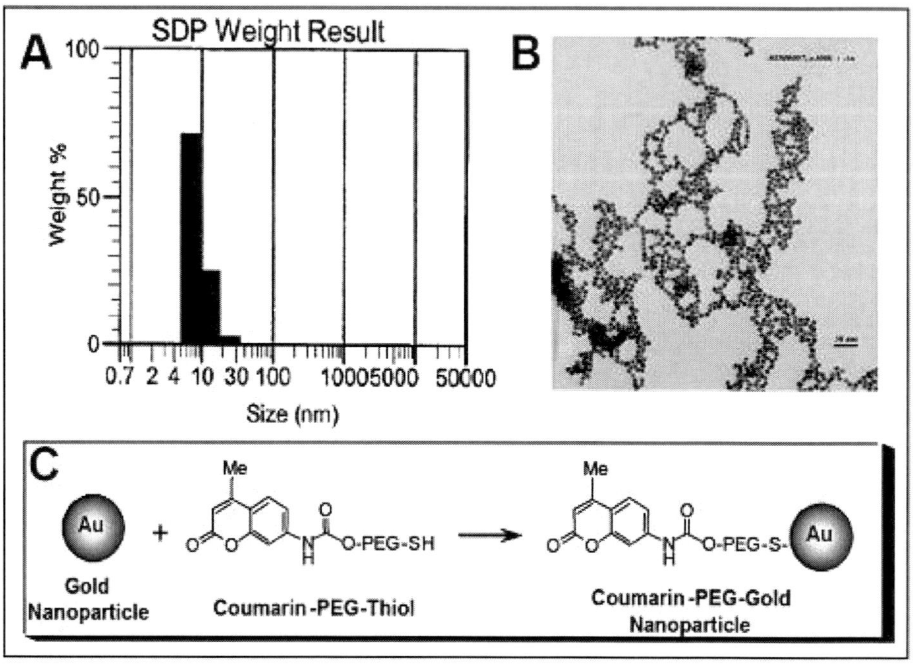

Figure 1. Coulter particle size distribution(A), the transmission electron microscopy image of gold nanoparticles(B), and schematic illustration of the functionalization of gold nanoparticles with coumarin-poly(ethylene glycol)-thiol(C).

Cellular cytotoxicy studies in Figure 2 show that neither gold nanoparticles, gold nanoparticles functionalized with methoxy-PEG-thiol (i.e., gold-PEG-OMe nanoparticles) or coumarin-PEG-thiol functionalized gold nanoparticles were cytotoxic at the concentrations studied. In contrast, polyethyleneimine, a known cytotoxic compound, did reduce cell viability to 40% of the control at the same concentrations. Although gold nanoparticles have been studied as very high efficient fluorescence quenchers [12], the results indicate that attachment of coumarin with the PEG spacer did significantly improve the fluorescence emission signal. This could be due to the fact that the dye was no longer in direct contact with the gold surface. Cellular uptake and distribution analysis using fluorescence confocal microscopy, as shown in Figure 3, show that the functionalized nanoparticles were efficiently internalized by non-specific endocytosis in mammalian tumor cells within one hour of incubation. The nanoparticles were found to have localized mainly in the cytosol and peri-nuclear region of the cells [13-18].

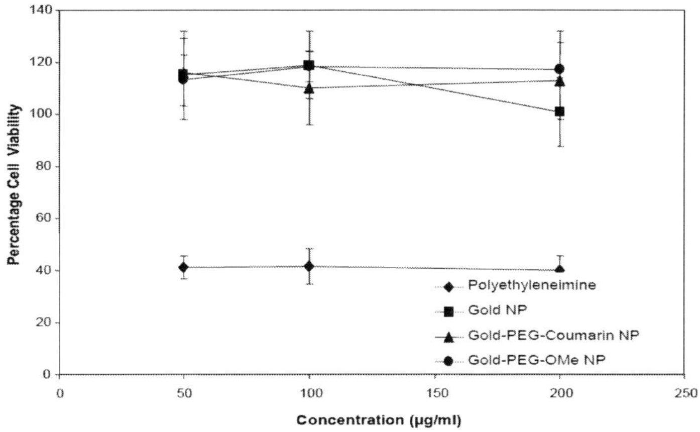

Figure 2. Cytotoxicity analysis of the controls and coumarin-poly (ethylene)-thiol functionalized gold nanoparticles in MDA-MB-231 breast cancer cells.

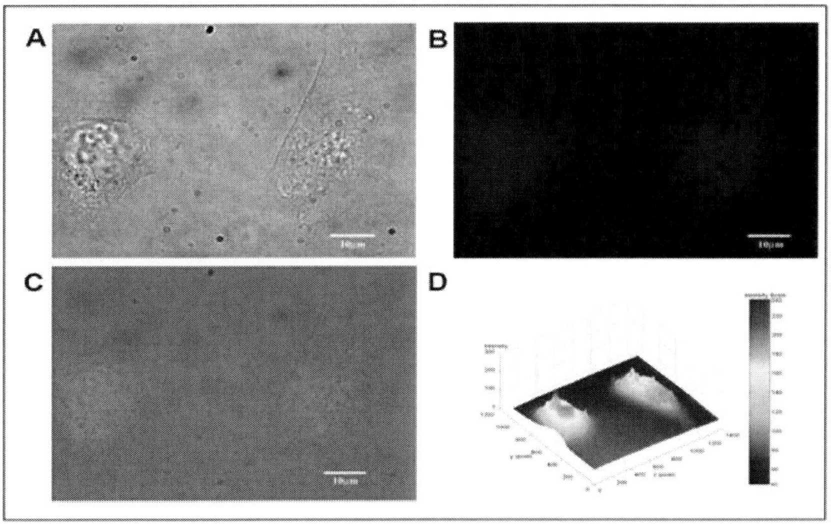

Figure 3. Cellular uptake and distribution of coumarin-poly (ethylene glycol)-thiol functionalized gold nanoparticles in MDA-MB-231 breast cancer cells.

3.2. Nanoparticle Composites for Coating Applications

During the past several years, advances in nanomaterials have allowed them to be formulated into numerous applications. The majority of these applications sought performance improvements that were previously unobtainable. Examples of such applications containing nanomaterials that have been commercialized include scratch/abrasion resistant transparent coatings, sunscreen lotions to provide visible transparent UV protection, polishing slurries to provide pristine surfaces for optics, and environmental catalysts to reduce pollution. The quest for improved scratch/abrasion resistant coatings is a goal for many coating formulators. Thousands of scratch resistant coating applications are present in our everyday lives. Examples of these applications include coatings for wood floors, safety glasses, electronic displays, automotive finishes, and polycarbonate panels. Improving the mar, scratch and/or abrasion in these transparent coating applications is a major challenge, particularly with regard to not affecting the other performance attributes of the coating. Incorporation of inorganic fillers into coatings to improve mechanical properties is well known. Drawbacks associated with this approach can include loss of transparency, reduced coating flexibility, loss of impact resistance, increase in coating viscosity, and the appearance of defects [19]. To overcome these defects a filler material should impart improved scratch resistance without causing the aforementioned detriments. Nanomaterials have the potential to overcome many of these detriments because of their inherent small size and particle morphology. Maintaining transparency in a coating containing inorganic filler particles is a challenge. Four properties dictate the degree of transparency in a composite material: film thickness, filler concentration, filler particle size, and the difference in refractive index between the bulk coating and the filler particle. Mie theory describes the relationship between particle size, concentration, refractive index, and light scattering for spherical particles dispersed in a bulk phase as shown in equation 1.

$$I_s = (Nd^6/\lambda^4)\ \{[(n_p/n_c)^2-1]/[(n_p/n_c)^2+2]\}(I_i)$$

I_s = Intensity of scattered light
N = Number of particles
d= Particle diameter
λ = Wavelength
n_p = Particle refractive index
n_c = Coating refractive index
I_i = Intensity of incident light

(1)

As is evident in equation 1, the magnitude of light scattering in a particle/coating composite is strongly influenced by the particle size. In addition, the greater the difference between the refractive indexes of the particle and that of the bulk coating, the greater the degree of light scattering. Silica particles, colloidal or fumed, and clays are among the most widely studied inorganic fillers for improving the scratch/abrasion resistance of transparent coatings [20-25]. These fillers are attractive from the standpoint that they do not adversely impact the transparency of coatings due to the fact that the refractive indices of these particles (fumed silica = 1.46, bentonite clay = 1.54) closely match those of most resin-based coatings. The drawback to silica-based fillers is that high concentrations of the particles are generally required to show a significant improvement in the scratch/abrasion resistance of a coating, and these high loadings can lead to various other formulation problems associated with viscosity, thixotropy, and film formation. The use of alumina particles in transparent coatings is much more limited even though alumina is significantly harder than silica-based materials, and as a scratch and abrasion-resistant filler, higher performance at lower loadings is often observed. For alumina particle sizes greater than 100-nm, the high refractive index (1.72) results in significant light scattering and a hazy appearance in most clear coatings. Currently, only high refractive index coatings, such as the melamine-formaldehyde resins used in laminate production, can use submicron alumina for scratch resistance and maintain transparency [26].

3.3. Nanoparticle Production

To use alumina as scratch-resistant filler in transparent coatings, the particle size must be sufficiently small to overcome its refractive index mismatch. Nanophase Technologies Corporation (NTC) has developed the Physical Vapor Synthesis (PVS) process that is capable of producing metal oxide nanoparticles via a bottom-up method starting from metallic feed. This process allows production of nonporous crystalline metal oxides having primary particle sizes less than 100 nm at economically viable rates with essentially no byproducts or waste streams [27].

NTC produces two grades of aluminum oxide using the PVS process: NanoTek™ and NanoDur™ alumina. Both grades feature a mixture of crystal phases and are spherical in shape, but the grades differ in terms of primary particle size. NanoTek™ alumina has a surface area of 35 m2/g corresponding to a mean particle size of 48 nm, whereas NanoDur™ alumina has a surface area of 45 m2/g with a mean particle size of 37 nm. A TEM image of NanoTek™ alumina is shown in Figure 4.

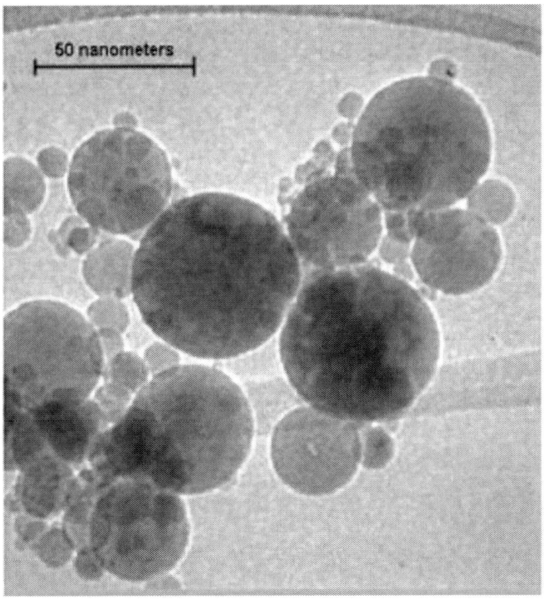

Figure 4. TEM image of NanoTek™ aluminium oxide.

3.3.1. Nanoparticle Dispersion

For nanoparticles to be of use in transparent coatings, it is critical that aggregates present in the powder be dispersible to their primary particle size in the coating formulation to avoid rapid settling and excessive light scattering. In addition, it is critical that the dispersed primary particles avoid re-aggregation during the coating curing process [28]. NTC has developed a proprietary particle dispersion stabilization process that involves specific surface treatments designed to yield nanoparticles that are compatible with a variety of different coating formulations. For example, stable dispersions of metal oxide nanoparticles can be prepared in solvents such as water, alcohols, polar and nonpolar hydrocarbons, plasticizers, and even directly in acrylate monomers with the appropriate surface treatment process. These surface treatments allow solids levels of up to 60 wt% to be dispersed and yet maintain a sufficiently low viscosity for ease of blending. The use of highly concentrated, non-aggregated, nanoparticle dispersions allows incorporation of the nanoparticles into a coating formulation without substantial dilution of the formulation with the dispersion liquid. This feature is particularly important in 100% solids coating formulations where the nanoparticle is dispersed in one of the reactive monomers [29-34].

3.3.2. Nanoparticle Composites

Using concentrated dispersions of aluminum oxide nanoparticles in various solvents and reactive monomers, composites were prepared for a variety of transparent coating formulations including emulsion-based polyurethanes and polyacrylates, solvent-based two-component polyurethanes and melamine-polyols, and 100% solids UV-curable coatings. Importantly, the alumina nanoparticles also featured surface-treatments designed specifically for the formulation chemistry of the coating type in which it was dispersed.

Nanoparticle coating composites were prepared using both the NanoTek™ and NanoDur™ grades of aluminum oxide in a solvent-based transparent melamine-polyol coating to compare the haze of the two different particle sizes. The haze results are shown in Figure 5.

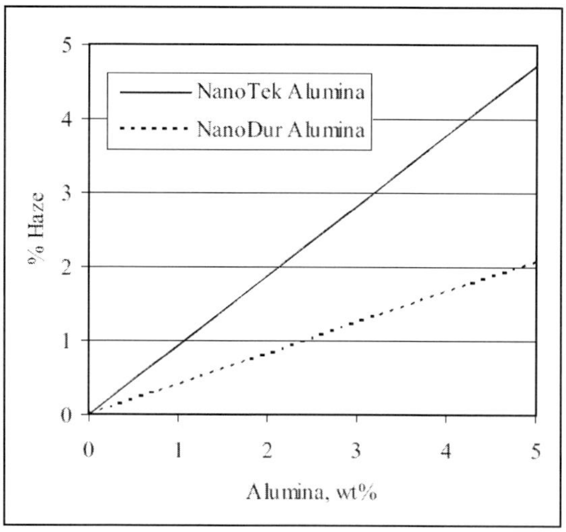

Figure 5. % Haze of alumina/melamine-polyol nanocomposite coatings thermally cured to 1 mil thickness.

Three aspects of the haze data in Figure 2 are noteworthy. (1) As expected, the larger particle size NanoTek™ alumina (48 nm, mean) results in higher haze in the melamine-polyol nano-composite compared to those made with NanoDur™ alumina (37 nm, mean), at a given alumina loading level. (2) If transparency in a clear coating is defined as < 1% haze, it is apparent that the particle size of alumina used in the composite should be no greater than 50 nm in order to maintain transparency. (3) Even with an alumina nanoparticle size grade of < 50 nm, improvement in scratch/abrasion resistance of the coating composite must be attainable with relatively low alumina loadings to stay within the transparency limit [35].

Extension of the alumina particle size haze study to a variety of coating formulations with different refractive indices allowed the derivation of the empirical formula shown in equation 2 to predict the haze resulting from alumina incorporation in any nanocomposite coating [36].

$$\% \text{ Haze} = (8.8\text{x}10^{-6} d^3 - 1.4154\ n_c + 2.02)(\text{wt }\%)(\text{mil})$$

d = Alumina particle diameter, nm
nc = Coating refractive index
wt % = Alumina loading level, wt % of total solids
mil = Cured coating thickness

(2)

3.3.3. Scratch-Resistant Nanocomposites

To evaluate the performance of alumina nanoparticles as scratch-resistant filler in a transparent coating, a nanocomposite was prepared with NanoDur™ alumina dispersed in a UV-curable coating formulation. The alumina nanoparticles were dispersed in 1,6-hexanedioldiacrylate, a reactive monomer, at 30 wt%, and this dispersion was blended with a UV-curable formulation to provide composite coatings with variable levels of alumina particles between 0.2 and 2.0 wt%. These composites were subjected to a scratch test involving 200 double rubs with a 0000 grade steel wool pad and the level of scratching was quantified by measuring the increase in haze due to the scratches. The performance of the alumina-containing composite coatings was compared with the neat coating without alumina particles. The performance of the alumina nanoparticles in Figure 6 is expressed as X times improvement in scratch resistance compared with the neat coating. It is evident the alumina nanoparticles significantly improve the performance of the UV-curable coating, up to a 9-fold improvement, even with very low levels of alumina incorporated in the composite [37-40].

The scratch resistance properties of alumina nanoparticles were also compared with silica particles at equivalent loading levels. The silica that was used was the commercially available Nanocryl silica from Hanse Chemie. The silica particles are surface-treated and dispersed in 1,6-hexanedioldiacrylate and were blended into the same UV-curable coating formulation as the alumina particles. The comparative performance is shown in Figure 7. As is evident in Figure 7, the alumina particles provide much better scratch resistance protection for the UV-curable coating compared to the silica particles at equivalent particle loadings. The much harder alumina particles are superior at preventing steel wool scratching compared to the softer silica particles [41-46].

The scratch resistance performance of alumina nanoparticles incorporated into a variety of other transparent coating compositions was also evaluated. A level of 1 wt% alumina particles was used in all cases, and the alumina particles were introduced in the coating formulations by dispersing the alumina at high concentration into the appropriate solvent used in the coating formulation, then blending into the formulation at a level to yield 1 wt% particles with respect to the total solids in the cured coating.

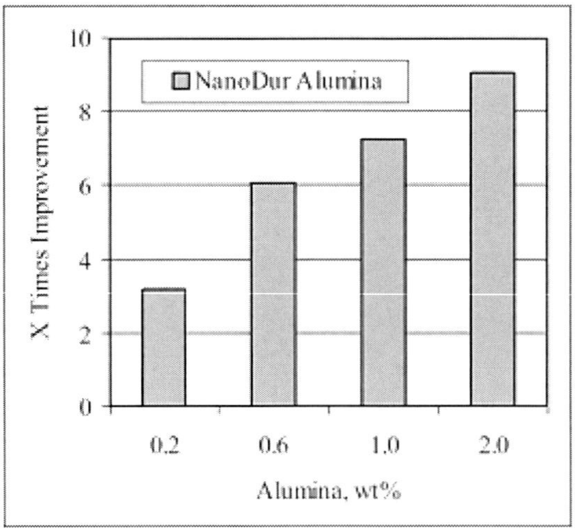

Figure 6. Scratch-resistance performance of NanoDur™ alumina particles in a UV-curable transparent coating.

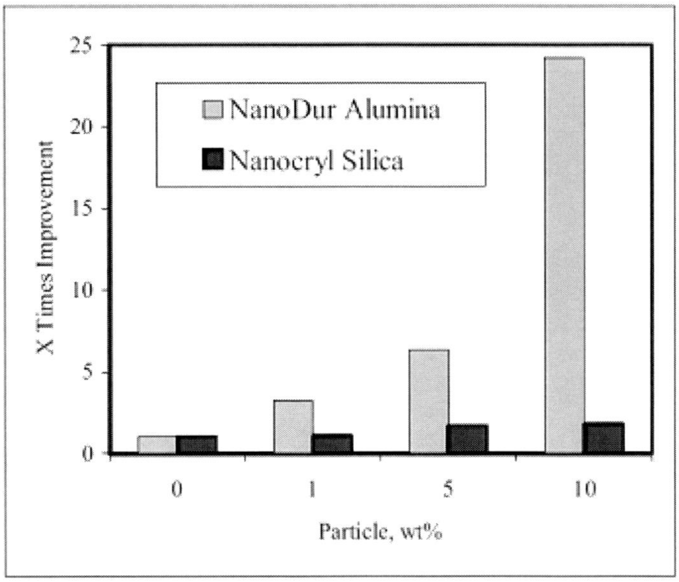

Figure 7. Comparison of the scratch-resistance performance of NanoDur™ alumina particles and Nanocryl™ silica particles in a UV-curable transparent coating.

The scratch resistance performance was measured using the same steel wool scratch test that was applied to the UVcurable coatings.

The results of this study are summarized in Figure 8. Several features can be noted by comparing the relative alumina performance in different transparent coating formulations in Figure 5. First, there is a range of scratch improvement dependent upon the particular formulation within the coating class. For example, the scratch resistance UV-curable coatings can be improved anywhere from 5 to 10 fold with the incorporation of 1 wt% alumina, depending on the type and concentration of reactive acrylate components used in the formulation. Performance ranges for the melamine-based coatings, 2°K polyurethane coatings, and emulsion-based coatings evaluated were also observed. Within a given coating class, those formulations that resulted in harder/stiffer coatings tended to show greater improvement with alumina incorporation than did those formulations that lead to softer/more thermoformable coatings [47].

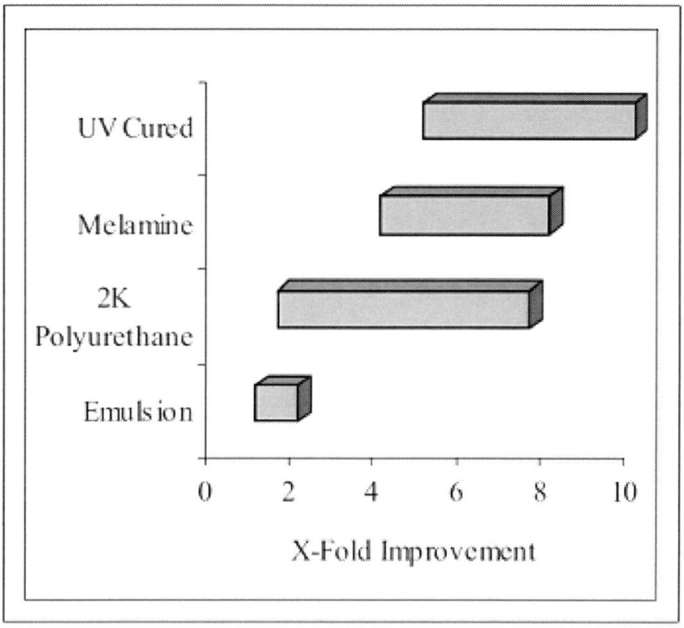

Figure 8. Scratch-resistance performance of NanoDur™ alumina particles at 1 wt% loading in various transparent coating formulations.

In addition, those transparent coating formulations which exhibit cross-linking upon curing, such as UV-curable, 2°K polyurethane, and melamine-based coatings, showed greater improvement in their scratch-resistance upon alumina nanoparticle incorporation compared to transparent coatings which do not cross-link but rather coalesce, such as emulsion-based coatings [48].

3.4. Applications of Magnetic Nanoparticles in Biomedicine

Magnetic nanoparticles offer some attractive possibilities in biomedicine. First, they have controllable sizes ranging from a few nanometers up to tens of nanometers, which places them at dimensions that are smaller than or comparable to those of a cell (10–100 μm), a virus (20–450 nm), a protein (5–50 nm) or a gene (2 nm wide and 10–100 nm long). This means that they can 'get close' to a biological entity of interest. Indeed, they can be coated with biological molecules to make them interact with or bind to a biological entity, thereby providing a controllable means of 'tagging' or addressing it. Second, the nanoparticles are magnetic, which means that they obey Coulomb's law, and can be manipulated by an external magnetic field gradient. This 'action at a distance', combined with the intrinsic penetrability of magnetic fields into human tissue, opens up many applications involving the transport and/or immobilization of magnetic nanoparticles, or of magnetically tagged biological entities[49]. In this way they can be made to deliver a package, such as an anticancer drug, or a cohort of radionuclide atoms, to a targeted region of the body, such as a tumor. Third, the magnetic nanoparticles can be made to resonantly respond to a time-varying magnetic field, with advantageous results related to the transfer of energy from the exiting field to the nanoparticle. For example, the particle can be made to heat up, which leads to their use as hyperthermia agents, delivering toxic amounts of thermal energy to targeted bodies such as tumors; or as chemotherapy and radiotherapy enhancement agents, where a moderate degree of tissue warming results in more effective malignant cell destruction. These, and many other potential applications, are made available in biomedicine as a result of the special physical properties of magnetic nanoparticles [50].

After reviewing some of the relevant basic concepts of magnetism, including the classification of different magnetic materials and how a magnetic field can exert a force at a distance, we will consider four particular applications: magnetic separation, drug delivery, hyperthermia treatments and magnetic resonance imaging (MRI) contrast enhancement. We will conclude with an outlook on future prospects in this area, especially with regard to the challenges faced in bringing current laboratory-tested technologies into mainstream use [50-55].

3.4.1. Basic Concepts

M–H curves

Figure 9 shows a schematic diagram of a blood vessel into which some magnetic nanoparticles have been injected. The magnetic properties of both the injected particles and the ambient biomolecules in the blood stream are illustrated by their different magnetic field response curves. To understand these curves better, we need to be aware of some of the fundamental concepts of magnetism, which will be recalled briefly here. Further details can be found in one of the many excellent textbooks on magnetism [56]. If a magnetic material is placed in a magnetic field of strength ***H***, the individual atomic moments in the material contribute to its overall response, the magnetic induction:

$$\boldsymbol{B} = \mu_0(\boldsymbol{H} + \boldsymbol{M}), \tag{1}$$

Where $\mu 0$ is the permeability of free space, and the magnetization $\boldsymbol{M} = \boldsymbol{m}/V$ is the magnetic moment per unit volume, where ***m*** is the magnetic moment on a volume V of the material. All materials are magnetic to some extent, with their response depending on their atomic structure and temperature. They may be conveniently classified in terms of their volumetric magnetic susceptibility, χ, where describes the magnetization induced in a material by ***H***.

$$\boldsymbol{M} = \chi \boldsymbol{H}, \tag{2}$$

In SI units, χ is dimensionless and both ***M*** and ***H*** are expressed in Am−1. Most materials display little magnetism, and even then only in the presence of an applied field; these are classified either as paramagnets, for which χ falls in the range 10^{-6}–10^{-1}, or diamagnets, with χ in the range -10^{-6} to -10^{-3}. However, some materials exhibit ordered magnetic states and are magnetic even without a field applied; these are classified as ferromagnets, ferrimagnets and antiferromagnets, where the prefix refers to the nature of the coupling interaction between the electrons within the material [57]. This coupling can give rise to large spontaneous magnetizations; in ferromagnets ***M*** is typically 104 times larger than would appear otherwise.

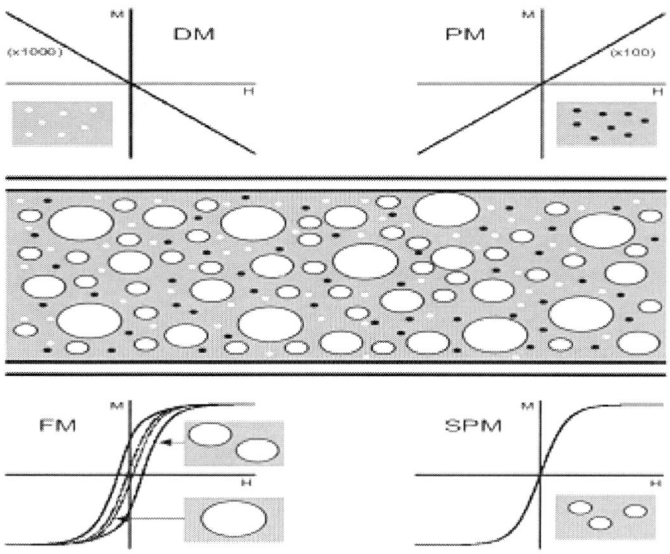

Figure 9. Magnetic responses associated with different classes of magnetic material.

The susceptibility in ordered materials depends not just on temperature, but also on ***H***, which gives rise to the characteristic sigmoidal shape of the *M–H* curve, with ***M*** approaching a saturation value at large values of ***H***. Furthermore, in ferromagnetic and ferrimagnetic materials one often sees hysteresis, which is a irreversibility in the magnetization process that is related to the pinning of magnetic domain walls at impurities or grain boundaries within the material, as well as to intrinsic effects such as the magnetic anisotropy of the crystalline lattice. This gives rise to open *M–H* curves, called hysteresis loops [58-65]. The shape of these loops are determined in part by particle size: in large particles (of the order micron size or more) there is a multi-domain ground state which leads to a narrow hysteresis loop since it takes relatively little field energy to make the domain walls move; while in smaller particles there is a single domain ground state which leads to a broad hysteresis loop. At even smaller sizes (of the order of tens of nanometres or less) one can see superparamagnetism, where the magnetic moment of the particle as a whole is free to fluctuate in response to thermal energy, while the individual atomic moments maintain their ordered state relative to each other. This leads to the anhysteretic, but still sigmoidal, *M–H* curve shown in Figure 10.

The underlying physics of superparamagnetism is founded on an activation law for the relaxation time τ of the net magnetization of the particle:

$$\tau = \tau_0 \exp\left(\frac{\Delta E}{k_\mathrm{B} T}\right), \tag{3}$$

where ΔE is the energy barrier to moment reversal, and $k_\mathrm{B}T$ is the thermal energy. For non-interacting particles the pre-exponential factor τ_0 is of the order 10^{-10}–10^{-12} s and only weakly dependent on temperature. The energy barrier has several origins, including both intrinsic and extrinsic effects such as the magnetocrystalline and shape anisotropies, respectively; but in the simplest of cases it has a uniaxial form and is given by $\Delta E = KV$, where K is the anisotropy energy density and V is the particle volume. This direct proportionality between ΔE and V is the reason that superparamagnetism—the thermally activated flipping of the net moment direction—is important for small particles, since for them ΔE is comparable to k_BT at, say, room temperature. However, it is important to recognize that observations of superparamagnetism are implicitly dependent not just on temperature, but also on the measurement time τ m of the experimental technique being used (see Figure 10). If τ << τ m, the flipping is fast relative to the experimental time window and the particles appear to be paramagnetic (PM); while if τ << τ m, the flipping is slow and quasi-static properties are observed—the so-called 'blocked' state of the system [66]. A 'blocking temperature' *T*B is defined as the mid-point between these two states, where τ = τ m. In typical experiments τ m can range from the slow to medium timescales of 102 s for DC magnetization and 10^{-1}–10^{-5} s for AC susceptibility, through to the fast timescales of 10^{-7}–10^{-9} s for 57Fe M¨ossbauer spectroscopy.

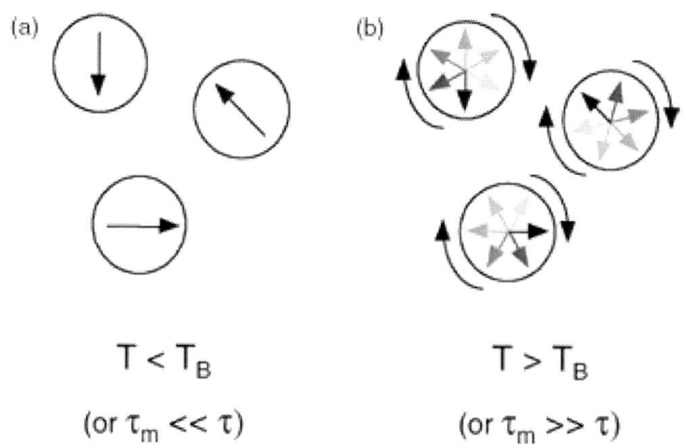

Figure 10. Illustration of the concept of superparamagnetism, where the circles depict three magnetic nanoparticles and the arrows represent the net magnetization direction in those particles.

All of the different magnetic responses discussed above are illustrated in Figure 9 for the case of ferromagnetic or ferrimagnetic nanoparticles injected into a blood vessel. Depending on the particle size, the injected material exhibits either a multi-domain, single-domain or superparamagnetic (SPM) *M–H* curve. The magnetic response of the blood vessel itself includes both a PM response—for example, from the iron-containing hemoglobin molecules, and a diamagnetic (DM) response—for example, from those intra-vessel proteins that comprise only carbon, hydrogen, nitrogen and oxygen atoms [67-70].

It should be noted that the magnetic signal from the injected particles, whatever their size, far exceeds that from the blood vessel itself. This heightened selectivity is one of the advantageous features of biomedical applications of magnetic nanoparticles. Returning to the hysteresis which gives rise to the open *M–H* curves seen for ferromagnets and antiferromagnets, it is clear that energy is needed to overcome the barrier to domain wall motion imposed by the intrinsic anisotropy and microstructural impurities and grain boundaries in the material. This energy is delivered by the applied field, and can be characterized by the area enclosed by the hysteresis loop. This leads to the concept that if one applies a time-varying magnetic field to a ferromagnetic or ferrimagnetic material, one can establish a situation in which there is a constant flow of energy into that material, which will perforce be transferred into thermal energy. This is the physical basis of hyperthermia treatments. Note that a similar argument regarding energy transfer can be made for SPM materials, where the energy is needed to coherently align the particle moments to achieve the saturated state.

Forces on Magnetic Nanoparticles

To understand how a magnetic field may be used to manipulate magnetic nanoparticles, we need to recall some elements of vector field theory. This is not always intuitive, and the reader is directed to recent reviews for further details. It is also important to recognize that a magnetic field gradient is required to exert a force at a distance; a uniform field gives rise to a torque, but no translational action. We start from the definition of the magnetic force acting on a point-like magnetic dipole $\boldsymbol{m}$:

$$\boldsymbol{F}_{\mathrm{m}} = (\boldsymbol{m} \cdot \nabla)\boldsymbol{B}, \tag{4}$$

which can be geometrically interpreted as differentiation with respect to the direction of $\boldsymbol{m}$. For example, if $\boldsymbol{m} = (0, 0, mz)$ then $\boldsymbol{m} \cdot \nabla = mz\ (\partial/\partial z)$ and a force will be experienced on the dipole provided there is a field gradient in $\boldsymbol{B}$ in the *z*-direction. In the case of a magnetic nanoparticle suspended in a weakly DM medium such as water, the total moment on the particle can be written $\boldsymbol{m} = V\mathrm{m}\boldsymbol{M}$, where Vm is the volume of the particle and $\boldsymbol{M}$ is its volumetric magnetization, which in turn is given by $\boldsymbol{M} = \_\chi\boldsymbol{H}$, where $\_\chi = \chi\mathrm{m} - \chi\mathrm{w}$ is the effective susceptibility of the particle relative to the water. For the case of a dilute suspension of nanoparticles in pure water, we can approximate the overall response of the particles plus water system by $\boldsymbol{B} = \mu 0\boldsymbol{H}$, so that equation (4) becomes:

$$\boldsymbol{F}_{\mathrm{m}} = \frac{V_{\mathrm{m}}\Delta\chi}{\mu_0}(\boldsymbol{B} \cdot \nabla)\boldsymbol{B}. \tag{5}$$

Furthermore, provided there are no time-varying electric fields or currents in the medium, we can apply the Maxwell equation $\nabla \times \boldsymbol{B} = 0$ to the following mathematical identity:

$$\nabla(\boldsymbol{B} \cdot \boldsymbol{B}) = 2\boldsymbol{B} \times (\nabla \times \boldsymbol{B}) + 2(\boldsymbol{B} \cdot \nabla)\boldsymbol{B} = 2(\boldsymbol{B} \cdot \nabla)\boldsymbol{B}, \tag{6}$$

to obtain a more intuitive form of equation (5):

$$\boldsymbol{F}_{\mathrm{m}} = V_{\mathrm{m}} \Delta\chi \nabla \left(\frac{B^2}{2\mu_0}\right) \quad \text{or}$$

$$\boldsymbol{F}_{\mathrm{m}} = V_{\mathrm{m}} \Delta\chi \nabla \left(\tfrac{1}{2}\boldsymbol{B} \cdot \boldsymbol{H}\right), \tag{7}$$

in which the magnetic force is related to the differential of the magnetostatic field energy density, 1 2$\boldsymbol{B}$ • $\boldsymbol{H}$. Thus, if $\Delta\chi > 0$, the magnetic force acts in the direction of steepest ascent of the energy density scalar field. This explains why, for example, when iron filings are brought near the pole of a permanent bar magnet, they are attracted towards that pole. It is also the basis for the biomedical applications of magnetic separation and drug delivery.

3.4.2. Magnetic Separation

Cell Labeling and Magnetic Separation

In biomedicine it is often advantageous to separate out specific biological entities from their native environment in order that concentrated samples may be prepared for subsequent analysis or other use. Magnetic separation using biocompatible nanoparticles is one way to achieve this. It is a two-step process, involving (i) the tagging or labeling of the desired biological entity with magnetic material, and (ii) the separating out of these tagged entities via a fluid-based magnetic separation device [71]. Tagging is made possible through chemical modification of the surface of the magnetic nanoparticles, usually by coating with biocompatible molecules such as dextran, polyvinyl alcohol (PVA) and phosopholipids—all of which have been used on iron oxide nanoparticles. As well as providing a link between the particle and the target site on a cell or molecule, coating has the advantage of increasing the colloidal stability of the magnetic fluid. Specific binding sites on the surface of cells are targeted by antibodies or other biological macromolecules such as hormones or folic acid. As antibodies specifically bind to their matching antigen this provides a highly accurate way to label cells. For example, magnetic particles coated with immunospecific agents have been successfully bound to red blood cells, lung cancer cells, bacteria, urological cancer cells and Golgi vesicles [72]. For larger entities such as the cells, both magnetic nanoparticles and larger particles can be used: for example, some applications use magnetic 'microspheres'—micron sized agglomerations of sub-micron sized magnetic particles incorporated in a polymeric binder.

The magnetically labeled material is separated from its native solution by passing the fluid mixture through a region in which there is a magnetic field gradient which can immobilize the tagged material via the magnetic force of equation (8). This force needs to overcome the hydrodynamic drag force acting on the magnetic particle in the flowing solution,

$$F_d = 6\pi \eta R_m \Delta v, \tag{8}$$

where η is the viscosity of the medium surrounding the cell (e.g., water), R_m is the radius of the magnetic particle, and $\Delta v = vm - vw$ is the difference in velocities of the cell and the water. There is also buoyancy force that affects the motion, but this is dependent on the difference between the density of the cell and the water, and for most cases of interest in biology and medicine, can be neglected. Equating the hydrodynamic drag and magnetic forces, and writing $Vm = 43\ \pi R^3_{m}$, gives the velocity of the particle relative to the carrier fluid as:

$$\Delta v = \frac{R_m^2 \Delta\chi}{9\mu_0\eta}\nabla(B^2) \quad \text{or} \quad \Delta v = \frac{\xi}{\mu_0}\nabla(B^2), \tag{9}$$

where ξ is the 'magnetophoretic mobility' of the particle—a parameter that describes how manipulable a magnetic particle is. For example, the magnetophoretic mobility of magnetic microspheres can be much greater than that of nanoparticles, due to their larger size. This can be an advantage, for example, in cell separations, where the experimental timeframe for the separations is correspondingly shorter. On the other hand, smaller magnetic particle sizes can also be advantageous, for example, in reducing the likelihood that the magnetic material will interfere with further tests on the separated cells [73].

Separator Design

Magnetic separator design can be as simple as the application and removal of a permanent magnet to the wall of a test tube to cause aggregation, followed by the removal of the supernatant (Figure 11(*a*)). However, this method can be limited by slow accumulation rates. It is often preferable to increase the separator efficiency by producing regions of high magnetic field gradient to capture the magnetic nanoparticles as they float or flow by in their carrier medium. A typical way to achieve this is to loosely pack a flow column with a magnetizable matrix of wire (e.g., steel wool) or beads and to pump the magnetically tagged fluid through the column while a field is applied (Figure 11(*b*)).

This method is faster than in the first case, although problems can arise due to the settling and adsorption of magnetically tagged material on the matrix. An alternative, rapid throughput method which does not involve any obstructions being placed in the column is the use of specifically designed field gradient systems, such as the quadrupolar arrangement shown in Figure 12 which creates a magnetic gradient radially outwards from the center of the flow column [74].

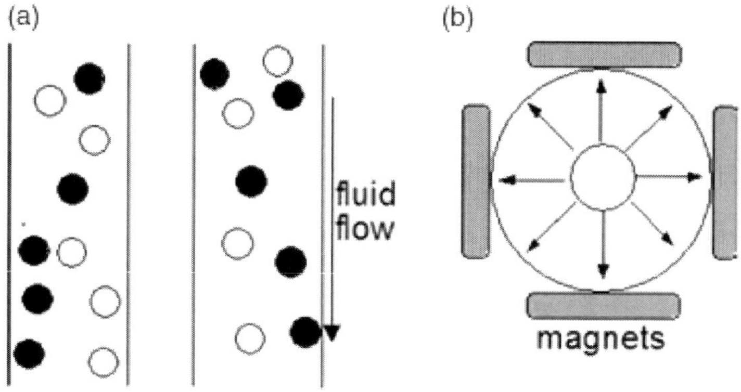

Figure 11. A rapid throughput method of magnetic separation, in which an annular column containing a flowing solution of magnetically tagged (•) and unwanted (○) biomaterials is placed within a set of magnets arranged in quadrature, (a) longitudinal cross-section of annular column, (b) transverse cross-section of four magnets with the resulting magnetic field lines.

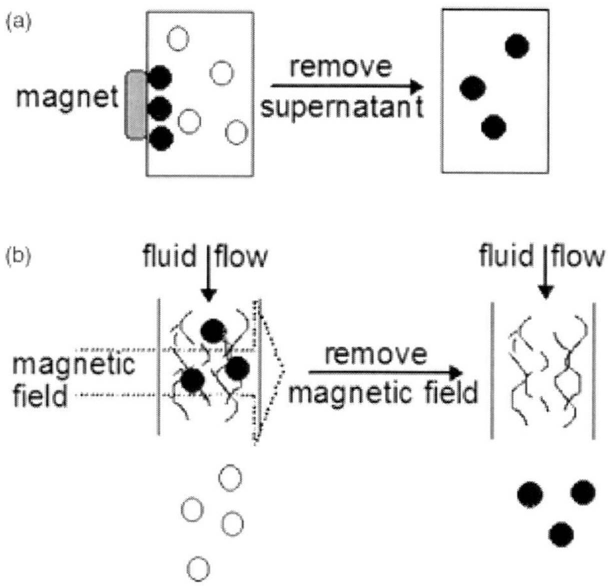

Figure 12. The standard methods of magnetic separation.

In addition to separating out the magnetically tagged material, the spatially varying magnitude of the field gradient can be used to achieve 'fluid flow fractionation'. This is a process in which the fluid is split at the outlet into fractions containing tagged cells or proteins with differing magnetophoretic mobilities. In a variant of this, the fluid is static while an applied magnetic field is moved up the container. The particles move up the container in the resulting field gradient at a velocity dependent on their magnetophoretic mobility. At the top of the container they enter a removable section and are held there by a permanent magnet. The bottom section of the container moves to the next section, a magnetic field with different strength to the first is applied and the process repeats. The result is a fractionation of the sample into aliquots of differing magnetophoretic mobility.

Applications

Magnetic separation has been successfully applied to many aspects of biomedical and biological research. It has proven to be a highly sensitive technique for the selection of rare tumor cells from blood, and is especially well suited to the separation of low numbers of target cells. This has, for example, led to the enhanced detection of malarial parasites in blood samples either by utilizing the magnetic properties of the parasite or through labeling the red blood cells with an immunospecific magnetic fluid. It has been used as a pre-processing technology for polymerase chain reactions, through which the DNA of a sample is amplified and identified. Cell counting techniques have also been developed. One method estimates the location and number of cells tagged by measuring the magnetic moment of the microsphere tags, while another uses a giant magnetoresistive sensor to measure the location of microspheres attached to a surface layered with a bound analyte.

In another application, magnetic separation has been used in combination with optical sensing to perform 'magnetic enzyme linked immunosorbent assays'. These assays use fluorescent enzymes to optically determine the number of cells labeled by the assay enzymes. Typically, the target material must first be bound to a solid matrix. In a modification of this procedure the magnetic microspheres act as the surface for initial immobilization of the target material and magnetic separation is used to increase the concentration of the material. The mobility of the magnetic nanoparticles allows a shorter reaction time and a greater volume of reagent to be used than in standard immunoassays where the antibody is bound to a plate. In a variation of this procedure, magnetic separation has been used to localize labeled cells at known locations for cell detection and counting via optical scanning. The cells are labeled both magnetically and fluorescently and move through a magnetic field gradient towards a plate on which lines of ferromagnetic material have been lithographically etched. The cells align along these lines and the fluorescent tag is used for optical detection of the cells.

3.4.3. Drug Delivery

Motivation and Physical Principles

The major disadvantage of most chemo-therapies is that they are relatively non-specific. The therapeutic drugs are administered intravenously leading to general systemic distribution, resulting in deleterious side-effects as the drug attacks normal, healthy cells in addition to the target tumor cells. For example, the side effects of anti-inflammatory drugs on patients who have chronic arthritis can lead to the discontinuation of their use. However, if such treatments could be localized, e.g., to the site of a joint, then the continued use of these very potent and effective agents could be made possible. Recognition of this led researchers in the late 1970s to propose the use of magnetic carriers to target specific sites (generally cancerous tumors) within the body.

The objectives are two-fold: (i) to reduce the amount of systemic distribution of the cytotoxic drug, thus reducing the associated side-effects; and (ii) to reduce the dosage required by more efficient, localized targeting of the drug. In magnetically targeted therapy, a cytotoxic drug is attached to a biocompatible magnetic nanoparticle carrier. These drug/carrier complexes—usually in the form of a biocompatible ferrofluid—are injected into the patient via the circulatory system. When the particles have entered the bloodstream, external, high-gradient magnetic fields are used to concentrate the complex at a specific target site within the body (Figure 13).

Once the drug/carrier is concentrated at the target, the drug can be released either via enzymatic activity or changes in physiological conditions such as pH, osmolality, or temperature, and be taken up by the tumor cells.

This system, in theory, has major advantages over the normal, non-targeted methods of cytotoxic drug therapy. The physical principles underlying magnetic targeting therapy are similar to those used in magnetic separation, and are derived from the magnetic force exerted on a SPM nanoparticle by a magnetic field gradient. The effectiveness of the therapy is dependent on several physical parameters, including the field strength, gradient and volumetric and magnetic properties of the particles.

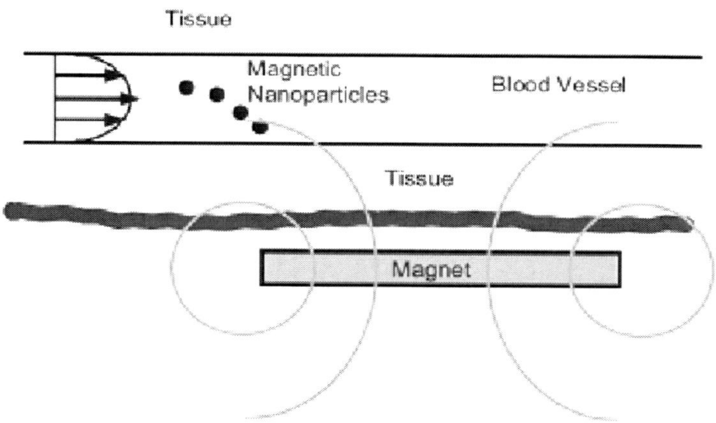

Figure 13. A hydrothetical magnetic drug delivery system shown in cross-section.

As the carriers (ferrofluids) are normally administered intravenously or intra-arterially, hydrodynamic parameters such as blood flow rate, ferrofluid concentration, infusion route and circulation time also will play a major role—as will physiological parameters such as tissue depth to the target site (i.e., distance from the magnetic field source), reversibility and strength of the drug/carrier binding, and tumor volume [75].

In general, larger particles (e.g., magnetic microspheres, ca 1 μm in diameter, comprising agglomerates of SPM particles) are more effective at withstanding flow dynamics within the circulatory system—particularly in larger veins and arteries. In most cases the magnetic field gradient is generated by a strong permanent magnet, such as Nd-Fe-B, fixed outside the body over the target site. Preliminary investigations of the hydrodynamics of drug targeting suggest that for most magnetite-based carriers, flux densities at the target site must be of the order of 0.2 T with field gradients of approximately 8 Tm^{-1} for femoral arteries and greater than 100Tm^{-1} for cartoid arteries. This suggests that targeting is likely to be most effective in regions of slower blood flow, particularly if the target site is closer to the magnet source. More recently, mathematical models have been developed to determine particle trajectories for a variety of field/particle configurations in two dimensions, including consideration of their motion as they approach the vessel wall. This is significant since near the walls, the particle motion is no longer governed by Stoke's law for the drag force, equation (8), and the hydrodynamic parameters are modified. Experimental work is underway to guide the development of these models.

Magnetic Drug Carriers

Since the first magnetic polymer carriers of the 1970s, a variety of magnetic nanoparticle and microparticle carriers have been developed to deliver drugs to specific target sites *in vivo*. The optimization of these carriers continues today. Generally, the magnetic component of the particle is coated by a biocompatible polymer such as PVA or dextran, although recently inorganic coatings such as silica have been developed. The coating acts to shield the magnetic particle from the surrounding environment and can also be functionalized by attaching carboxyl groups, biotin, avidin, carbodi-imide and other molecules. As shown in Figure 14, these molecules then act as attachment points for the coupling of cytotoxic drugs or target antibodies to the carrier complex. The carriers typically have one of two structural configurations: (i) a magnetic particle core (usually magnetite, Fe_3O_4, or maghemite, $\gamma - Fe_2O_3$) coated with a biocompatible polymer or (ii) a porous biocompatible polymer in which magnetic nanoparticles are precipitated inside the pores.

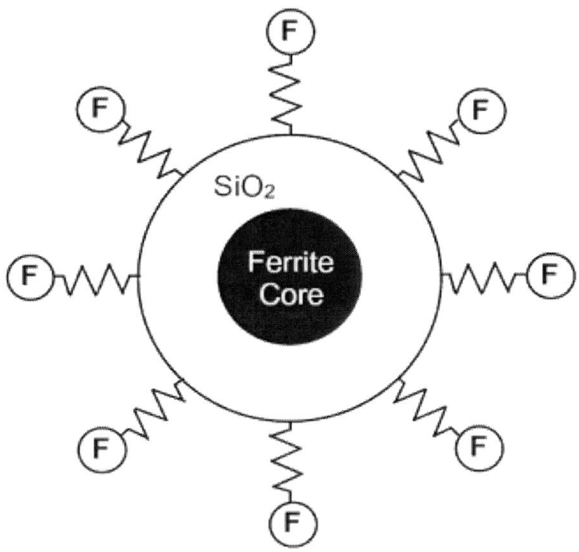

Figure 14. Schematic diagram of a functionalized magnetic nanoparticle showing a core/shell structure with a shell of silica and functionalized groups attached to the shell.

Recent developmental work on carriers has largely focused on new polymeric or inorganic coatings on magnetite/maghemite nanoparticles, although noble metal coatings such as gold are also being considered. Research also continues into alternative magnetic particles, such as iron, cobalt, nickel or yttrium aluminium iron garnet. Cobalt/silica carriers are currently being investigated for their use in eye surgery to repair detached retinas.

Targeting Studies

Magnetic carriers were first used to target cytotoxic drugs (doxorubicin) to sarcoma tumors implanted in rat tails. The initial results were encouraging, showing a total remission of the sarcomas compared to no remission in another group of rats which were administered with ten times the dose but without magnetic targeting. Since that study, success in cytotoxic drug delivery and tumor remission has been reported by several groups using animal models including swine, rabbits and rats. Kubo et al. recently offered a variation on these techniques. They implanted permanent magnets at solid osteosarcoma sites in hamsters and delivered the cytotoxic compounds via magnetic liposomes. They reported a four-fold increase in cytotoxic drug delivery to the osteosarcoma sites when compared with normal intravenous (non-magnetic) delivery, as well as a significant increase in anti-tumor activity and the elimination of weight-loss as a side effect.

This technique has also been employed to target cytotoxic drugs to brain tumors. Brain tumors are particularly difficult targets due to the fact that the drug must cross the blood–brain barrier. Pulfer and Gallo demonstrated that particles as large as 1–2 μm could be concentrated at the site of intracerebral rat glioma-2 (RG-2) tumors. Though the concentration of the particles in the tumor was low, it was significantly higher than non-magnetic particles. In a later study the group demonstrated that 10–20 nm magnetic particles were even more effective at targeting these tumors in rats. Electron microscopic analysis of brain tissue samples revealed the presence of magnetic carriers in the interstitial space in tumors but in normal brain tissue, they were only found in the vasculature. Mykhaylyk and others recently had less success using magnetite–dextran nanoparticles but were able to target rat glial tumors by disrupting the blood–brain barrier immediately prior to particle injection.

Studies of magnetic targeting in humans were rare up to now. A Phase I clinical trial conducted by Lübbe and others demonstrated that the infusion of ferrofluids was well tolerated in most of the 14 patients studied. In addition, the authors reported that the ferrofluid was successfully directed to the advanced sarcomas without associated organ toxicity. More recently, FeRx Inc. was granted fast-track status to proceed with multi-center Phases I and II clinical trials of their magnetic targeting system for hepatocellular carcinomas (a type of liver tumor). This currently appears to be the most promising clinical application.

As promising as these results have been, there are several problems associated with magnetically targeted drug delivery. These limitations include (i) the possibility of embolization of the blood vessels in the target region due to the accumulation of the magnetic carriers, (ii) difficulties in scaling up from animal models due to the larger distances between the target site and the magnet, (iii) once the drug is released, it is no longer attracted to the magnetic field, and (iv) toxic responses to the magnetic carriers. Recent pre-clinical and experimental results indicate, however, that it is still possible to overcome these limitations and use magnetic targeting to improve drug retention and also address safety issues.

Radionuclide and Gene Delivery

One way to overcome the limitation caused by the release of the drug from the carrier is to use a system in which the therapeutic agent remains coupled to the magnetic carrier throughout the duration of the treatment. Based on this idea, the possibility of targeting radionuclides via magnetic carriers has been investigated. The advantage these complexes have over cytotoxic drug/magnetic carrier complexes is that the effectiveness of the radionuclide does not require the tumor cells to actually take up the agent. If the radionuclide is targeted to a region near the tumor site and held there, the radiation will affect the surrounding tumor tissue while it is still attached to the magnetic carrier. This type of system was tested in 1995 using *in vitro* and mouse models.

In both cases, targeting of a magnetic carrier coupled to a β-emitter (Y-90) was effective at concentrating radiation to the desired site. In the mouse tumor model, there was a significant increase in radioactivity at the tumor site compared to using the same complex without a magnetic field: 73 ± 32% vs. 6 ± 4%. Since this study, H¨afeli and others have demonstrated the effectiveness of this technique in both animal and cell culture studies using both yttrium-90 and rhenium-188.

Investigations also have begun with an objective of using magnetic carriers for gene therapy. In this case, a viral vector carrying the therapeutic gene is coated onto the magnetic carrier's surface. By holding the carrier at the target site via external magnetic fields, the virus is in contact with the tissue for a longer period of time, increasing the efficiency of gene transfection and expression. New magnetic carriers are being developed specifically for these applications and is an area showing great promise.

REFERENCES

[1] Jiles D 1991 Introduction to Magnetism and Magnetic Materials (London: Chapman and Hall)

[2] Morrish A H 2001 The Physical Principles of Magnetism (New York: IEEE Press) Néel L 1949 *Ann. Geophys.* 5 99–136.

[3] Brown W F Jr 1963 Thermal fluctuations of a single-domain particle. *Phys. Rev.* 130 1677–86.

[4] Gerber R 1994 Magnetic separation Applied Magnetismed G Asti (Dordrecht: Kluwer) pp 165–220.

[5] Zborowski M 1997 Physics of magnetic cell sorting Scientific and Clinical Applications of Magnetic Carriersed M Zborowski (New York: Plenum) pp 205–31.

[6] Hatch G P and Stelter R E 2001 Magnetic design considerations for devices and particles used for biological high-gradient magnetic separation (HGMS) systems. *J. Magn. Magn. Mater.* 225 262–76.

[7] Molday R S and MacKenzie D 1982 Immunospecific ferromagnetic iron–dextran reagents for the labeling and magnetic separation of cells. *J. Immunol. Methods* 52 353–67.

[8] Sangregorio C, Wiemann J K, O'Connor C J and Rosenzweig Z 1999 A new method for the synthesis of magnetoliposomes. *J. Appl. Phys.* 85 5699–701.

[9] Pardoe H, Chua-anusorn W, St Pierre T G and Dobson J 2001 Structural and magnetic properties of nanoscale iron oxide particles synthesized in the presence of dextran or polyvinyl alcohol. *J. Magn. Magn. Mater.* 225 41–6.

[10] Högemann D, Josephson L, Weissleder R and Basilion J P 2000 Improvement of MRI probes to allow efficient detection of gene expression. *Bioconjugate Chem.* 11 941–6.

[11] Levy L, Sahoo Y, Kim K S, Bergey E J and Prasad P N 2002 Nanochemistry: synthesis and characterization of multifunctional nanoclinics for biological applications. *Chem. Mater.* 14 3715–21.

[12] Zhang Y, Kohler N and Zhang M 2002 Surface modification of superparamagnetic magnetite nanoparticles and their intracellular uptake. *Biomaterials* 23 1553–61.

[13] Tibbe A, de Grooth B, Greve J, Liberti P, Dolan G and Terstappen L 1999 Optical tracking and detection of immunomagnetically selected and aligned cells. *Nature Biotechnol.* 17 1210–13.

[14] Kularatne B Y, Lorigan P, Browne S, Suvarna S K, Smith M O and Lawry J 2002 Monitoring tumour cells in the peripheral blood of small cell lung cancer patients. *Cytometry* 50 160–7.

[15] Morisada S, Miyata N and Iwahori K 2002 Immunomagnetic separation of scum-forming bacteria using polyclonal antibody that recognizes mycolic acids. *J. Microbiol. Methods* 51 141–8.

[16] Zigeuner R E, Riesenberg R, Pohla H, Hofstetter A and Oberneder R 2003 Isolation of circulating cancer cells from whole blood by immunomagnetic cell enrichment and unenriched immunocytochemistry in vitro. *J. Urol.* 169 701–5.

[17] Mura C V, Becker M I, Orellana A and Wolff D 2002 Immunopurification of Golgi vesicles by magnetic sorting. *J. Immunol. Methods* 260 263–71.

[18] Ugelstad J, Prestvik W S, Stenstad P, Kilaas L and Kvalheim G 1998 Selective cell separation with monosized magnetizable polymer beads Magnetism in Medicine ed H Nowak (Berlin: Wiley-VCH) pp 471–88.

[19] Hancock J P and Kemshead J T 1993 A rapid and highly selective approach to cell separations using an immunomagnetic colloid. *J. Immunol. Methods* 164 51–60.

[20] Owen C S 1983 Magnetic cell sorting Cell Separation: *Methods and Selected Applications* (New York: Academic)

[21] Rheinl¨ander T, K¨otitz R, Weitschies W and Semmler W 2000 Magnetic fractionation of magnetic fluids. *J. Magn. Magn. Mater.* 219 219–28.

[22] Moore L, Rodeiguez A, Williams P, McCloskey B, Nakamura M, Chalmers J and Zborowski M 2001 Progenitor cell isolation with a high-capacity quadrapole magnetic flow sorter. *J. Magn. Magn. Mater.* 225 277–84.

[23] Rheinl¨ander T, Roessner D, Weitschies W and Semmler W 1999 Comparison of size–selective techniques for the fractionation of magnetic nanospheres. *Presented at Frontiers in Magnetism* (Stockholm, Sweden).

[24] Todd P, Cooper R, Doyle J, Dunn S, Vellinger J and Deuser M 2001 Multistage magnetic particle separator. *J. Magn. Magn. Mater.* 225 294–300.

[25] Liberti P A, Rao C G and TerstappenLWMM2001 Optimization of ferrofluids and protocols for the enrichment of breast tumor cells in blood. *J. Magn. Magn. Mater.* 225 301–7.

[26] Paul F, Melville D, Roath S and Warhurst D 1981 A bench top magnetic separator for malarial parasite concentration IEEE Trans. *Magn. MAG*-17 2822–4.

[27] Seesod N, Nopparat P, Hedrum A, Holder A, Thaithong S, Uhlen M and Lundeberg J 1997 An integrated system using immunomagnetic separation, polymerase chain reaction, and colorimetric detection for diagnosis of Plasmodium Falciparum. *Am. J. Tropical Med. Hygiene* 56, 322–8.

[28] Hofmann W-K, de Vos S, Komor M, Hoelzer D, Wachsman W and Koeffler H P 2002 Characterization of gene expression of CD34+ cells from normal and myelodysplastic bone marrow. *Blood* 100 3553–60.

[29] Delgratta C, Dellapenna S, Battista P, Didonato L, Vitullo P, Romani G and Diluzio S 1995 Detection and counting of specific cell populations by means of magnetic markers linked to monoclonal antibodies. *Phys. Med. Biol.* 40 671–81.

[30] Edelstein R L, Tamanaha C R, Sheehan P E, Miller M M, Baselt D R, Whitman L J and Colton R J 2000 The BARC biosensor applied to the detection of biological warfare agents. *Biosensors Bioelectron.* 14 805–13.

[31] Kala M, Bajaj K and Sinha S 1997 Magnetic bead enzyme-linked immunosorbent assay (ELISA) detects antigen-specific binding by phage-displayed scFv antibodies that are not detected with conventional. *ELISA Anal. Biochem.* 254 263–6.

[32] Yazdankhah S P, Hellemann A-L, Ronningen K and Olsen E 1998 Rapid and sensitive detection of Staphylococcus species in milk by ELISA based on monodisperse magnetic particles. *Veterinary Microbiol.* 62 17–26.

[33] Widder K J, Senyei A E and Scarpelli D G 1978 Magnetic microspheres: a model system for site specific drug delivery in vivo. *Proc. Soc. Exp. Biol. Med.* 58 141–6.

[34] Senyei A, Widder K and Czerlinski C 1978 Magnetic guidance of drug carrying microspheres. *J. Appl. Phys.* 49 3578–83.

[35] Mosbach K and Schr¨oder U 1979 Preparation and application of magnetic polymers for targeting of drugs. FEBS Lett. 102 112–6.

[36] Alexiou C, Arnold W, Klein R J, Parak F G, Hulin P, Bergemann C, Erhardt W, Wagenpfeil S and Lubbe A S 2000 Locoregional cancer treatment with magnetic drug targeting. *Cancer Res.* 60 6641–8.

[37] L¨ubbe A S, Bergemann C, Brock J and McClure D G 1999 Physiological aspects in magnetic drug-targeting. *J. Magn. Magn. Mater.* 194 149–55.

[38] Voltairas P A, Fotiadis D I and Michalis L K 2002 Hydrodynamics of magnetic drug targeting. *J. Biomech.* 35 813–21.

[39] Cummings L J, Richardson G and Hazelwood L 2000 Drug delivery by magnetic microspheres. *Proc. Mathematics in Medicine Study Group* (Nottingham, UK: University of Notingham) pp 69–83.

[40] Dobson J, Gross B, Exley C, Mickhailova A, Batich C and Pardoe H 2001 Biomedical applications of biogenic and biocompatible magnetic nanoparticles Modern Problems of Cellular and Molecular Biophysics ed S Ayrapetyan and A North (Yerevan: Noyan Tapan/UNESCO) pp 121–30.

[41] Mehta R V, Upadhyay R V, Charles S W and Ramchand C N 1997 Direct binding of protein to magnetic particles. *Biotechnol. Techn.* 11 493–6.

[42] Koneracka M, Kopcansky P, Antalk M, Timko M, Ramchand C N, Lobo D, Mehta R V and Upadhyay R V 1999 Immobilization of proteins and enzymes to fine magnetic particles. *J. Magn. Magn. Mater.* 201 427–30.

[43] Koneracka M, Kopcansky P, Timko M, Ramchand C N, de Sequeira A and Trevan M 2002 Direct binding procedure of proteins and enzymes to fine magnetic particles. *J. Mol. Catalysis B Enzymatic* 18 13–8.

[44] Hans M L and Lowman A M 2002 Biodegradable nanoparticles for drug delivery and targeting. *Curr. Opin. Solid State Mater. Sci.* 6 319–27.

[45] Gr¨uttner C and Teller J 1999 New types of silica-fortified magnetic nanoparticles as tools for molecular biology applications. *J. Magn. Magn. Mater.* 194 8–15.

[46] Rudge S R, Kurtz T L, Vessely C R, Catterall L G and Williamson D L 2000 Preparation, characterization, and performance of magnetic iron–carbon composite microparticles for chemotherapy. Biomaterials 21 1411–20.

[47] Arias J L, Gallardo V, Gomez-Lopera S A, Plaza R C and Delgado A V 2001 Synthesis and characterization of poly(ethyl-2-cyanoacrylate) nanoparticles with a magnetic core *J. Control. Release* 77 309–21.

[48] Gomez-Lopera S A, Plaza R C and Delgado A V 2001 Synthesis and characterization of spherical magnetite/biodegradable polymer composite particles. *J. Colloid Interface* Sci. 240 40–7.
[49] Mornet S, Grasset F, Portier J and D E 2002 Maghemite/silica nanoparticles for biological applications. *Eur. Cells Mater.* 3 110–3.
[50] Deng Y, Wang L, Yang W, Fu S and Elaissari A 2003 Preparation of magnetic polymeric particles via inverse microemulsion polymerization process. *J. Magn. Magn. Mater.* 257 69–78.
[51] Santra S, Tapec R, Theodoropoulou N, Dobson J, Hebard A and Tan W 2001 Synthesis and characterization of silica-coated iron oxide nanoparticles in microemulsion: the effect of nonionic surfactants. *Langmuir* 17 2900–6.
[52] Carpenter E E 2001 Iron nanoparticles as potential magnetic carriers. *J. Magn. Magn. Mater.* 225 17–20.
[53] Stevenson J P, Rutnakornpituk M, Vadala M, Esker A R, Charles S W, Wells S, Dailey J P and Riffle J S 2001 Magnetic cobalt dispersions in poly(dimethylsiloxane) fluids J. Magn. Magn. Mater. 225 47–58.
[54] Connolly J, St Pierre T G, Rutnakornpituk M and Riffle J S 2002 Silica coating of cobal nanoparticles increases their magnetic and chemical stability for biomedical applications. *Eur. Cells Mater.* 3 106–9.
[55] Sun X, Gutierrez A, Yacaman M J, Dong X and Jin S 2000 Investigations on magnetic properties and structure for carbon encapsulated nanoparticles of Fe, Co, Ni. *Mater. Sci. Eng.* 286 157–60.
[56] Grasset F, Mornet S, Demourgues A, Portier J, Bonnet J, Vekris A and Duguet E 2001 Synthesis, magnetic properties, surface modification and cytotoxicity evaluation of Y3Fe5−xAlxO12 garnet submicron particles for biomedical applications. *J. Magn. Magn. Mater.* 234 409–18.
[57] Dailey J P, Phillips J P, Li C and Riffle J S 1999 Synthesis of silicone magnetic fluid for use in eye surgery. *J. Magn. Magn. Mater.* 194 140–8.
[58] Rutnakornpituk M, Baranauskas V, Riffle J S, Connolly J, St Pierre T G and Dailey J P 2002 Polysiloxane fluid dispersions of cobalt nanoparticles in silica spheres for use in ophthalmic applications. *Eur. Cells Mater.* 3 102–5.
[59] Widder K J, Morris R M, Poore G A, Howard D P and Senyei A E 1983 Selective targeting of magnetic albumin microspheres containing low-dose doxorubicin—total remission in Yoshida sarcoma-bearing rats. *Eur. J. Cancer Clin. Oncol.* 19 135–9.
[60] Goodwin S, Peterson C, Hob C and Bittner C 1999 Targeting and retention of magnetic targeted carriers (MTCs) enhancing intra-arterial chemotherapy. *J. Magn. Magn. Mater.* 194 132–9.
[61] Goodwin S C, Bittner C A, Peterson C L and Wong G 2001 Single-dose toxicity study of hepatic intra-arterial infusion of doxorubicin coupled to a novel magnetically targeted drug carrier. *Toxicol. Sci.* 60 177–83.
[62] Pulfer S K and Gallo J M 1999 Enhanced brain tumor selectivity of cationic magnetic polysaccharide microspheres. *J. Drug Targeting* 6 215–28.
[63] Pulfer S K, Ciccotto S L and Gallo J M 1999 Distribution of small magnetic particles in brain tumor-bearing rats. *J. Neuro-Oncol.* 41 99–105.
[64] Zandonella, C. *Nature* 423: 10-12 (2003).
[65] West, J.L. and N.J. Halas. *Curr. Opinion Biotech.*, 11: 215-217 (2000).

[66] Sahoo, S.K. and V. Labhasetwar. *Drug Disc. Today.*, 8: 1112-1120 (2003).
[67] LaVan, D.A., D.M. Lynn, and R. Langer. *Nature Revs.,* 1: 77-84 (2002).
[68] Goodman, C.M., C.D. McCusker, T. Yilmaz, and V.M. Rotello, *Bioconj. Chem.*, 15: 897-900 (2004).
[69] Hirsch, L.R., R.J. Stafford, J.A. Bankson, S.R. Shersen, B. Rivera, R.E. Price, J.D. Hazle, N.J. Halas, and J.L. West. *Proc. Natl. Acad. Sci., USA* 100: 13549-13553 (2003).
[70] Hainfeld, J.F., D.N. Slatkin, and H.M. Smilowitz, *Phys. Med. Biol.*, 49: N309-N315 (2004).
[71] Thomas, M and A. Klibanov. *Proc. Natl. Acad. Sci., USA* 100: 9138-9143 (2003).
[72] J-M Nam, C.S. Thaxton, and C.A. Mirkin, *Science*, 301: 1884 (2003).
[73] Potineni, A., D.M., Lynn, R. Langer, and M.M. Amiji. *J. Controlled Rel.*, 86: 223-234 (2003).
[74] Bouzide, A., and G. Sauve. *Org. Lett.*, 4: 2329-2332 (2002).
[75] Fan, C., S. Wang, J.W. Hong, G.C. Bazan, K.W. Plaxco, and A.J. Heeger. *Proc. Natl.* Acad.

Chapter 4

pH-SENSITIVE DRUG DELIVERY HYDROGELS AND NANOGELS

4.1. INTRODUCTION

In recent years, extensive efforts have been devoted to the use of potential pharmaceutical devices such as novel drug delivery systems (DDS), since they provide a suitable means of site-specific and/or time-controlled delivery of therapeutic agents. Among various kinds of polymeric systems, which have been used as drug containers or release rate controlling barriers, hydrogels have gained considerable interest. Hydrogels are cross-linked, three-dimensional hydrophilic networks that swell but not dissolve when brought into contact with water. Hydrogels can be formulated in a variety of physical forms, including slabs, microparticles, nanoparticles, coatings, and films. As a result, they are commonly used in clinical practice and experimental medicine for a wide range of applications, including biosensors, tissue engineering and regenerative medicine, separation of biomolecules or cells and barrier materials to regulate biological adhesions. Among these applications, hydrogel-based drug delivery devices have become a major area of research interest. Hydrogels can protect drugs from hostile environments, e.g., the presence of enzymes and low pH in the stomach.

Their porosity permits loading of drugs into the gel matrix and subsequent drug release at a pre-designed rate. Hydrogels can also be designed to control drug release by changing the gel structure in response to environmental stimuli, such as pH, temperature, ionic strength, and electric field. Numerous articles provide the synthesis and application of hydrogels in the pharmaceutical field, hydrogels characterization and their use as intelligent carriers for novel pharmaceutical formulations. Swelling behavior of hydrogels, different mechanisms of solvent diffusion into and drug release from hydrogels and mathematical description of these phenomena are also presented.

4.2. HYDROGELS

Hydrogels are water swollen polymer matrices, with a strong tendency to absorb water. Their ability to swell, under physiological conditions, makes them an ideal material for biomedical applications. The hydrophilicity of the network is due to the presence of chemical groups such as hydroxylic, carboxylic, primary, secondary amides, sulphonic and others that can be found on the polymer backbone or as lateral chains. It is also possible to produce hydrogels containing a significant portion of hydrophobic polymers, by blending or copolymerizing hydrophilic and hydrophobic polymers, or by producing interpenetrating networks (IPN) or semi-interpenetrating polymer networks (s-IPN) of hydrophobic and hydrophilic polymers. Hydrogels can be classified as neutral or ionic, based on the nature of the side groups. In neutral hydrogels, the driving force for swelling is the water-polymer thermodynamic mixing contribution to the overall free energy, along with the elastic polymer contribution. Hydrogels can be classified as affine or phantom, based on their mechanical and structural characteristics. Hydrogels are also classified as homopolymeric or copolymeric, based on the starting materials. They can also be classified based on the physical properties of the network as amorphous, semi-crystalline, hydrogen bonded structures, super-molecular structures and hydrocolloidal aggregates. An important class of hydrogels is the stimuli responsive gels. These gels show swelling behavior dependent on their physical environment, which allows for their usage in a number of applications. Hydrogels can be prepared from natural or synthetic polymers (Table 1). Several techniques have been reported for the synthesis of biomedical hydrogels. Chemically cross-linked gels have ionic or covalent bonds between polymer chains. Copolymerization, suspension polymerization, polymerization by irradiation, chemical reaction of complementary groups, and crosslinking using enzymes are some common examples. Chemically cross-linked gels imply the use of a cross-linking agent which is often toxic. This requires that the unreacted cross-linking agent be removed from the gel, which can affect the gel integrity. For these reasons, physically cross-linked gels are now coming into prominence. Several cross-linking methods such as ionic interactions, crystallization, hydrogen bonds protein interaction, and crosslinking by hydrophobic interactions have been investigated exploring the preparation of physically cross-linked gels [1].

Table 1. Natural polymers and synthetic monomers used in hydrogel fabrication

Natural polymers	Synthetic monomers/polymers
Chitosan	Hydroxyethylmethacryate (HEMA)
Alginate	*N*-(2-Hydroxy propyl)methacrylate (HPMA)
Fibrin	*N*-Vinyl-2-pyrrolidone (NVP)
Collagen	*N*-Isopropylacrylamide (NIPAMM)
Gelatin	Vinyl acctate (VAc)
Hyaluronic acid	Acryolic acid (AA)
Dextran	Methacrylic acid (MAA)
	Polyethylene glycol acrylate/methacrylate (IPEGA/PEGMA)
	Polyethylene glycol diacrylate/dimethacrylate (PEGDA/PEGDMA)

4.3. Swelling Behavior of Hydrogels

When a hydrogel in its initial dry state is placed in an aqueous solution, water molecules will penetrate into the polymer network. The entering molecules are going to occupy some space, and as a result some meshes of the network will start expanding, allowing other water molecules to enter within the network. Evidently, swelling is not a continual process. The elasticity of the covalently or physically cross-linked network will counter-balance the infinite stretching of the network to prevent its destruction. Thus, by balancing these two opposite forces, a net force, known as the swelling pressure (Psw) is produced, which is equal to zero at equilibrium obtained with pure water, and that can be expressed as:

$$P_{sw} = k \times C^{n} \tag{1}$$

where k and n are constants. At the equilibrium there is no additional swelling. Swelling can be described as the increase in weight, volume, or length. Thus, the amount of water up-taken by the hydrogel (m_w) is given by:

$$m_w = \frac{m_{HG,w} - m_{HG,d}}{m_{HG,w}} \tag{2}$$

where, $m_{HG,w}$ and $m_{HG,d}$ are the wet and dry hydrogel weights, respectively. In addition, the percent of swelling does not exceed 100. Then, the degree of swelling (Dsw) is given by:

$$D_{sw} = \frac{m_{HG,w}}{m_{HG,d}} \qquad D_{sw} \geq 1 \tag{3}$$

whereas the swelling ratio (Rsw) is:

$$R_{sw} = D_{sw} \frac{\rho_0}{\rho_{sw}} = \frac{V_s}{V_r} \tag{4}$$

where ρ_0 is the density of the hydrogel in the dry state, ρ_{sw} is the density of the swollen gel, whereas V_s and V_r are the volumes of the hydrogel in the wet and dry states, respectively.

4.3.1. Thermodynamics of Gel Swelling

The structure of hydrogels that do not contain ionic moieties can be analyzed by the Flory-Rehner theory. The combination of thermodynamic and elasticity theories states that a cross-linked polymer gel that is immersed in a fluid and allowed to reach equilibrium with its surroundings is subject only to two opposing forces: the thermodynamic force of mixing and the retractive force of the polymer chains. The change of chemical potential due to mixing can be expressed using heat and the entropy of mixing. The change in chemical potential due to the elastic retractive forces of the polymer chains can be determined from the theory of rubber elasticity. In the case of ionic polymers, the swelling equilibrium of the polymeric matrix is more complicated since it heavily depends also on the ionic strength (Figure 1). The free energy change of ionic hydrogel, corresponding to the volume change during swelling, ΔG, is the sum of the contributions due to the mixing of pure solvent with an initially pure, amorphous, unstrained gel network, $\Delta G1$, due to configurational changes of the gel structure, $\Delta G2$, and due to the mixing of ions with solvent, $\Delta G3$. An ionic gel is subjected to a swelling pressure, π, which is expressed as the sum of three components corresponding to each contribution to ΔG.

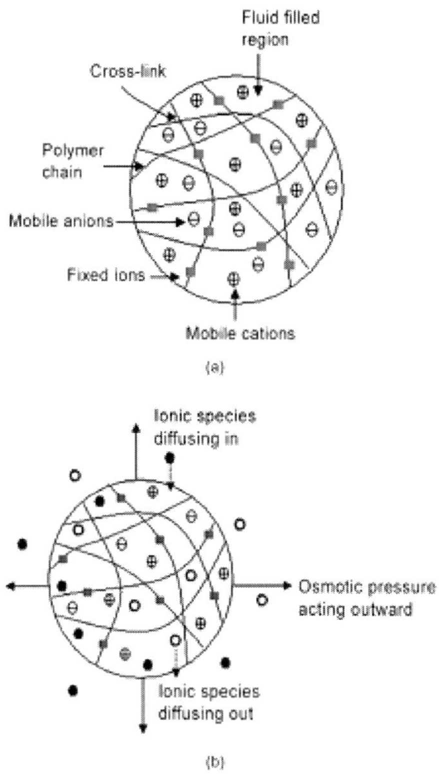

Figure 1. Schematic of an ionic polymeric hydrogel (a), and the swelling phenomena of an ionic hydrogel in a buffered pH solution (b).

A molecular theory of swollen gels by including free volume as a third component in the binary mixture of solvent and polymer was developed. This theory could predict the phase behavior such as near-critical lower consolute boundaries, low-temperature upper consolute boundaries and closed-loop miscibility gaps. The theory was further applied to correlate the experimental data of poly (isopropylacrylamide) and poly (diethylacrylamide) gels. The predicted values by the theory agree closely with those measured experimentally. In 2001, a multi-component, one and two-dimensional model for the transient and equilibrium swelling of polyelectrolyte gels was developed [2]. The model accounts for the effect of network stress, osmotic pressure, and electrical potential on the species diffusive flux. The osmotic pressure and the network stress were derived from the Helmholtz free energy of the system. They try to simulate one and two dimensional swelling in unconstrained and constrained geometries for a salt-solvent polymer system. Through solving the mathematical model numerically, one can obtain the transient and equilibrium fields of electrical potential, concentrations, deformation, and stress. The swelling behavior of cross-linked ionic hydrogels has been studied [3] by using a quasi-chemical thermodynamic model. The calculated volume transition temperature of PNIPAm gel is 0.8°C lower than the experimental value and the predicted solvent volume fraction in the collapsed and swollen gel states are about 2% larger than the corresponding experimental data measured at the transition point. Applying the same energy parameters obtained from regressing PNIPAm gel swelling pressure data, their model has also been capable of correctly representing the major features found in the swelling behavior of linear poly (*N-tert*-butylacrylamide) and poly (*N-tert*-butylacrylamide) gels, after the model parameters that characterize the molecular structure were changed in accordance with each polymer repetitive unit.

Yang et al. developed a lattice model for the binary polymer solutions based on statistical associating theory [4]. A new molecular thermodynamic model for thermo-sensitive hydrogels was proposed by combining Yang et al.'s lattice model and an elastic term. When the energy parameter is expressed as a quadratic of inverse temperature, this model can describe the swelling behavior of neutral thermo-sensitive hydrogels. The experimental swelling curves of two kinds of polyacrylamide based gels were correlated. The swelling equilibrium of IPAAm/NaMA hydrogels in aqueous solutions of the single salts, NaCl and Na2HPO4, was investigated experimentally and described mathematically by developing a new thermodynamic model [5]. This new model combines an extension of Pitzer's model for aqueous electrolyte solutions for the excess Gibbs energy of an aqueous phase with an extension of the phantom network theory. The studies indicated that such an extension is necessary to account for the large volume changes observed when ionic hydrogels are dissolved in water and salt solutions at low salt concentrations. A study assumed that the swelling mechanism results from the equilibrium of different forces and can be triggered by chemical, thermal or electrical stimulations [6]. The chemical field could be described by a convection-diffusion equation, while the electric field was directly obtained by solving the Poisson equation in the gel and solution domain. The mechanical field was formulated by the momentum equation. To model the global macroscopic behavior they used the statistical theory which is capable to describe the overall swelling ratio. By refining the scale, the mesoscopic coupled multi-field theory was applied. Their mesoscopic formulation is capable of capturing the whole gel and solution domain for chemical, thermal and electrical stimulations without prescribing any jumps of concentration or electric potential at the interface.

A new molecular thermodynamic model for describing the swelling behavior of thermo-sensitive hydrogels was developed by Huang et al. They assumed that a neutral hydrogel system contains only polymer network and water. The Gibbs energy inside the hydrogel is defined as: $\Delta G = \Delta Gm + \Delta Ge$, where ΔGm is the mixing term of hydrogel network and water and ΔGe is the elastic term deriving from the network elasticity. Their model consists of two terms: the contribution of the mixing of hydrogel network and water, which is dependent on the local polymer concentration and the interaction between polymer segment and solvent, and the elastic contribution derived from the network elasticity, which is dependent on the cross-linking degree of the gel network. Their important parameters are ε (energy parameter, reflecting the interaction between water and gel network) and V^* (a size parameter which represents the crosslinking degree of the hydrogel). When the energy parameter is expressed as a quadratic of inverse temperature, this model can describe the swelling equilibrium behavior of neutral thermo-sensitive hydrogels very well.

4.3.2. Kinetics of Gel Swelling

It is well known that sorption processes for polymer solvent systems frequently do not conform to the behavior expected from the classical theory of diffusion. The slow reorientation of polymer molecules can lead to a wide variety of anomalous effects for both permeation and sorption experiments. The four basic categories of the sorption phenomena in polymers may be described as follows. Fickian or Case I transport is characterized by the single parameter D, the diffusion coefficient. Penetrants in rubbery polymers and at low activities in glassy polymers typically exhibit Fickian behavior. Molecular relaxation is either faster than diffusion (well above the glass transition temperature, Tg) or so slow that is not observed on the time scale of the experiment (well below Tg). In slab samples, Case I diffusion is characterized by a linear increase of polymer weight gain as a function of the square root of sorption time. It asymptotically approaches a fixed equilibrium value.

Case II transport is characterized by the single parameter V, the velocity of the advancing penetrant front. Diffusion is very rapid compared to relaxation, with relaxation occurring at an observation rate. Here, the rate of mass uptake is directly proportional to time. Diffusion behavior which is intermediate between that of Case I and Case II is regarded as Non-Fickian behavior. Therefore, non-Fickian or anomalous transport is observed when the diffusion and relaxation rates are comparable. Super Case II diffusion is characterized by an acceleration of the absorption rate towards the end of the front penetration process. This change is attributed to the expansion forces exerted by the swollen gel on the glassy core.

The role of Case II diffusion as another limiting case of diffusion may be debated because of the controversy over the existence of Super Case II diffusion, and the fact that Case I diffusion generally precedes the onset of Case II diffusion. In addition, Case I behavior may reoccur in Case II diffusion when the sample dimension is large. Therefore, Case II diffusion may be more appropriately named Case II transport or Case II sorption. Case II sorption is uniquely characterized by a constant movement of the penetration front after a glassy polymer is immersed in a non-dissolving liquid penetrant. Numerous mathematical models have been proposed describing the kinetics of hydrogel swelling. The models may be divided into three categories.

The Fickian diffusion models apply Fick's laws to the distribution of solvent in a gel sample during swelling or collapse. These models predict that the fractional approach to equilibrium increases linearly with the square root of time up to roughly 0.4 and that the swelling curve, the fractional approach to equilibrium vs. square root of time, is not sigmoidal even if the diffusion coefficient is a function of composition. The collective diffusion models, developed by Tanaka et al., treat the swelling of a gel as the expansion of a network driven by a gradient of stress. These models describe small volume changes, but they fail to predict the sigmoidal swelling curves resulting from large volume change. Sigmoidal experimental swelling curves are often taken to indicate non-Fickian behavior. Deviations from the fixed boundary Fickian behavior are usually attributed to some of the following phenomena: (i) variable surface concentration, (ii) a history dependent diffusion coefficient, (iii) stresses between parts of the gel swollen to different extents and (iv) polymer relaxation. The first three have been discussed, while the last has been modeled [7].

However, it has been shown that the sigmoidal swelling behavior can be well described by Fickian diffusion when the movement of the gel surface is taken into account correctly. Although these models predict the swelling curves for large volume changes reasonably well, they are subject to three problems: (i) they do not allow for the movement of the gel boundary, (ii) they require three or more parameters to fit experimental data, and (iii) the diffusion coefficients may show unusual composition dependence, e.g., a maximum at an intermediate composition. Singh and Fan developed a generalized mathematical model for the simultaneous transport of a drug and a solvent in a planar glassy polymer matrix. The swelling behavior of the polymer is characterized by a stress-induced drift velocity term 'v'. The change of volume due to the relaxation phenomenon is assumed to be instantaneous [8]. The model incorporates convective transport of the two species induced by volume expansion and by stress gradient. Vasheghani-Farahani has already indicated that the NIPAm gels exhibit Fickian behavior upon swelling, but their shirking at 35°C is non-Fickian. The rate of swell is also slower than the rate of collapse. He has developed a Fickian mathematical model, which describes the kinetics of swelling and collapse of ionic hydrogels, through the use of a material coordinate and a chemical potential driving force. In this way, explicit relationships between diffusion coefficients in polymer material coordinate and in laboratory coordinate for cylindrical and spherical geometries have been obtained. Swelling of superabsorbent acrylamide/sodium acrylate hydrogels prepared using multifunctional cross-linkers have been studied by Karadag et al.

The research involved the study of the influence of cross-linkers and the relative content of sodium acrylate on swelling, initial swelling rate, swelling rate constant, swelling coefficient and diffusional behavior of water in the hydrogel. Results indicate that acrylamide-sodium acrylate hydrogels showed greater swelling in water compared to acrylamide hydrogels and the water intake of hydrogels followed non-Fickian type diffusion. Swelling behavior and drug release kinetics of ionic and non-ionic IPNs of PNIPAm and calcium alginate were studied by Khorram et al.. They used calcium alginate as a mould to obtain uniform, large size, macroporous spherical beads of NIPPAm hydrogels [9-12]. Their swelling observations indicated that the equilibrium swelling degree of homopolymer gel increased after calcium alginate extraction, while it had no effect on the lower critical solution temperature (LCST).

In addition, equilibrium swelling degree of copolymer composite hydrogels containing Na^+ and Ca^{2+} cations were greater than that of the extracted hydrogel containing only monovalent cations. It was also observed that, swelling kinetics of hydrogels as well as their drug release followed Fickian behavior. Bouquerand et al. developed a model of flavor release from encapsulated flavour particles immersed in water by considering a pseudo-steady state condition in the swollen region of the particles. Their model predicts a very different release with time from the encapsulated flavour if the particle develops a hydrogel at the surface (swelling) compared to gradual erosion. But, the proposed model contains some errors which are corrected [13-20].

Osmotic swelling and de-swelling studies have been performed on gelatin suspended in water-ethanol marginal solvent at room temperature, where the alcohol concentration was changed from 0 to 100% (v/v). The osmotic pressure of polymer-solvent mixing (π m) is much smaller than the osmotic pressure due to network elasticity (π e). In addition, the osmotic pressure arising from ionic contributions, π ion was found to play a significant role in controlling volume phase transitions. For gelatin hydrogels, total swelling pressure of gel, π tot could be related to gelatin volume fraction (ϕ2), relaxed volume of network (V0), and cross-link density (ve) while it is independent of gel pH and swelling time [21-25].

4.4. Biomedical Applications of Hydrogels

Hydrogels have been successfully used in biomedical fields due to their high water content and the consequent biocompatibility. Potential applications of hydrogels in tissue engineering, synthetic extracellular matrix (ECM) and three dimensional scaffolds are well highlighted in a recent work. The proliferation and differentiation of mesenchymal stem cells (MSC) in a three dimensional (3-D) network of nanofibers formed by self-assembly of peptide-amphiphile (PA) molecules was investigated by Hosseinkhani et al. A 3-D network of nanofibers was formed by mixing cell suspensions in media with dilute aqueous solution of PA. In another work, a hybrid scaffold consists of two biopolymers, a hydrogel formed through self-assembly of peptide-amphiphile with cell suspensions in media and a collagen sponge reinforced with poly(glycolic acid) fiber incorporation, were used successfully to enhance bone formation. A novel injectable 3-D scaffolds with encapsulated growth factor was formed by the mixing of PA aqueous solution with basic fibroblast growth factor (bFGF) suspension. Hosseinkhani et al. showed that when aqueous solution of PA was subcutaneously injected together with bFGF suspension into the back of mice, a transparent 3-D hydrogel was formed at the injected site and induced significant angiogenesis around the injected site, in marked contrast to bFGF injection alone or PA injection alone. Materials controlling the activity of enzymes, phospholipids bilayer destabilizing agents, materials controlling reversible cell attachment, nanoreactors with precisely placed reactive groups in three-dimensional space, smart microfluidics with responsive hydrogels, and energy-conversion systems are the promising applications of hydrogels in biomedical and pharmaceutical areas [26-35]. The soft and hydrophilic nature of hydrogels makes them particularly suitable as novel drug delivery systems.

Control of hydrogel swelling properties can be used as a method to trigger drug release. Through proper design, hydrogels can be used in a variety of applications including sustained, targeted, or stealth biomolecule delivery. Hydrogels can be engineered to exhibit bioadhesiveness to facilitate drug targeting, especially through mucus membranes, for non-invasive drug administration.

Hydrogels offer an important "stealth" characteristic in vivo owing to their hydrophilicity which increases the in vivo circulation time of the delivery device by evading the host immune response and decreasing phagocytic activities. Dinarvand's group developed poly (lactide*co*-glycolide)-based hydrogel nanoparticles with modified surface properties to increase the blood circulating half-life of drug carriers [36].

4.5. Intelligent Hydrogels in DDS

The intelligent response of environmentally responsive hydrogels allows for release that is controlled by the conditions of the environment. Temperature responsive and pH-responsive hydrogels have been widely used to create drug delivery systems that exhibit a pulsatile release in response to temperature or pH changes. By incorporating enzymes within environmentally responsive hydrogels, researchers have created drug delivery systems that are responsive to biological analytes. Another area of drug delivery where hydrogels have proven beneficial is in systems where molecular recognition is utilized for enhanced residence times, sustained delivery, and/or targeted drug delivery [37-40].

4.5.1. Temperature Responsive Hydrogels in DDS

Temperature sensitive hydrogels can be classified as negatively thermosensitive, positively thermosensitive, and thermally reversible hydrogels. Negatively thermosensitive hydrogels are those showing lower critical solution temperature (LCST) behavior, while positively thermosensitive gels are known to have an upper critical solution temperature (UCST). The LCST polymers exhibit a hydrophilic-to-hydrophobic transition with increasing temperature, whereas the UCST systems undergo the opposite transition. Thermally reversible hydrogels are those that can experience cyclic phase transitions (sol-gel transition), such as poloxamers, gelatin and other natural polymers [41].

4.5.2. Negatively Thermosensitive Hydrogels

The LCSTs of several typical negatively thermosensitive polymers are listed in Table 2. Negatively thermosensitive hydrogels tend to shrink or collapse as the temperature is increased above the LCST, and swell upon lowering the temperature below the LCST.

The change in the hydration state, which causes the volume phase transition, reflects competing hydrogen bonding properties, where intra- and inter-molecular hydrogen bonding of the polymer molecules are favored compared to a solubilization by water. Thermodynamics can explain this with a balance between entropic effects due to the dissolution process itself and due to the ordered state of water molecules in the vicinity of the polymer. Enthalpic effects are due to the balance between intra- and intermolecular forces and due to solvation, e.g., hydrogen bonding and hydrophobic interaction. The transition is then accompanied by coil-to-globule transition. By controlling the polymer composition and topology, the coil-to-globule transition could be kinetically and thermodynamically controlled. Control of deswelling kinetics was achieved by using graft copolymer structure. The comb-type grafted hydrogels of cross-linked P (NIPAAm) grafted with oligo- NIPAAm and poly (NIPAAm-*g*-PEG) exhibited fast response to temperature changes [42-45]. Grafted short chains of oligo-NIPAAm in the former case contributed to fast dehydration, whereas in the latter case the hydrophilic PEG chains provided a water channel for a fast deswelling mechanism. Copolymerization of NIPAAm with hydrophobic butylmethacrylate decreases the LCST of aqueous copolymer solution and copolymerization with hydrophilic comonomers, such as acrylic acid or hydroxy ethyl methacrylate, results in an increase in LCST.

Table 2. Lower critical solution temperatures (LCST) of several typical thermosensitive polymers

Polymer	LCSTs (°C)
Poly(*N*-isopropylacrylamide) (PNIPAM)	32
Poly(*N,N*-diethylacrylamide) (PDEAM)	25
Poly(*N*-ethylmethacrylamide) (PNEMAM)	58
Poly(methyl vinyl ether) (PMVE)	34
Poly(2-ethoxyethyl vinyl ether) (PEOVE)	20
Ploy(*N*-vinylcaprolactam (PNVCa)	30-35
Poly(*N*-vinylisobutyramide) (PNVIBAM)	39
Poly(*N*-vinyl-*n*-butyramide) (PNVIBAM)	32

Bulmus et al. used the PNIPAAm polymers for on/off control of avidin-biotin binding. Below the transition temperature of 32°C, NIPAAm copolymers are in a fully extended conformation in water because of favorable polymer-water interactions. The PNIPAAm with fully extended conformation interferes with the biotin-binding site on the avidin, whereas above the transition temperature, the polymers are collapsed and cannot interfere with the binding sites [46-48].

Cellulose nitrate and cellulose acetate monolayer membranes containing *n*-heptyl-cyanobiphenyl were developed as thermoresponsive barriers for drug permeation. Methimazole and paracetamol as hydrophilic and hydrophobic drug models were used, respectively. Atyabi's group found that upon changing the temperature of the system around 41.5°C, both cellulose membranes without cyanobiphenyl showed no temperature sensitivity to drug permeation, whereas the results for cyanobiphenyl entrapped membranes exhibited a distinct jump in permeability when the temperature was raised to above the 41.5°C for both drug models. Drug permeation through the membranes as reversible, reproducible and followed zero order kinetics. The pattern of on/off permeation through these membranes was more distinguished for methimazole compared to that of paracetamol, seemingly due to its lower molecular weight [49-51].

4.5.3. Positively Thermosensitive Hydrogels

Certain hydrogels formed by IPNs show positive thermosensitivity, i.e., swelling at high temperature and shrinking at low temperature. IPNs of poly (acrylic acid) and polyacrylamide (PAAm) or P(AAm-*co*-BMA) perform as positively thermosensitive hydrogels. Increasing the BMA content shifted the transition temperature to a higher temperature. The swelling of those hydrogels was reversible, responding to stepwise temperature changes. This resulted in reversible changes in the release rate of a model drug, ketoprofen, from a monolithic device. Another UCST system was fabricated based on a combination of acrylamide (AAm) and acrylic acid (AAc). The IPNs obtained were very stable at 70°C in an aqueous solution. Dissociation temperatures of the hydrogels shifted to higher values with increasing AAm content. These IPNs showed limited swelling ratios between their swelling transition temperatures and lower swelling ratios above these transition temperatures. Transition temperatures shift to higher values with increasing AAm content. Reversible and pulsatile solute release, reflecting the "on" state at higher temperatures and the "off" state at lower temperatures, was recorded. A positively thermosensitive drug-release microcapsule was fabricated. The microcapsule had a core layered with carbazochrome sodium sulphonate (a water-soluble model drug) particles and a thermosensitive coat composed of an ethylcellulose matrix containing nano-sized thermosensitive hydrogels (Figure 2). The hydrogel particles consisted of newly synthesized composite latex with a PNIPAAm shell that could reversibly change the shell thickness in water with response to an environmental temperature change. This microcapsule demonstrated a positively thermosensitive on-off pulsatile drug release around 32°C, suggesting that the shrinkage of PNIPAAm shells most likely created many voids in the coat and thereby imparted the higher water-permeability to the coat [52-54].

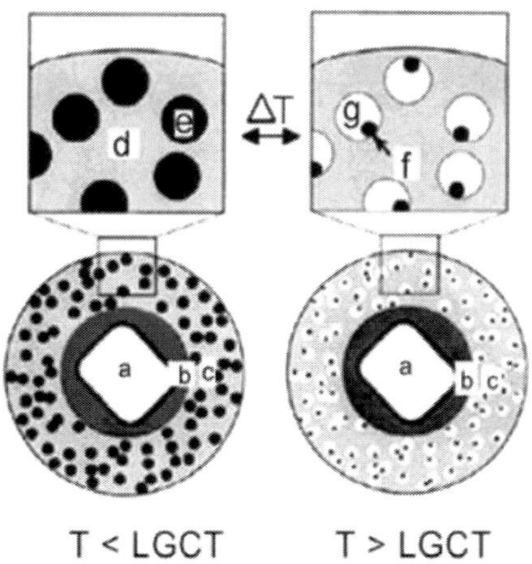

Figure 2. The ideal structure of microcapsules with positively thermosensitive drug release.

A novel family of monodisperse thermosensitive core-shell microspheres has been developed by Xiao et al.. The microspheres were fabricated in a three-step process. First, monodisperse poly (acrylamid- *co*-styrene) seeds were prepared by emulsifier free emulsion polymerization. Then, poly (acrylamide) or poly [acrylamid-*co* (butylate)] shells were fabricated on the microsphere seeds by free radical polymerization.

Finally, the core-shell microspheres with poly (acrylamide)/poly (acrylic acid) based IPN shells were finished by a method of sequential IPN synthesis. The swelling ratio of the microspheres could be tuned by controlling the hydrophobic monomer (BMA) or cross-linker (MBA) dosage in the PAAM shell fabrication and IPN synthesis.

4.5.4. Thermally Reversible Hydrogels

Aqueous solutions of some polymers undergo sol to gel transition in response to a certain stimulus (Figure 3). Among them, thermally reversible hydrogels are of most interest and will be discussed in this section. Polymers with hydrophobic domains can crosslink in aqueous environments via reverse thermal gelation. Hydrophobicity-driven gelation often occurs via the mechanism shown in Figure 4. The hydrophobic segment is coupled to a hydrophilic polymer segment by post-polymerization grafting or by directly synthesizing a block copolymer to create a polymer amphiphile. Such amphiphiles are water soluble at low temperatures. As the temperature is increased, hydrophobic domains aggregate to minimize the hydrophobic surface area contacting the bulk water, reducing the amount of structured water surrounding the hydrophobic domains and maximizing the solvent entropy.

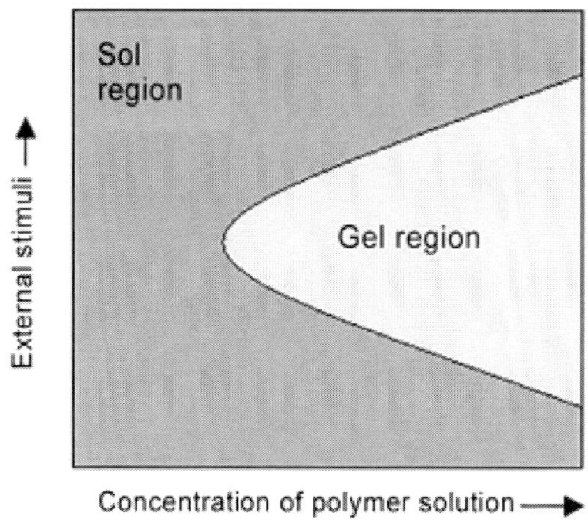

Figure 3. Schematic representation of sol-to-gel transition in stimuli-sensitive polymers.

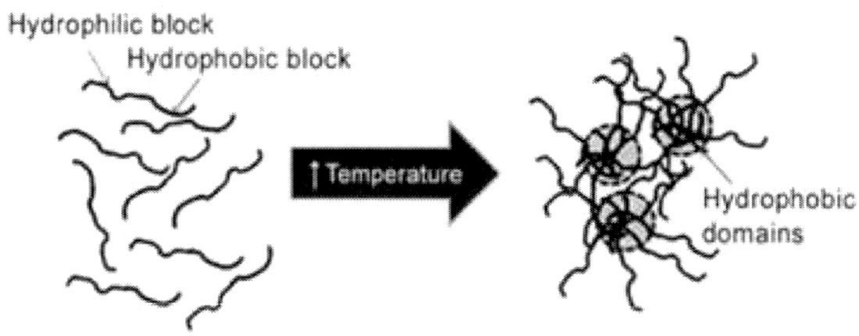

Figure 4. Mechanism of in situ physical gelation driven by hydrophobic interactions.

The chemical structures of some common hydrophobic blocks which can undergo reverse thermal gelation at or near physiological temperature are shown in Figure 5.

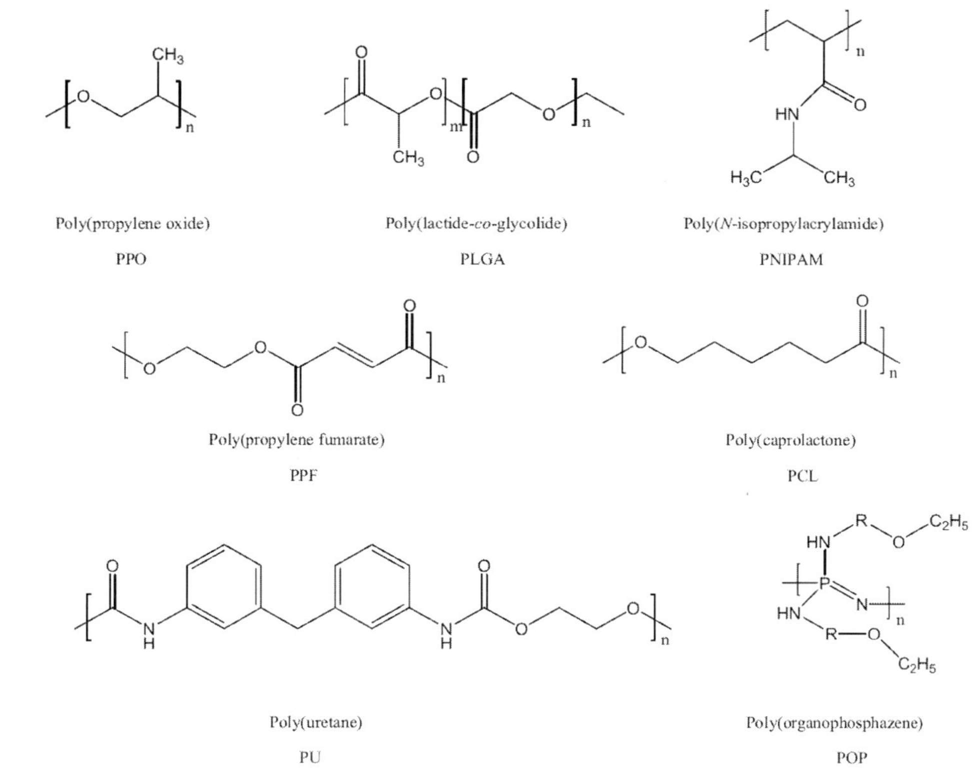

Figure 5. Chemical structures and abbreviations of common thermogelling hydrophobic blocks.

Triblock copolymers of poly (ethylene oxide)-poly (propylene oxide)-poly (ethylene oxide) (PEOPPO-PEO, the Poloxamers/Pluronics) are the most widely used thermally reversible hydrogels. Aqueous solution of poloxamers demonstrates phase transitions from sol to gel at 5-30°C and gel to sol at 35-50°C with the temperature increasing monotonically over the polymer concentration range of 20 to 30 wt%. The transition mechanism from gel-to-sol is related to the shrinkage of PEO corona of the micelles because of temperature effects on PEO solubility and interaction of PEO chains with the PPO hard core. A recent small angle neutron scattering study proposed the transition of micelle structure from spherical to cylindrical, thus releasing micellepacking constraints, as the cause of the high temperature gel-soltransition.

A sol-gel transition behavior was also observed for an altered triblock structure in which poly (butylenes oxide) (PBO) was used in place of PPO in the middle block or with PEO-PBO diblock copolymers. Poloxamers can also be modified by adding an additional polymer block at each chain terminus, forming a multi-block copolymer with improved properties for drug delivery. PL (G)A-based gelators typically exhibit better biodegradability, higher gelation temperatures, and longer periods of sustained drug release compared to poloxamer systems.

Release of hydrophilic compounds from PLGA-PEG-PLGA copolymers was found to be diffusion controlled, while release of hydrophobic compounds showed an initial diffusion-controlled stage followed by a prolonged polymer degradation-controlled stage. Some natural polymers also undergo reverse thermal gelation. Chitosan-glycerol phosphate-water system is an interesting example, which has being investigated for protein delivery, gene delivery and tissue engineering applications. Recently, novel di-block copolymers of chitosan and PEG were synthesized by block copolymerization of monomethoxy-PEG onto chitosan backbone, using potassium persulphonate as a free radical initiator. The hydrogels that are obtained undergo a thermosensitive transition from a free flowing solution at room temperature to a gel around body temperature. Their gelation time varied from 6 to 11 min. Figure 6 illustrates viscosity of a 2% w/v of chitosan-*b*-PEG copolymer in an aqueous solution versus temperature.

A sharp increase in viscosity around 35°C indicates the beginning of the gelation process. It is also found that solutions with high polymer concentrations and low PEG content gel faster than those with low polymer concentrations or high PEG content. Similar trends were observed by Bhattarai et al. for their thermosensitive PEG-grafted chitosan copolymers. Their hydrogel is synthesized by grafting an appropriate amount of PEG onto the chitosan backbone and used for drug release of bovine serum albumin (BSA) as a model protein. Chitosan was first modified with a PEG-aldehyde to yield an imine (Schiff base) that was subsequently converted into PEG-*g*-chitosan through reduction with sodium cyanoborohydride (NaCNBH3). The required time for gelation of their PEG-*g*-chitosan copolymers varied from 10 minutes to 1 hour, depending on polymer concentration. They have found that the required amount of grafted PEG for an injectable thermosensitive copolymer is approximately 36-55 wt%. Below the 36 wt% of grafted PEG, the obtained copolymers were found to be hardly soluble in water.

Grafting PNIPAM linear chains onto natural polymers can also convert those polymers into physically cross-linkable hydrogels. For example, PNIPAM grafted hyaluronic acid formed a gel in vivo which showed a burst release of riboflavin for 12 hours and sustained release thereafter; NIPAM-grafted chitosan has also been used to control the release of 5-fluorouracil [55-57].

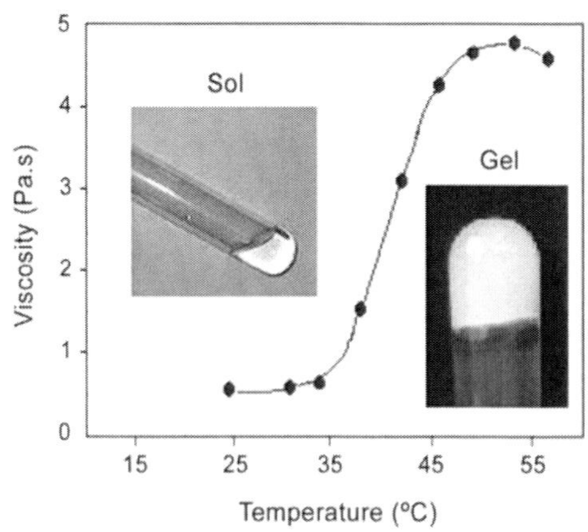

Figure 6. Viscosity of chitosan-b-PEG copolymer versus temperature for 2% w/v polymer solution.

4.5.5. pH Sensitive Hydrogels in DDS

Since the pH change occurs at many specific or pathological body sites, it is one of the important environmental parameters for drug delivery systems. For nonionic hydrogels, the degree of swelling only depends on the chemical compositions of the polymers. However, for ionic hydrogels the swelling depends not only on the chemical composition but also on the pH of the surrounding medium. Therefore, the pHsensitive polymers show dramatic changes on the pH and on the compositions of the external solutions.

The pH-sensitive polymers can be classified as acidic weak polyelectrolytes and basic weak polyelectrolytes according to the method of ionization i.e., donating or accepting protons. Anionic hydrogels deprotonate and swell more when the external pH is higher than pKa of the ionizable groups bonded on polymer chains, while cationic hydrogels protonate and swell more when external pH is lower than the pKb of the ionizable groups. Depending on the ionic monomers used to fabricate the hydrogel, the pH dependent swelling curves exhibit one or more inflection points near the pKa/pKb of the ionizable groups as shown in Figure 7 [58].

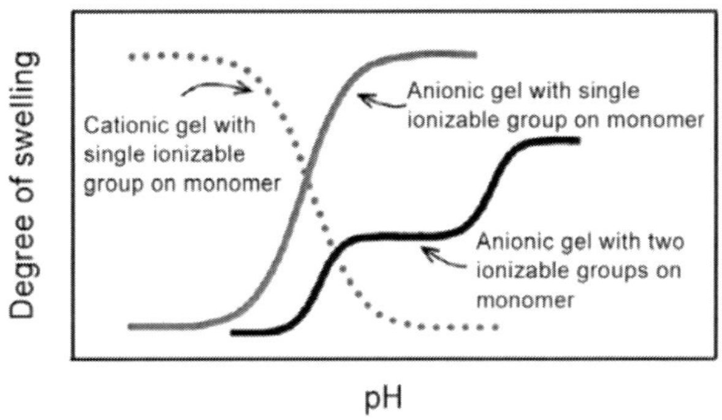

Figure 7. Schematic of relative ionic hydrogel swelling as a function of pH.

Typical acidic pH-sensitive polymers for drug delivery are based on the polymers containing carboxylic groups, such as poly (acrylic acid), poly (methacrylic acid), poly (*L*-glutamic acid) , and polymers containing sulphonamide groups . Typical examples of the basic polyelectrolytes include poly (tertiary amine methacrylate) such as poly (2-(dimethylamino) ethyl methacrylate) and poly (2-(diethylamino) ethyl methacrylate), poly (2-vinylpyridine) and biodegradable poly (β -amino ester).

A pH-sensitive polymeric carrier has been synthesized in which a poly (vinylpyrrolidone-*co*dimethyl maleic anhydride) (PVD) was conjugated to doxorubicin (DOX), that could gradually release free drug in response to changes in pH, from near neutral to slightly acidic pH. The superior anticancer activity of PVD-DOX conjugate is mainly due to the controlled release and enhanced tumor accumulation of the drug. Alginate-*N,O* carboxymethyl chitosan (NOCC) gel beads coated by chitosan were prepared for a colon specific drug delivery system and their swelling behavior was investigated. Their studies showed that swelling degree at pH 7.4 was considerably higher than that at pH 1.2, which indicates the pH sensitivity of these networks. The swelling degree of beads also decreased by chitosan coating and the presence of NOCC due to the hydrogen bond formation and ionic interaction of functional groups of the polymer chains. The synthesis and swelling behavior of a superabsorbent hydrogel based on starch (St) and polyacrylonitrile (PAN) are investigated. The absorbency of the hydrogels indicated that the swelling ratios decreased with increasing ionic strength. The hydrogels exhibited a pH-responsive swelling-deswelling behavior at pHs 2 and 8. This on/off switching behavior provides the hydrogel with the potential to control delivery of bioactive agents. Release profiles of ibuprofen from the hydrogels were studied under both simulated gastric and intestinal pH conditions. The release was much quicker at pH 7.4 than at pH 1.2. The swelling rates of the hydrogels with various particle sizes were investigated as well. A biodegradable pH-sensitive hydrogel for potential colon-specific drug delivery was synthesized. Their composite hydrogel, based on a methacrylated and succinic derivative of dextran, and a methacrylated and succinic derivative of poly (*N*-2-hydroxyethyl)-*DL*-aspartamide was produced by photo cross-linking polymerization. In vitro drug release studies, performed using 2-methoxyestradiol as a model drug, show that obtained hydrogel is able to release the drug in simulated intestinal fluid, due to its pH-sensitive swelling and enzymatic degradability [59-60].

4.5.6. Temperature/pH Sensitive Hydrogels in DDS

More recently, studies have been conducted to fabricate and characterize hydrogels with dual-sensitivities. This was accomplished by copolymerizing a temperature-sensitive monomer, usually *N*-isopropylacrylamide, and a pH-sensitive monomer such as acrylic acid or methacrylic acid. Stayton's group has investigated a series of copolymers containing propylacrylic acid (PAA) and NIPAAm pendant chains as pH- and thermosensitive moieties, respectively. This new class of copolymers can sense environmental changes in the physiological range and has found usefulness in intracellular drug delivery in which subtle pH differences across the endosomal membrane triggers the delivery of protein or DNA [61-63].

Hollow beads of IPN of PNIPAAm were synthesized using Ca-alginate as the polymerization mould. Drug loading was carried out using injection of drug solution into the cores of hollow beads. Acetaminophen and diltiazem hydrochloride were used as low-water-soluble and water-soluble model drugs. Drug loading into this hydrogel based reservoir controlled release system indicated its advantage for loading of sparingly soluble drug in aqueous solutions. The swelling behavior of poly ((2-dimethyl amino) ethyl methacrylate-*co*-BMA) was investigated. These hydrogels demonstrated dual sensitivity to both pH and temperature. It was shown that the pH-sensitive or temperature-sensitive phase transition behavior of the gels can be changed by changing the temperature or pH of the swelling medium at constant hydrogel composition. Increasing the temperature decreased the transition pH, while increasing the pH of the surrounding medium decreased the transition temperature of the temperature-sensitive phase transition. Increasing the BMA content reduced the transition pH and temperature of the pH and temperature-sensitive phase transition, respectively. The results of equilibrium swelling and compression-strain measurements were used to calculate the polymer-solvent interaction parameters of these hydrogels using the Flory-Rehner equation of equilibrium swelling [65-67].

The pH and temperature double responsive poly (*N*-isopropylacrylamide-*co*-IA) hydrogels were prepared by copolymerization in mixed solvents of water and dimethylsulphoxide (DMSO). The results indicated that only hydrogels prepared in the highest DMSO/water ratio media exhibited improved properties such as higher swelling ratios, faster deswelling and reswelling kinetics compared with traditional P (NIPAAm-*co*-IA) hydrogels. They have proposed that there may be an energy barrier that existed for conformation transitions in the process of removal of DMSO by water. Only the gels which could overcome the energy barrier exhibited expanded network structures and improved properties while, the other gels maintained their contracted network structures and poor properties.

4.5.7. Bioresponsive Hydrogels in DDS

Bioresponsive hydrogels, which undergo structural and/or morphological changes in response to a biological stimulus, have been investigated for numerous applications in drug delivery, tissue regeneration, bioassays/biosensors, and biomimetic systems. A considerable amount of work on bioresponsive hydrogels for drug delivery relates to the release of insulin in response to raised blood sugar levels as a potential autonomous treatment of insulin-dependent diabetes. Glucose oxidase molecules are immobilized onto a basic polymeric carrier. Following the enzyme reaction, glucose is converted to gluconic acid and therefore the pH of the hydrogel is temporarily lowered. In this situation, the basic groups on the polymer are protonated and induced swelling of the gel which enhances the release profile of insulin (Figure 8).

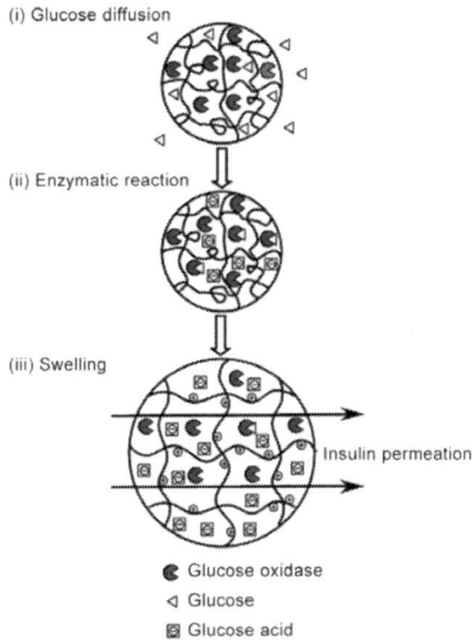

Figure 8. Schematic representation of the glucose-sensitive hydrogel membrane consisting of poly (amine) and glucose-oxidase-loaded membrane.

This system works as a feedback loop: upon release of insulin the sugar levels drop, resulting in a pH increase that stops the release of further insulin. Bioresponsive hydrogels can be designed in a degradable form in response to external stimuli such as enzymes. Such systems deliver physically entrapped guest molecules, held freely within the carrier, and do not require chemical modification for targeted delivery. Plunkett et al. have developed a procedure to synthesize hydrogels with cross-links composed of different enzyme-cleavable peptides. Law et al. described self-assembled peptide sequences that release therapeutic payloads upon specific interaction with disease-associated proteases. The core peptide sequence consists of a protease cleavable region flanked by two self-assembly motifs. Successful enzyme cleavage results in drug release. However, the extent of cleavage is limited by the degree of cross-linkage required to form a suitable gel. A biochemically and stimulus responsive triblock copolymer is developed. The polymer forms a micellar, dithiol cross-linked NIPAAm gel at 37°C that can be degraded by glutathione via cleavage of a central disulphoide bond. Thus, the system offers the possibility of payload release from a loaded gel following glutathione-stimulated degradation. Kumashiro et al. have synthesized a temperature responsive hydrogel that only allows enzyme-triggered polymer degradation above a lower critical solution temperature and below a higher critical solution temperature. They anticipate that this technique will allow the release of drug molecules depending on both enzyme selectivity and changes in body temperature [68-70].

Kim et al. have prepared the PNIPAm-*co*-AAc hydrogels by photo-polymerization with the peptide cross-linker, which provides for the enzyme degradable capability of the hydrogels. In the presence of collagenase the peptide cross-linked hydrogels were successfully degraded in dependence on the concentration of the enzyme and the initial cross-linking density. By conjugating a bio-affinity pair to the hydrogels, a new group of bioresponsive hydrogels was prepared, which has been used as a convenient construct for post-functionalization. One interesting example is in targeting cells as potential drug carriers, where PNIPAm core-shell hydrogel nanoparticles are conjugated with the folic acid that is a ligand for targeting cancer cells. When the folic acid labeled hydrogel particles are incubated with cancer cells that over express the folate receptors, the hydrogel particles are incorporated into the cancer cells via receptor- mediated endocytosis.

4.6. Kinetics of Drug Release From Hydrogels

Hydrogels can be used in many different types of controlled release systems, as diffusion-controlled systems, swelling-controlled systems and chemically controlled systems. In this section, the mechanism of drug release in each type of system is described.

4.6.1. Diffusion-controlled Delivery Systems

Diffusion-controlled is the most widely applicable mechanism for describing drug release from hydrogels, divided into two major types: reservoir devices and matrix devices (Figure 9). Reservoir systems consist of a polymeric membrane surrounding a core containing the drug. In matrix devices, the drug is dispersed throughout the three-dimensional structure of the hydrogel. Drug release from each type of system occurs by diffusion through the macromolecular mesh or through the water filled pores. Fick's law of diffusion is commonly used in modeling diffusion controlled release systems. For a reservoir system where the drug depot is surrounded by a polymeric hydrogel membrane, Fick's first law of diffusion can be used to describe drug release through the membrane:

$$J_i = -D_{ip}\frac{dC_i}{dX} \tag{5}$$

Where Ji is the molar flux of the drug (mol/cm2s), Ci is the concentration of drug, and Dip is the diffusion coefficient of the drug in the polymer. For the case of a steady-state diffusion process, i.e., constant molar flux, and constant diffusion coefficient, eqn (5) can be integrated to give the following expression:

$$J_i = K\frac{D_{ip}\Delta C_i}{\delta} \tag{6}$$

Where, δ is the thickness of the hydrogel and K is the partition coefficient, defined as the ratio of drug concentration in the gel per drug concentration in solution. To maintain a constant release rate or flux of drug from the reservoir, the concentration difference must remain constant. This can be achieved by designing a device with excess solid drug in the core. Under these conditions, the internal solution in the core will remain saturated. This type of device is an extremely useful device, allowing for time-independent or zero-order release. For a matrix system where the drug is uniformly dispersed throughout the matrix, unsteady-state drug diffusion in a one-dimensional slap-shaped matrix can be described by the Fick's second law:

$$\frac{\partial C_i}{\partial t} = \frac{\partial}{\partial X}\left(D_{ip}\frac{\partial C_i}{\partial X}\right) \tag{7}$$

This form of the equation is for one-dimensional transport with non-moving boundaries and can be evaluated for the case of constant diffusion coefficients and concentration-dependent diffusion coefficients. For the case of concentration independent diffusion coefficients, eqn (6) can be analyzed by application of the appropriate boundary conditions. Upon application of perfect-sink condition, the solution can be written in terms of the amount of drug released at a given time, Mt, normalized to the amount released at infinite times, M∞

$$\frac{M_t}{M_\infty} = 4\left(\frac{D_{ip}t}{\delta}\right)^{1/2}\left[\frac{1}{\pi^{1/2}} + 2\sum_{n=0}^{\infty}(-1)^n \, ierfc\frac{n\delta}{2(D_{ip}t)^{1/2}}\right] \tag{8}$$

And at short times, this solution can be approximated as:

$$\frac{M_t}{M_\infty} = 4\left(\frac{D_{ip}t}{\pi\delta^2}\right)^{1/2} \tag{9}$$

Fu et al. obtained an analytical solution of Fick's law for cylindrical geometry considering mass transfer in three dimensions:

$$\frac{M_t}{M_\infty} = 1 - \frac{8}{h^2 r^2}\sum_{m=1}^{\infty}\alpha_m^{-2}\, exp(-D\alpha_m^{-2}t)\times$$

$$\sum_{m=1}^{\infty}\beta_n^{-2}\, exp(-D\beta_n^2 t) \tag{10}$$

where M_3 and $M\infty$ are the amounts of drug released at time t and infinite, time respectively; and h denotes the half-length, and r the radius of the cylinder; D is the constant diffusivity; β n = (2n+1) π /2h and J0(r α) = 0, such that J0 a zero order Bessel function and m and n are integers. This model is applicable to tablets that range from the shape of a flat disk (radius > thickness) to that of a cylindrical rod (radius < thickness).

The exact solution for the release kinetics of a solute from an infinite spherical and rectangular reservoir with the burst effect initial condition into a finite external volume was developed. The governing diffusional equation for transport in the membrane is solved by the time Laplace transform method. Abdekhodaie concluded that the approximate solution is very accurate at later times but at early times deviations from the exact solution increase. His release kinetics results indicate that as the external fluid volume increases, the cumulative release at any time and the releasable amount of the solute at infinite time increase. Based on the obtained models, the fractional release of model drugs decreases when the polymeric coating thickness increases.

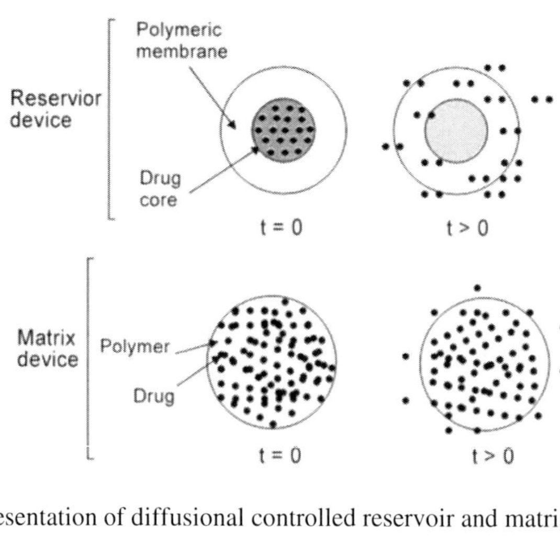

Figure 9. Schematic representation of diffusional controlled reservoir and matrix devices.

Experimentally, cumulative release profiles of theophylline microspheres coated with ethylene vinyl acetate copolymer into different external volumes agreed with the mathematical predictions. In most systems, the drug diffusion coefficient is dependent on the drug concentration as well as the concentration of the swelling agent. In order to analyze the diffusive behavior of drug delivery systems in this case, one must choose an appropriate relationship between the diffusion coefficient and the drug concentration. Based on theories that account for the void space in the gel structure, known as free-volume, researchers have proposed relationships between the diffusion coefficient and to the gel property. One of the most widely used equations, that have been proposed, relates the drug diffusion coefficient in the gel to the drug concentration in the following manner [71-73]:

$$D_{ip} = D_{iw} \exp\left[-\beta(C_s - C_i)\right] \tag{11}$$

where, Diw is the diffusion coefficient in the pure solution, β is a constant dependent on the system, and Co is the concentration of drug in the solution. Additionally, a similar equation was written to relate the diffusion coefficient to the concentration of the swelling agent (Cs) and the drug in the gel:

$$D_{ip} = D_{iw} \exp\left[-\beta(C_i - C_o)\right] \tag{12}$$

The structure and morphology of a polymer network will also significantly affect the ability of a drug to diffuse through a hydrogel. For porous hydrogels, when pore sizes are much larger than the molecular dimensions of the drug, the diffusion coefficient can be related to the porosity and the tortuosity of the hydrogels:

$$D_{eff} = D_{iw} \frac{K_p \varepsilon}{\tau} \tag{13}$$

where Deff is the effective diffusion coefficient, Diw is the diffusion coefficient of the solute in the pure solvent, Kp is the partition coefficient, ε is network porosity and τ is its tortuosity. In 1961, Higuchi developed a mathematical model to describe drug release from a matrix system. This model is based on the several hypotheses, as: (1) initial drug concentration in the matrix is much higher than drug solubility, (2) drug diffusion takes place only in one dimension, (3) drug particles are much smaller than system thickness, (4) matrix swelling and dissolution is negligible, (5) drug diffusivity is constant and (6) perfect sink conditions are always attained in the release environment. Accordingly, model expression is given by:

$$M_t = A\sqrt{D(2C_0 - C_s)C_s t} \qquad C_0 > C_s \tag{14}$$

Where Mt is the amount of drug released until time t, A is the release area, D is the drug diffusion coefficient, C0 is the initial drug concentration in the matrix while Cs is drug solubility. Interestingly, this model shows an Mt square root dependence on time as exact Fick's solution predicts when the amount released is less than the 60 percent. Another simple and useful empirical equation is the so-called power law equation:

$$\frac{M_t}{M_\infty} = kt^n \tag{15}$$

The constants, k and n, are characteristics of the drug polymer system. Peppas et al. were the first to give an introduction to the use and the limitations of these equations. The diffusional exponent, n, is dependent on the geometry of the device as well as the physical mechanism for release (Table 3). Based on the diffusional exponent, the drug transport in slab geometry is classified as Fickian diffusion, Case II transport, non-Fickian or anomalous transport and Super Case II transport. For systems exhibiting Case II transport, the dominant mechanism for drug transport is due to polymer relaxation as the gels swells.

These types of devices, known as swelling-controlled release systems, will be described in more detail later. Anomalous transport occurs due to a coupling of Fickian diffusion and polymer relaxation. For the case of anomalous transport, Peppas and Sahlin developed the following model to describe the release behavior of dynamically swelling hydrogels:

$$\frac{M_t}{M_\infty} = k_1 t + k_2 t^{1/2} \tag{16}$$

This expression describes the release rates in terms of relaxation-controlled transport process, k1t, and the diffusion-controlled process, k2t1/2. Shang et al. have shown that the partition coefficient can be controlled by immobilizing a certain amount of drug molecules in the membrane while maintaining diffusivity of the free drug of the same type. One of the potential applications of this control scheme is for hydrogel-based pulsatile DDS, which requires a sharp change in the drug permeation rate when the device is elicited on and off to better mimic the physiological release profile of certain hormones. Based on their hypothesis, the amount of immobilization needed for a required permeation rate can be calculated from:

$$K_{eff} = \left[\frac{exp((\mu_{D,0} - \mu_{M,0})/RT)\gamma_{D,free}}{C_{D,free}^{\alpha-1}\, \gamma_{M,free}(bC_{M,immob})^{\beta}} \right]^{1/\alpha}$$

$$for\ C_{M,immob} > 0 \tag{17}$$

where Keff is the partition coefficient of free drugs; $\mu_{D,0}$ and $\mu_{M,0}$ are the reference chemical potentials in the donor and membrane, respectively; $\gamma_{D,free}$ and $\gamma_{M,free}$ are the activity coefficients of free drugs in the donor and membrane, respectively; CD, free is the free drug concentration in the donor and CM, immob is the concentration of immobilized drug in the membrane; R is the gas constant and T is the temperature; α, β and b are fitting parameters. This partition control scheme offers an opportunity to modify the drug release profile of reservoir drug delivery systems with greater flexibility, i.e., the difference in drug permeation rates between the 'off-state' and the 'on-state' can be increased and the drug release profile can be modified after its membrane is synthesized [74-77].

4.6.2. Swelling-Controlled Delivery Systems

Swelling-controlled release occurs when diffusion of drug is faster than hydrogel swelling. The modeling of this mechanism usually involves moving boundary conditions where molecules are released at the interface of rubbery and glassy phases of swollen hydrogels. The rate of drug release is controlled by the velocity and position of the front dividing the glassy (dry) and rubbery (swelled) portions of the polymer. Drug diffusion time and polymer chain relaxation time are two key parameters determining drug delivery from polymeric matrices. In diffusion-controlled delivery systems, the time-scale of drug diffusion, t, (where t = δ (t) 2/D and δ (t) is the time-dependent thickness of the swollen phase) is the rate-limiting step while in swelling-controlled delivery systems the time-scale for polymer relaxation (λ) is the rate limiting step. The Deborah number (De) is used to compare these two time-scales:

$$D_e = \frac{\lambda}{t} = \frac{\lambda D}{\delta_{(t)}^{\ 2}} \tag{18}$$

In diffusion-controlled delivery systems (De≤1), Fickian diffusion dominates the molecule release process and diffusion equations described in the previous section can be used to predict molecule release. In swelling-controlled delivery systems (De≥1), the rate of molecule release depends on the swelling rate of polymer networks. This type of transport is known as Case II transport and results in zero-order release kinetics. A dimensionless swelling interface number, Sw, correlates the moving boundary phenomena to hydrogel swelling:

$$S_w = \frac{V\delta(t)}{D} \tag{19}$$

Where, V is the velocity of the hydrogel swelling front and D is the drug diffusion coefficient in the swollen phase. For a slab system when Sw≤1, drug diffusion is much faster than the movement of glassy rubbery interface and thus a zero-order release profile is expected.

Among the first models intended to describe drug release from a swellable matrix are those presented. The considerable volume extension due to matrix swelling is accounted by introducing a moving boundary diffusion problem and chemical potential for swelling of ionizable networks. A good agreement between experimental results and model calculations was obtained for drug concentration profiles within the polymer in a system of KCl distributed in HPMC matrix tablets.

Table 3. Drug transport mechanisms and diffusional exponents for hydrogel slabs

Type of transport	Diffusional exponent (n)	Time dependence
Fickian diffusion	0.5	$t^{1/2}$
Anomalous transport	0.5 < n < 1	t^{n-1}
Case II transport	1	Time independent
Super case II transport	n > 1	t^{n-1}

A research group developed a mathematical model to describe swelling-controlled release. They introduced additional boundary conditions derived from a volume balance and accounted for two dimensional movement of the swelling front in the radial or axial directions. This model assumes a homogeneous mixture of drug and polymer at t = 0, perfect sink conditions, and geometrical symmetry of the tablet. Model predictions were verified using compressed poly (ethylene oxide) (PEO) hydrogel tablets with different molecular weights. The results of water uptake, swelling and dissolution of PEO matrices as well as drug release are shown to agree well with the mathematical model. Vasheghani et al. developed mathematical models for drug release from swelling controlled systems. The swelling process from a rubbery state was described by taking into account the movement of interface position R(t) (Figure 10) as:

$$\frac{dR(t)}{dt} = -\frac{D_m}{\varphi_p^\infty}\frac{\partial \varphi_p}{\partial r}\bigg|_{r=R(t)} \tag{20}$$

Where, Dm is diffusion coefficient of solvent in rubbery polymer, r is distance and $\phi p\ \infty$ is volume fraction of polymer at interface. It was shown that the model developed is relatively accurate for describing simultaneous drug release and hydrogel dimensional change. In addition, modeling of swelling controlled drug release from glassy polymers was developed by using Fick's law and considering movement of the interface of hydrogel surrounding solution and the moving front of glassy gel interface, R1(t), as:

$$\frac{dR_1(t)}{dt} = -\frac{D_m}{(1-\varphi_p^*)}\frac{\partial \varphi_p}{\partial r}\bigg|_{r=R_1(t)} \tag{21}$$

Where, the volume fraction of polymeric gel at diffusion front ϕp *, is defined by:

$$\varphi_p^* = \frac{1}{C^*(\rho_p/\rho_s)+1} \tag{22}$$

Where, C* is the concentration of solvent in terms of g/g polymer at diffusion front and ρ p and ρ s are the density of polymer and solvent, respectively. The proposed model accurately described the dimensional changes of the polymeric network (polymer-solvent and glassy-gel interfaces) and the dichlofenac release from HPMC discs [78-86].

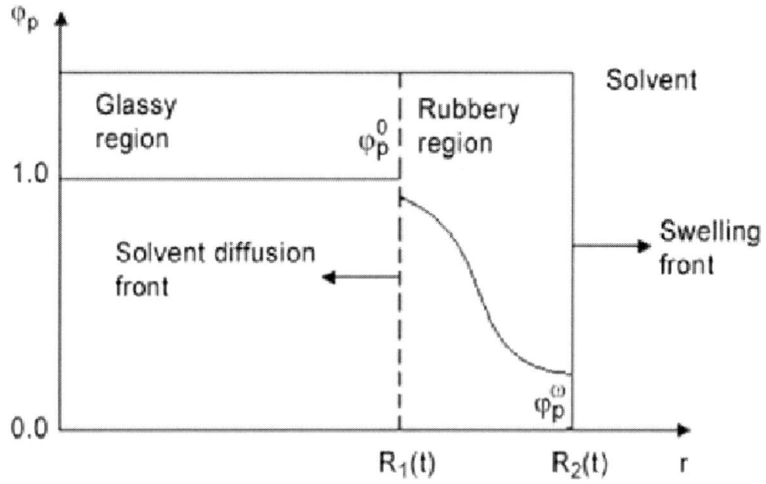

Figure 10. Schematic diagram representing the swelling of the hydrogels with the inward movement of diffusing front of the solvent

4.6.3. Chemically-Controlled Delivery Systems

There are two major types of chemically-controlled release systems; erodible drug delivery systems and pendant chain systems. In erodible systems, drug release occurs due to degradation or dissolution of the hydrogel. In pendant chain systems, the drug is affixed to the polymer backbone through degradable linkages. As these linkages degrade, the drug is released.

4.6.4. Erodible Drug Delivery Systems

In erodible drug delivery systems (either matrix or reservoir), also known as degradable or absorbable release systems, drug release is mediated by the rate of surface erosion. In reservoir devices, an erodible membrane surrounds the drug core. In matrix devices, the drug is dispersed within the three-dimensional structure of the hydrogel. Drug release is controlled by drug diffusion through the gel or erosion of the polymer. In true erosion controlled devices, the rate of drug diffusion will be significantly slower than the rate of polymer erosion and the drug is released as the polymer erodes. One of the earliest mathematical models, where the release mechanism only depends on matrix erosion rates, was developed by Hopfenberg:

$$\frac{M_t}{M_\infty} = 1 - \left(1 - \frac{k_a t}{C_0 a_0}\right)^n \tag{23}$$

Where a_0 is the radius of a spherical or cylindrical geometry or half-thickness for slab geometry and C_0 is the drug concentration in the surface eroding device. In this equation, n is 1, 2, or 3 for slab, cylinder, or sphere, respectively. It is clear that only for a slab-shaped device,, the drug release must be a zero order profile.

A general mathematical model for heterogeneous eroding networks was developed, accounting for different radial and vertical erosion rate constants for a flat tablet:

$$\frac{M_t}{M_\infty} = 1 - \left(1 - \frac{k_a t}{C_0 a_0}\right)^2 \left(1 - \frac{2K_b t}{C_0 b_0}\right) \tag{24}$$

Where ka and kb are radial and vertical degradation constants. Here, a_0 is the initial radius of the tablet and b_0 is the thickness of the tablet. By changing the radius to thickness ratio of the device, one can easily obtain various drug release rates. It is important to note that swelling of the matrices is either not considered or is assumed to occur prior to erosion and drug release in the two above models. In addition, these models only focused on the surface erosion of the hydrophobic polymers.

The research developed a generalized statistical *co*-kinetic model to predict the degradation behaviors of acrylated poly (vinyl alcohol) (PVA) hydrogels. In this model, a statistical approach was used to predict the different configurations of the cross-linking molecules and kinetic chains. It also accounts for the probability of an intact degradable linkage. The model was verified by experimental observation of gel swelling, mass loss and compressive modulus.

Monte Carlo simulation is used to predict protein release from cross-linked dextran microspheres. Monte Carlo simulation is good for describing network morphological changes. However, it does not provide any information regarding molecule release. Diffusion equations (Fick's law) must be incorporated in order to link the network degradation to molecule diffusion.

Based on Vlugt-Wensink et al. studies, a kinetic Monte Carlo scheme for the degradation of a small domain inside the dextran microsphere was developed. The only processes included in their model are diffusion and degradation. The general effects of diffusion, cross-link density, protein loading, and clustering of proteins on the release were also studied in their report. However, swelling of the hydrophilic microspheres and changes in swelling with matrix degradation were not accounted for in the described model.

4.6.5. Pendant Chain Systems

In a pendant chain system, the drug molecule is chemically linked to the backbone of a polymer. In the presence of enzymes or fluids, chemical or enzymatic hydrolysis occurs with concomitant release of the drug at a controlled rate. The drug may be linked directly to the polymer or can be linked via a "spacer" group. In the biodegradable system, polymers gradually decompose and bring about a controlled release of drug. The drug is dispersed uniformly throughout the polymer and is slowly released as the polymer disintegrates. The release of covalently attached drugs is determined by the degradation rate of the polymer drug linkage. Generally, these linkages degrade by hydrolysis allowing degradation and also release rates to be illustrated by simple first-order kinetic relationships. However, in specific applications, the drug polymer linkages may be designed to be enzymatically degraded which leads to more complex release kinetics.

A statistical-*co*-kinetic model has been developed to predict the effects of hydrolytic or enzymatic degradation on the macroscopic properties of hydrogels. Dubose et al. covalently incorporated fluorescently labeled probe molecules within the three-dimensional network structure of PEG-based hydrogels. Bond cleavage kinetics, microstructural network characteristics and detailed analysis of degradation products are the important parameters accounted for in this study. They demonstrated that hydrolytic degradation of covalent bonds as well as the cleavage of immobilized probe molecules resulted in a biphasic release profile in which a constant molecular release profile is obtained prior to gel dissolution and an almost instantaneous burst release following gel dissolution [87-100].

Targeting of drugs specifically to the colon is advantageous in the treatment of diseases such as amoebiasis, Crohn's disease, ulcerative colitis, and colorectal cancer. In addition, it has shown great potential in the oral delivery of therapeutic peptides and proteins, which are unstable in the upper part of the gastrointestinal (GI) tract. The colonic region is recognized as having less diversity and intensity of enzymatic activities than the stomach and small intestine. Various strategies are available for targeting drug release selectively to the colon. The designing of prodrugs is based on the concept of preventing the release of drugs in the stomach and small intestine and drug release is triggered by the use of specific property at the target site such as altered pH or high activity of certain enzymes in comparison to nontarget tissues. Since it is known that azo function can be reduced in the colon, many novel polymers containing azo groups either in the polymeric backbone or in the crosslinks have been synthesized. To promote further selective degradation in the vicinity of the colonic environment, delivery systems have been designed that contain both pH-sensitive acidic monomers and degradable azo aromatic crosslinks. Chitosan is a functional linear polymer derived from chitin, which is the most abundant natural polysaccharide on the earth after cellulose and is not digested in the upper GI tract by human digestive enzymes. Chitosan is a copolymer consisting of 2-amino-2-deoxy-D-glucose and 2-acetamido-2-deoxy-D-glucose units links with β- (1-4) bonds. It should be susceptible to glycosidic hydrolysis by microbial enzymes in the colon because it possesses glycosidic linkages similar to those of other enzymatically depolymerized polysaccharides. The polysaccharide, on reaching the colon, undergoes assimilation by microorganisms, degradation by enzymes or break down of the polymer back bone leading to a subsequent reduction in molecular weight. There is also a loss of mechanical strength and it is unable to hold the drug any longer. Chitosan has drawn attention for its potential to achieve site-specific delivery to the colon. Chitosan, a natural linear polyamine with a high ratio of glucosamine to acetyl-glucosamine units, is a weak base and carries a positive charge. Its solubility is pH-dependent, and it reacts readily with negatively charged surfaces and materials, including polymers and DNA. Ionic gelation, complex coacervation, emulsion cross-linking, and spray-drying are methods commonly used for the preparation of chitosan nanoparticles. Among those methods, ionic gelation and complex coacervation are mild processes occurring in a pure aqueous environment and are ideal for maintaining the in-process stability of drugs. Ionic gelation and complex coacervation are very similar except that the former involves the gelation of chitosan using an electrolyte such as tripolyphosphate (TPP), whereas the latter employs an oppositely charge dionic polymer such as alginate. A new type of chitosan (CS) hydrogel, using dextran sulfate (DS) as a polyanionic polymer, was developed to achieve complex coacervation for the incorporation and controlled release of an anti-angiogenesis hexapeptide. This was the first report describing the use of DS to formulate CS-based hydrogels. Although there have been investigations of how the properties of CS and formulation variables such as CS molecular weight (MW), concentrations of CS and 5-ASA, and formulation pH affect the formation and encapsulation capability of hydrogels, to our knowledge, no attempts have been made to study how the ratio of CS to the oppositely charged polymer influences the formulation and properties of hyrogel. In this study, 5-aminosalicylic acid (5-ASA) was chosen as a model drug. These hydrogels had good spherical geometry. It is obvious that the surface of the hydrogel shrank and a densely cross-linked gel structure was formed. This may explain the greater retardation of drug release from matrices of higher cross linked content. The average drug entrapment was found to be 81.21%±1.86%in the hydrogel. Results of drug content and EE demonstrated that drug content increased from 10.23±0.50 mg/100mg to 26.24±0.43 mg/100mg of hydrogel with increasing the amount of drug from 5% to 20% wt/wt. No significant increase in drug content was observed on further increasing the amount of drug, i.e., above 15% wt/wt , which could be due to the limited solubility of the drug in DMF and that is endorsed from the presence of drug particles on the surface of the hydrogel prepared with a 20% drug concentration. The percent EE was increased up to 86.32%±0.20% with increasing polymer concentrations to 4%. Concentration of the cross-linking agent exhibited no significant effect on the percent EE. FESEM images of superparamagnetic CS-DS hydrogel (Figure 11) show that hydrogels have a solid and near consistent structure [101-120].

Furthermore, the incorporation of 5-ASA into the hydrogel produced a smooth surface and compact structure. The particle size observed in FESEM is smaller than that measured by the Zetasizer. This is because dried hydrogel was used in FESEM, whereas particles in the liquid dispersion were analyzed by the Zetasizer. CS-DS hydrogels are hydrophilic and would be expected to swell in water, thus producing a large hydrodynamic size when measured by the Zetasizer (Figure 12).

5-aminosalicylic acid (5-ASA) – loaded hydrogel was obtained spontaneously upon the mixing of the DS aqueous solution (0.1%wt/vol) with the CS solution (0.1% wt/vol) under magnetic stirring, with 5-ASA dissolved in CS-DS solution. The incorporation of 5-ASA into the CS-DS hydrogel resulted in a sharp increase in the particle size of the nanoparticle dispersion. The significant increases in particle size give a good induction of the incorporation of 5-ASA into CS-DS hydrogel. A study was undertaken to investigate the effect of the order of 5-ASA mixing with CS and DS. The data obtained show that the order of 5-ASA mixing had no effect on the size, entrapment efficiency, and yield of 5-ASA-loaded hydrogel.

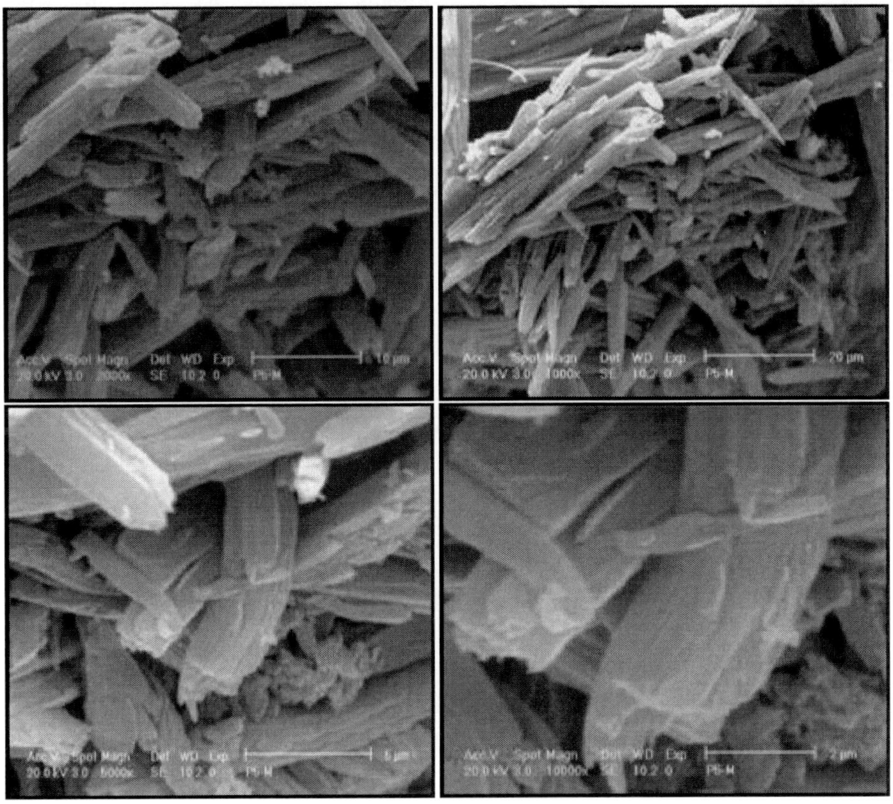

Figure 11. Morphology of chitosan-dextran sulfate (CS-DS) hydrogel by emission scanning electron microscopy field.

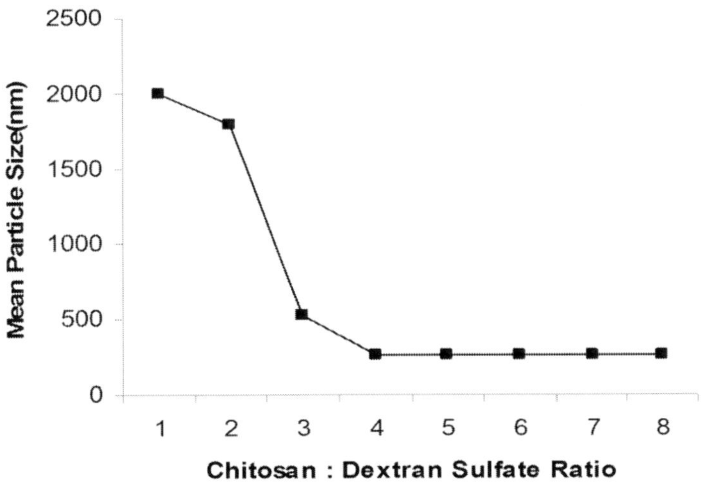

Figure 12. Influence of the Chitosan: Dextran sulfate ratio on the particle size.

The effect of drug concentration, chitosan-dextran sulfate concentration, GST concentration, and cross-linking time on in vitro drug release was also observed. In vitro drug release after 5 hours was 86.3%±4.0% in the case of hydrogel having 15% drug, while it was 88.2%±3.1% for hydrogel with 20% drug. The effect of chitosan-dextran sulfate on the release of drug was found to be limited. It is also observed that the amount of drug released from hydrogel decreased with increasing cross-linking time. These properties are probably explained by the promotion of cross-links between chitosan-dextran sulfate chains and GST. Freeze-drying of the samples resulted in larger and more porous hydrogel compared with air-dried hydrogel. Freeze-drying had the advantage of avoiding drug extraction by immediately freezing and removing the water present within the hydrogel and a burst effect during the dissolution study (Figure 13).

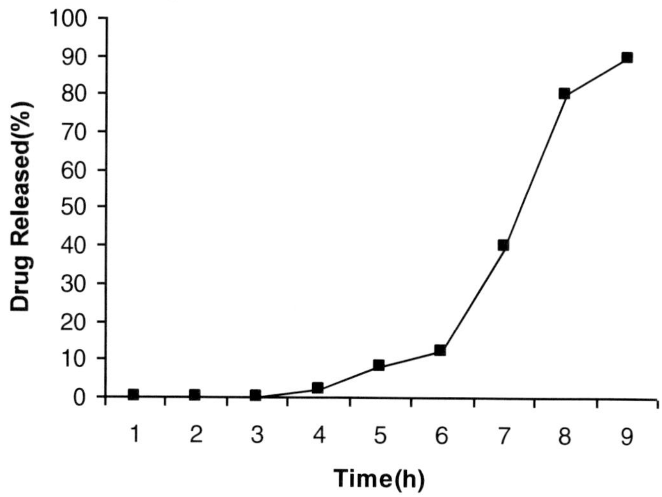

Figure 13. The *in-vitro* release profile of 5-aminosalicylic acid from chitosan-dextran sulfate hydrogel in various simulated gastrointestinal fluids (n=3).

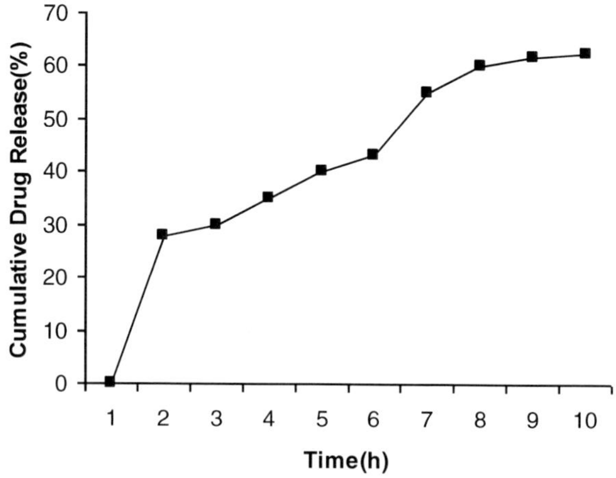

Figure 14. *In vitro* drug release from chitosan-dextran sulfate hydrogel in SCF (pH 7.0) in the presence of 5-ASA.

Conventional dissolution testing is less likely to accurately predict in vivo performance of colon delivery systems triggered by bacteria residing in the colon (because aspects of the colon's environment, i.e., scarcity of fluid, reduced motility, and presence of microflora cannot be simulated in conventional dissolution methods). Hence, release studies were performed in an alternate release medium. The release profiles of 5-ASA-loaded hydrogel was evaluated in water and a phosphate buffer, which was either in a different ionic strength or with saline, in order to study the underpinning mechanisms for drug release. The greatest release for 5-ASA –loaded hydrogel occurred in the release media of a high ionic strength. By contrast, a small portion of 5-ASA was released in water over the release study period. The burst release was observed with hydrogel, and it may have arisen from desorption of loosely attached 5-ASA from the surface of the matrix polymers (Figure 14) [121-125].

The opposite charges of CS-DS hydrogel were responsible for the formation of nanoparticles. The charge ratio between the negatively charged sulfate groups (N) in DS and the positively charged amine groups (P) in CS was determined. Under the experimental conditions (pH=3-4), DS carries ~74 sulfate groups per mole, equivalent to 5.78×10^{-3} negatively charged groups per gram of DS ; CS has ~2073 amino groups per gram of CS. Table 2 shows the conductivity of dispersion of hydrogel. Natural polymers have potential pharmaceutical applications because of their low toxicity, biocompatibility, and excellent biodegradability. In recent years, biodegradable polymeric systems have gained importance for the design of surgical devices, artificial organs, drug delivery systems with different routes of administration, carriers of immobilized enzymes and cells, biosensors, ocular inserts, and materials for orthopedic applications. These polymers are classified as either synthetic (polyesters, polyamides or polyanhydrides) and natural (polyamino acids or polysaccharides). Polysaccharide-based polymers represent a major class of biomaterials, which includes agarose, alginate, carageenan, dextran, and chitosan. Chitosan [β(1,4)2-amino-2-D-glucose] is a cationic biopolymer produced by alkaline N-deacetylation of chitin, which is the main component of the shells of crab, shrimp, and krill. Chitosan is a functional linear polymer derived from chitin, the most abundant natural polysaccharide on the earth after cellulose, and it is not digested in the upper GI tract by human digestive enzymes. Chitosan is a copolymer consisting of 2-amino-2-deoxy-D-glucose and 2-acetamido-2-deoxy-D-glucose units links with β- (1-4) bonds. It should be susceptible to glycosidic hydrolysis by microbial enzymes in the colon because it possesses glycosidic linkages similar to those of other enzymatically depolymerized polysaccharides. Among diverse approaches that are possible for modifying polysaccharides, the grafting of a synthetic polymer is a convenient method for adding new properties to a polysaccharide with minimum loss of its initial properties. Graft copolymerization of vinyl monomers onto polysaccharides using free radical initiators, has attracted the interest of many scientists. Up to now, considerable work has been devoted to the grafting of vinyl monomers onto the substrates, especially starch and cellulose. Existence of polar functionally groups as carboxylic acid are needed not only for bioadhesive properties but also for the pH-sensitive properties of the polymer. The increase of MAA content in the hydrogels provides more hydrogen bonds at low pH and more electrostatic repulsion at high pH. A part of our research program is chitosan modification to prepare materials with pH-sensitive properties for uses as drug delivery. The free radical graft copolymerization polymethacrylic acid onto Chitosan was carried out at 70°C with bis-acrylamide as a cross-linking agent and persulfate as an initiator. Polymer bonded drugs usually contain one solid drug bonded together in a matrix of a solid polymeric binder. They can be produced by polymerizing a monomer such as methacrylic acid (MAA), mixed with a particulate drug, by means of a chemical polymerization catalyst, such as AIBN or by means of high-energy radiation, such as x-ray or γ-rays. The modified hydrogel and satranidazole as a model drug was converted to nanoparticles by the freeze-drying method. The equilibrium swelling studies and in vitro release profiles were carried out in enzyme-free simulated gastric and intestinal fluids [SGF (pH 1) and SIF (pH 7.4) respectively)]. The influences of different factors, such as the content of MAA in the feed monomer and swelling were studied. In the present study, the beads were prepared by the classical method, which involves spreading a uniform layer of polymer dispersion followed by a drying step for removal of the solvent system. Since the methodology of bead preparation involved a heating step, it may have had a detrimental effect on the chemical stability of drug. Hence, the stability assessment of satranidazole impregnated in bead was done using the stability-indicating method. For this purpose, satranidazole was extracted from bead and analyzed by HPLC. A single peak at 21.2 minutes representing satranidazole (with no additional peaks) was detected in the chromatogram, suggesting that the molecule was stable during preparation of beads. Satranidazole was extracted from different regions of the N-sulfonato-N,O-carboxymethylchitosan/PMAA bead using acetonitrile: ETOH(1:1 vol/vol) solvent system. After normalization of the satranidazole concentration on the weight basis of the bead, the variation in the distribution of satranidazole in different regions of bead was <16%. The composition of the polymer defines its nature as a neutral or ionic network as well as its hydrophilic/hydrophobic characteristics. Ionic hydrogels, which could be cationic, containing basic functional groups or anionic and acidic functional groups, have been reported to be very sensitive to changes in the environmental pH. The swelling properties of the ionic hydrogels are unique due to the ionization of their pendent functional groups. The equilibrium swelling behavior of ionic hydrogels containing acidic and/or basic functional groups is illustrated in Figure 15 [126-130].

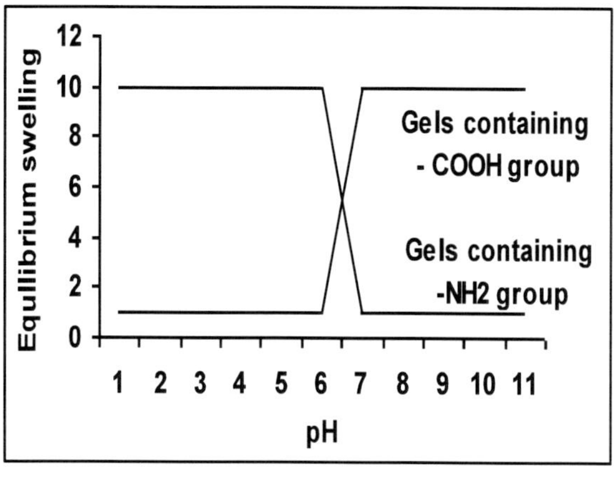

Figure 15. Equilibrium degree of swelling in response to pH.

Hydrogels containing basic functional groups are found to have increased swelling activity in acidic conditions and reduced swelling in basic conditions. The pH sensitive anionic hydrogels show low swelling activity in acidic mediums and very high activity in basic mediums. As shown in Figure 16, an increase in the content of MAA in the feed monomer mixtures resulted in less swelling in simulated gastric fluid but greater swelling in and simulated intestinal fluids.

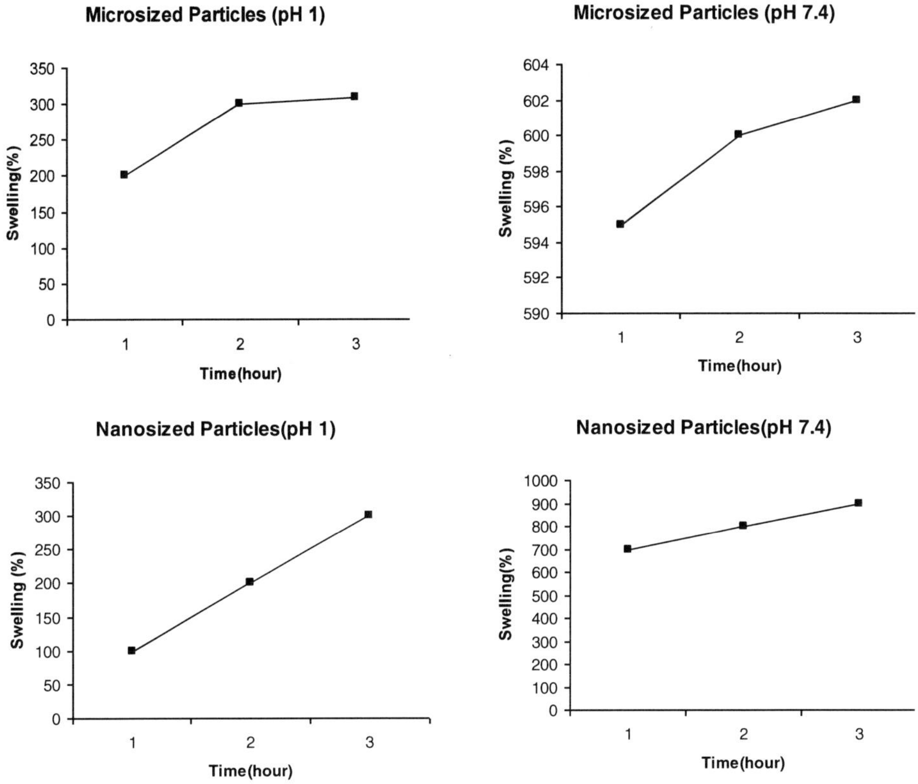

Figure 16. Time-dependent swelling behavior of micro and nano carriers for satranidazole drug model as a function of time at 37°C.

This is because the increase of MAA content in the hydrogels provides more hydrogen bonds at low pH and more electrostatic repulsion at high pH. Figure 17 shows the scanning electron microscope (SEM) of a graft NOCCS copolymer with polymethacrylic acid and nano polymer bonded drug, respectively.

Nano and micro polymer bonded drugs (50 mg) were poured into 3 mL of aqueous buffer solution (SGF: pH 1 or SIF: pH 7.4) (Figure 18). The mixture was introduced into a cellophane membrane dialysis bag. The bag was closed and transferred to a flask containing 20 mL of the same solution maintained at 37°C. The external solution was continuously stirred, and 3 mL samples were removed at selected intervals. The removed volume was replaced with SGF or SIF.

The triplicate samples were analyzed by UV spectrophotometer, and the quantity of satranidazole was determined using a standard calibration curve obtained under the same conditions.

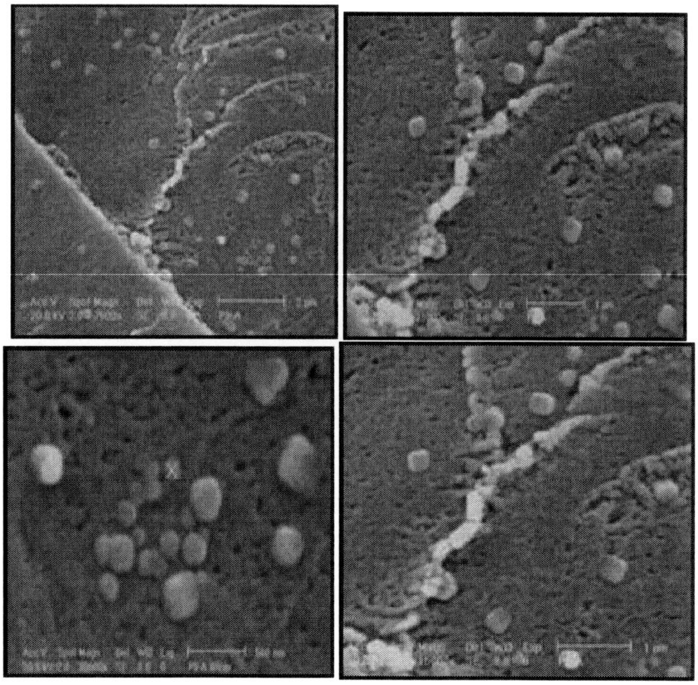

Figure 17. SEM of NOCCS-Polymethacrylic acid hydrogel beads with satranidazole.

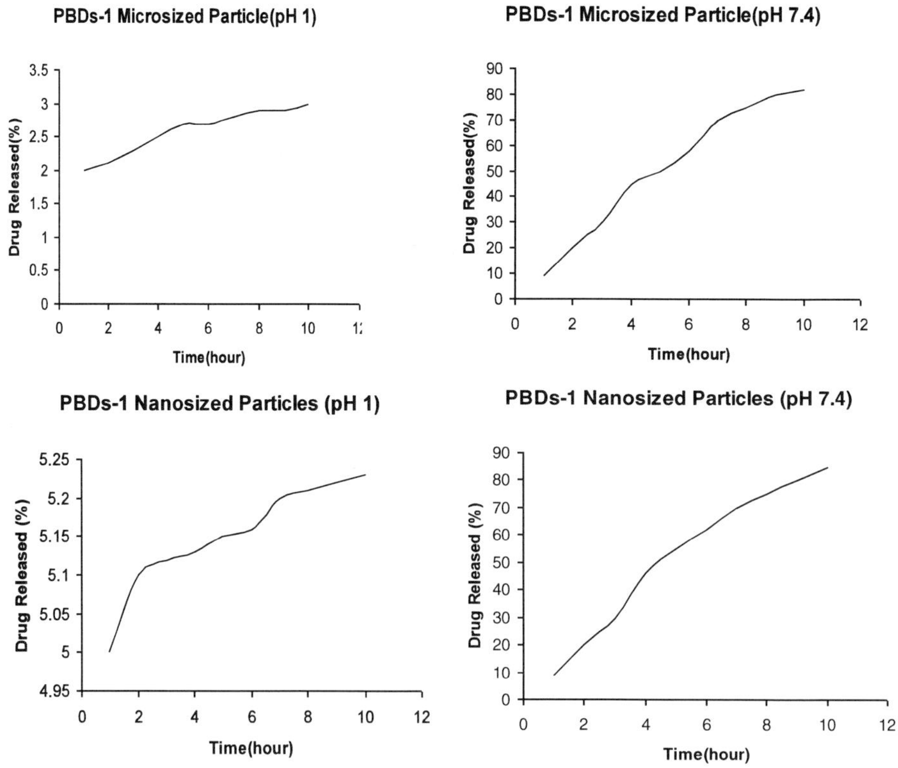

Figure 18. Release of satranidazole drug from micro and nano polymeric carriers as a function of time at 37ºC.

The thermal behavior of a polymer is important for controlling the release rate in order to have a suitable drug dosage form. The glass transition temperature (Tg) was determined from the DSC thermograms. The values are given in Table 4. The higher Tg values are probably related to the introduction of crosslinks, which would decrease the flexibility of the chains and the ability of the chains to undergo segmental motion, which would increase the Tg values. On the other hand, the introduction of a strongly polar carboxylic acid group can increase the Tg value because of the formation of internal hydrogen bonds between the polymer chains. X-ray diffraction is a proven tool to study crystal lattice arrangements and yields very useful information on the degree of crystallinity. The X-ray diffraction pattern of hydrogel with satranidazole has several high-angle bead at the following 2θ values: 21, 22, 25, 26.5, 28 and 32. The 26.5° 2θ peak had the highest intensity, and the hump in the baseline occurred from 7 ° to 45 ° 2θ, as observed for chitosan bead (Figure 19). In order to develop a local delivery system for satranidazole, a biodegradable N-sulfonato-N, O-carboxymethylchitosan-PMAA bead was attempted, which to our knowledge is the first effort of its kind.

Table 4. DSC data and composition of copolymer

Polymer Samples	Molar composition of monomers in the feed				Degree of Substitution (DS)	T_g (° C)
	NOCCS (gr)	MAA (gr)	CA (gr)	IN (gr)		
P- 1	1	3	0.05	0.05	0.52	135
P- 2	1	2	0.05	0.05	0.49	142

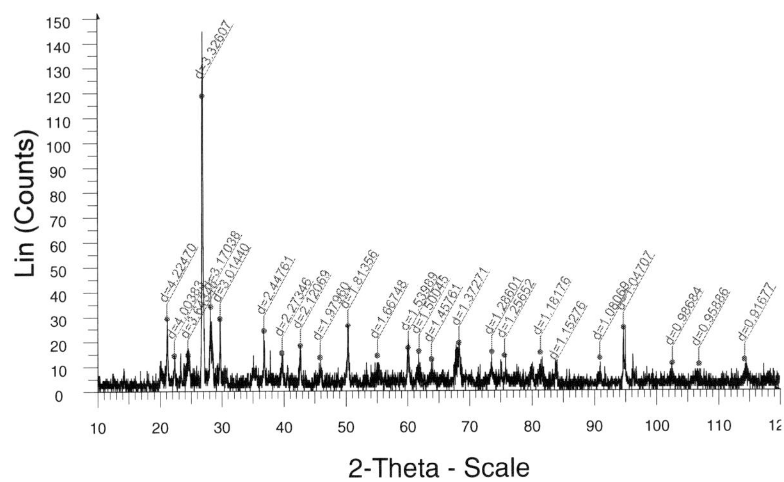

Figure 19. XRD patterns of NOCCS-Polymethacrylic acid hydrogel beads.

The potential formulation problems were anticipated since chitosan is only soluble in aqueous acidic solutions, whereas satranidazole, being a hydrophobic drug, is insoluble under similar conditions. In the early stages of the formula optimization studies, the satranidazole was incorporated into the bead .The primary mechanism for release of drugs from matrix systems in vitro are swelling, diffusion, and disintegration. In vitro degradation of chitosan beads were prepared by the solution casing method occurred less rapidly as the degree 73%

deacetylated showed slower biodegradation. Since the grade of chitosan used in the present study was of high molecular weight with a degree of deacetylation ≥85%, significant retardation of the release of satranidazole from beads is attributed to the polymer characteristics. In addition, diffusion of satranidazole may have been hindered by increased tortuosity of the polymer accompanied by a swelling mechanism. As observed from SEM photomicrographs, the crystals of satranidazole have a different appearance than recrystallized satranidazole. These nanoparticles do not have clearly defined crystal morphological features in the SEM photomicrographs. Hence, it appears that the irregularly shaped particles are surface deposited with poloxamer, which gives them an appearance resembling that of coated particles. An x-ray diffraction technique is also used to study the degree of crystallinity of pharmaceutical drugs and excipients. A lower °2θ value indicates larger d-spacings, while an increase in the number of high-angle reflections indicates higher molecular state order. In addition, broadness of reflections, high noise and low peak intensities are characteristic of a poorly crystalline material. A broad hump in the diffraction pattern of chitosan extending over a large range of °2θ suggests that chitosan is present in an amorphous state in the bead. The existence of polar functionally groups as carboxylic acid are needed not only for their bioadhesive properties but also for the pH-sensitive properties of polymer. The increase of MAA content in the hydrogels provides more hydrogen bonds at low pH and more electrostatic repulsion at high pH. It is as a part of our research program on chitosan modification to prepare materials with pH-sensitive properties for uses as drug delivery. In novel research the absorption of paclitaxel could be enhanced by administration with chitosan-polymethacrylic acid nanoparticles because of their greater permeability properties. In the present study, nanoparticles were prepared by the classical method, which involves spreading a uniform layer of polymer dispersion followed by a drying step for removal of the solvent system. Since the bead preparation methodology involved a heating step, it may have had a detrimental effect on the chemical stability of drug. Hence, the stability assessment of paclitaxel impregnated in bead was done using the stability-indicating method. For this purpose, paclitaxel was extracted from bead and analyzed by HPLC. A single peak at 21.2 minutes representing paclitaxel (with no additional peaks) was detected in the chromatogram, suggesting that the molecule was stable during the preparation of the beads. In order to develop a local delivery system for paclitaxel, a biodegradable N- chitosan-PMAA bead was attempted, which to our knowledge is the first effort of its kind. Potential formulation problems were anticipated since chitosan is only soluble in aqueous acidic solutions, whereas paclitaxel, being a hydrophobic drug, is insoluble under similar conditions. In the early stages of formula optimization studies, it was observed that paclitaxel was incorporated into bead. The primary mechanisms for release of drugs from matrix systems in vitro are swelling, diffusion, and disintegration. In vitro degradation of chitosan-PMAA nanoparticles prepared by a solution casing method occurred less rapidly as the degree 73% deacetylated showed slower biodegradation. Since the grade of chitosan used in the present study was of high molecular weight with a degree of deacetylation ≥85%, significant retardation of the release of paclitaxel from nanoparticles is attributed to the polymer characteristics. In addition, diffusion of paclitaxel may have been hindered by increased tortuosity of polymer accompanied by a swelling mechanism. As observed from SEM photomicrographs, the crystals of paclitaxel have a different appearance than recrystallized paclitaxel. These nanoparticles do not have clearly defined crystal morphological features in the SEM photomicrographs. Hence, it appears that the irregularly shaped particles are surface deposited with poloxamer, which gives them an appearance resembling that of coated particles. X-ray diffraction is also used to study the degree of crystallinity of pharmaceutical drugs and excipients. A lower °2θ value indicates larger d-spacings, while an increase in the number of high-angle reflections indicates higher molecular state order. In addition, broadness of reflections, high noise, and low peak intensities are characteristic of a poorly crystalline material. A broad hump in the diffraction pattern of chitosan extending over a large range of 2θ suggests that chitosan is present in an amorphous state in the nanoparticles. Recent studies have been reported investigation of the effect of 4.0 generation (G) PAMAM dendrimer present in an anionic phospholipids composition, consisting of hydrogenated soyaphosphotidylcholine, cholesterol, diacetyl phosphate and poly (ethylene glycol)-derivatized phosphotidylethanolamine, on the hydration and liquid crystalline structure formation. In the novel study, the 4.0 generation PAMAM dendrimer was exhaustively studied as controlled – release systems for parenteral administration of a model drug 5-aminosalicylic acid and analyzed using various release kinetic studies. This study gives us an insight about the biodistribution pattern of the drug and it's localization at the site of inflammation with PAMAM dendrimer. In the present study, films were prepared by the classical method, which involves spreading a uniform layer of polymer dispersion followed by a drying step for removal of the solvent system. Since the film preparation methodology involved a heating step, it may have had a detrimental effect on the chemical stability of the drug. Hence, stability assessment of 5-ASA impregnated in film was done using the stability-indicating method. For this purpose, 5-ASA was extracted from film. 5-ASA was extracted from different regions of Ac-PAMAM-CS film using acetonitrile: ETOH (1:1 vol/vol) solvent system. After normalization of the amount of 5-ASA on weight basis of film, the results indicated the variation in distribution of 5-ASA in different regions of film was <18%. The mechanical strength of films is described in terms of tensile strength, and brittle films are characterized by a decrease in the percentage of elongation at break. The area under the curve is related to the energy required to break polymeric material, and tough polymers have larger areas requiring large amounts of energy for rupture. In order to understand the arrangement of polymer chains in the presence of 5-ASA, force-time profiles of films were generated as shown in Figure 20. Although the force of elongation at break is slightly lowered in the presence of 5-ASA in film, the area under the profile has been increased. For convenience of interpretation, each profile is further described in terms of ascending and descending portions. The time to plateau of the ascending portion of 5-ASA–Ac-PAMAM-CS film was greater in comparison with the control film. As brittleness is reflected in the plateau and greater time is indicative of lack of brittleness of film. In order to confirm in vivo degradation of film, a time-dependent visual monitoring of film texture and integrity following implantation in mice was performed. On days 2, 10, 20 and 50, mice were humanely killed by cervical dislocation, and an incision was made in the site for examination of the film.

After 2 days of implantation, it was observed that film turned white and opaque when compared with the control film, which was initially yellowish and translucent. With progression of the implantation period, the film turned soft and delicate (on the 10th day) with loss of mechanical strength. In a period of 30 to 40 days, the film began to get embedded into the surrounding tissues, and after 50 days it was completely embedded. Although this experiment was qualitative in nature, the film's loss of integrity and identity confirms that the implant was amenable to biodegradation. The tissue responses to implanted film were studied by microscopic examination of tissues in the implanted area (Figure 21).

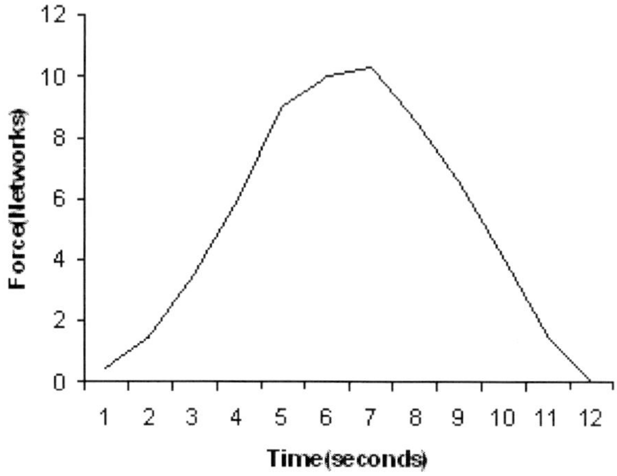

Figure 20. Force – time profiles of hydrogel film to study the influence of formulation approach on mechanical properties of film.

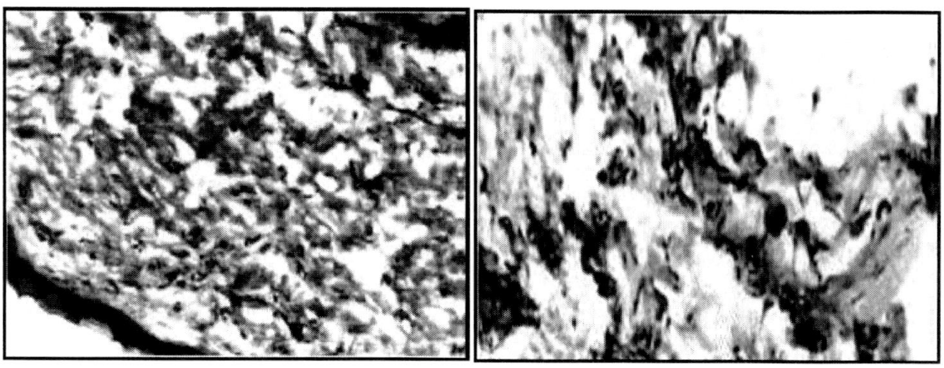

Figure 21. Histopathological examination for generalized inflammatory response to assess biocompatibility of subcutaneous implanted hydrogel film in mice.

The lack of an increase in macrophages at the site of implantation suggests that inflammatory responses were either minimal or absent. It has been reported previously that the degree of deacetylation of CS is a key factoring determining not only release rate but also inflammatory responses. Typically, CS gels with low degrees of deacetylation have short residence times and higher degrees have longer residence time and minimal or no inflammation. In the present study, the grade of Ac-PAMAM-CS used for fabrication of film was ≥85% deacetylated, hence, the high degree of deacetylation of CS may be the reason for reducing local inflammatory responses following implantation. X-ray diffraction is a proven tool to study crystal lattice arrangements and yields very useful information on the degree of sample crystallinity. The x-ray diffraction patterns of 5-ASA-Ac-PAMAM-CS hydrogel film were obtained and compared, which revealed marked differences in the molecular state of 5-ASA (Figure 22).

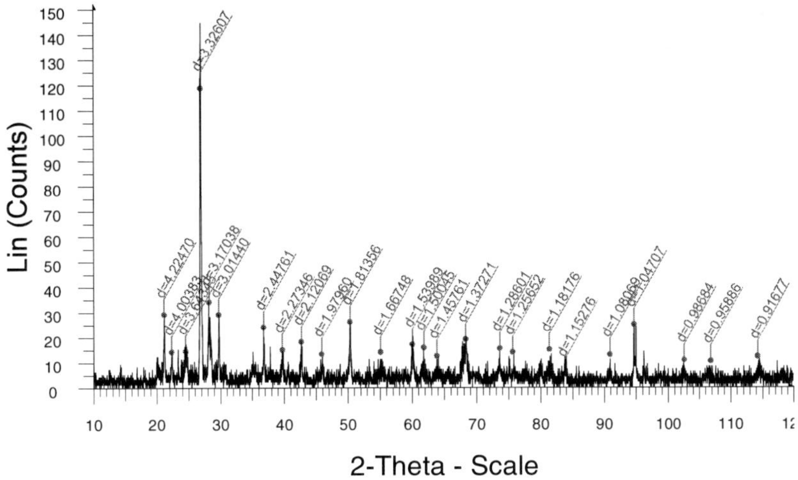

Figure 22. X-ray diffraction patterns of hydrogel film at ambient temperature to examine solid –state features of 5-aminosalicylic acid in film.

An X-diffractogram of 5-ASA-Ac-PAMAM-CS hydrogel film shows several high-angle diffraction peaks were observed at the following 2θ values: 21.3°,22.5°,24.1°,27.4° ,28.2° ,37.4° ,43.3° , 51.6°.The 27.4° 2θ peak had the highest intensity as observed for hydrogel film. Transmission infrared spectra of hydrogel powder were acquired to draw information on the molecular state of hydrogel (Figure 23).

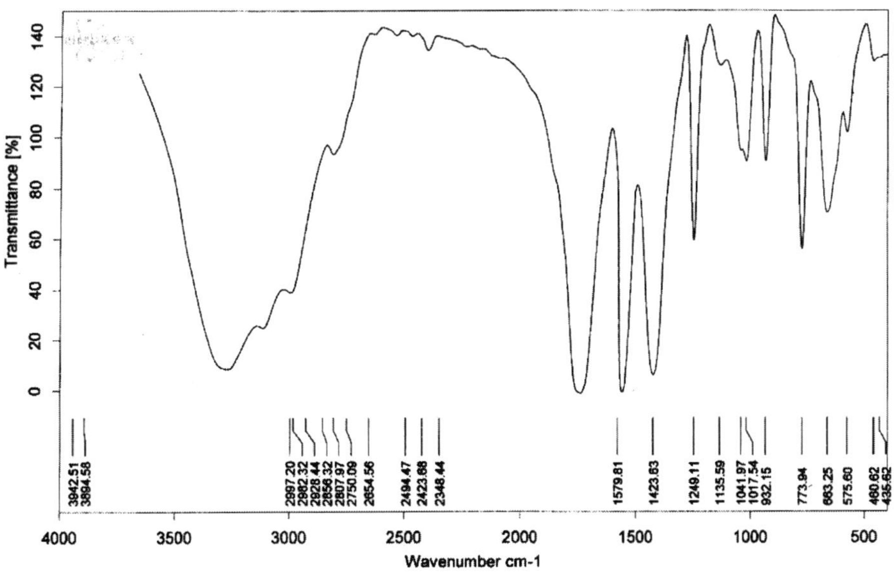

Figure 23. FTIR spectra of Hydrogel film to examine the alterations in molecular vibrations of 5-aminosalicylic acid.

CS is an amino glucose characterized by a small proportion of amide groups via an amide linkage with acetic acid. In the infrared spectrum, powder CS hydrogel exhibited a broad at 3431cm^{-1}, which is assigned to the N-H and hydrogen bonded O-H strength vibration frequencies, while a sharp peak at 3600 cm^{-1} is that of free O-H bond stretch of glucopyranose units. Further, in the C-H strength region of FT-IR spectrum, the higher intensity peak at 2920 cm^{-1} is assigned to the asymmetric and the lower intensity peak at 2835 cm^{-1} is assigned to the symmetric modes of CH_2. In addition, the characteristic band due to CH_2 scissoring, which usually occurs at 1423 cm^{-1} was also present in the sample. Since the grade of CS hydrogel used in the present study was ≥85% deactylated, an amide bond peak was present in the spectra and the C=O strength of amide bond was observed at 1679 cm^{-1}. The effect of drug concentration, CS-Ac-PAMAM concentration, GST concentration, and cross-linking time on in vitro drug release was also observed. In vitro drug release after 5 hours was 86.3%±4.0% in the case of hydrogel having 15% drug, while it was 89.2%±3.1% for hydrogel with 20% drug. The effect of CS- Ac-PAMAM dendrimer on the release of the drug was found to be limited. It is also observed that the amount of released drug from hydrogel decreased with increasing crosslinking time (Figure 24).

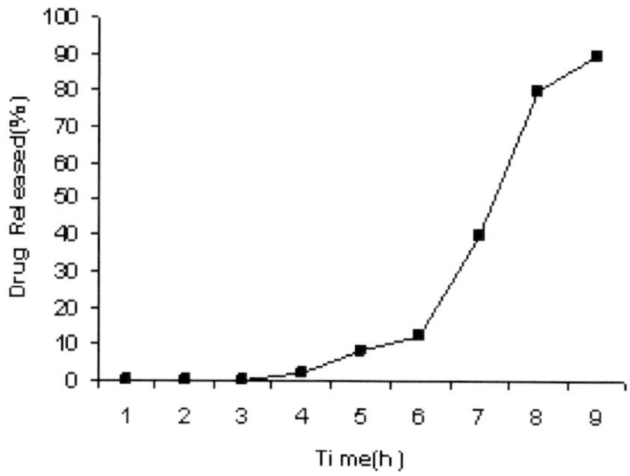

Figure 24. The in-vitro release profile of 5-aminosalicylic acid from chitosan-PAMAM hydrogel in various simulated gastrointestinal fluids (n=3).

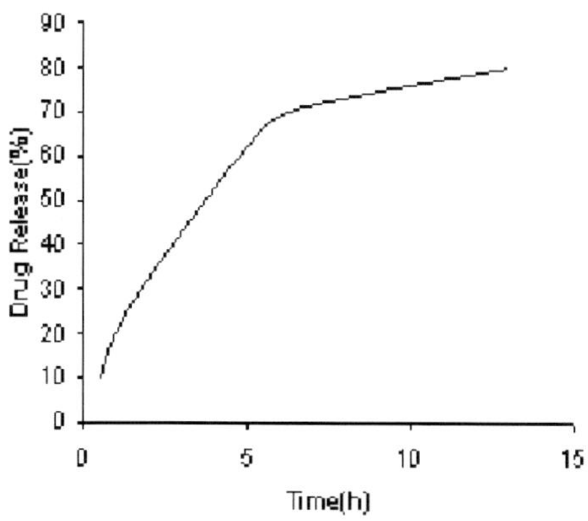

(a)

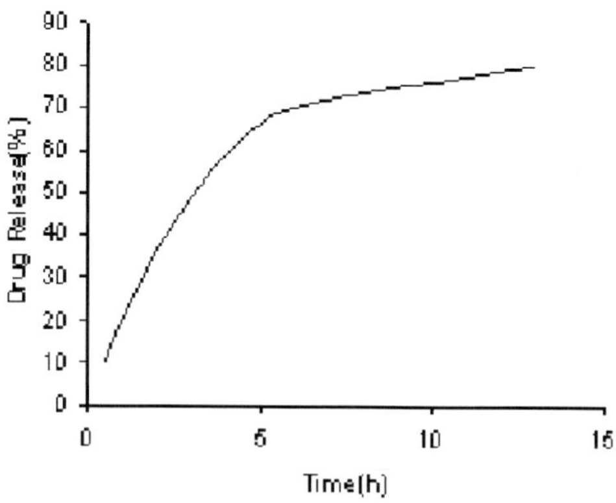

(b)

Figure 25. Release of drug (5-ASA) from (a) micro and (b) nano polymeric carriers as a function of time at 37°C.

The concentration of 5-ASA in the decanted Tris HCl buffer and 2 washing solutions was determined by measuring the absorbance at 235 nm using a GBS Cintra 10-UV-Visible Spectrophotometer (Shimadzu, Japan). The determinations were made in triplicate, and the results were averaged (Figure 25).

TCS which are gaining popularity because of their high mucoadhesiveness and extended drug release properties. The derivatization of the primary amino groups of CS with coupling reagents bearing thiol functions leads to the formation of TCS. The use of MET, which inhibits estrogen biosynthesis, is an attractive treatment for postmenopausal women with hormone – dependent breast cancer. Since the early 1980s, the concept of mucoadhesion has gained considerable interest in pharmaceutical technology. It might open the door for novel, highly efficient dosage forms especially for oral drug delivery.

The most important goal of cancer chemotherapy is to minimize the exposure of normal tissues to drugs while maintaining their therapeutic concentration in tumors. Interestingly, nanoparticles (NPs) exhibit a significant tendency to accumulate in a number of tumors after intravenous injection. The uptake and consequently bioavailability of the drug may be increased and the frequency of dosing reduced with the result that patient compliance is improved. Various natural and synthetic polymers have been discovered as mucoadhesive excipients. Their mucoadhesive properties can be explained by their interaction with the glycoproteins of the mucus, based mainly on non-covalent bonds such as ionic interactions, hydrogen bonds and van der Waals forces. The biopolymer CS is obtained by alkaline deacetylation of chitin which is one of the most abundant polysaccharides in nature. Shell wastes of shrimp, lobster and crab are the main industrial sources of chitin. The primary amino group accounts for the possibility of relatively easy chemical modification of CS and salt formation with acids. At acidic pH, the amino groups are protonated, which promotes solubility, whereas CS is insoluble at alkaline and neutral pH. Because of its favorable properties, such as enzymatic biodegradability, non-toxicity and biocompatibility CS has received considerable attention as a novel excipient in drug delivery systems, and has been included in the European Pharmacopoeia since 2002. The administration of nanoparticles will also provide the advantage of facilitating their injection through standard infiltration needles. So far, there was one published study on MET nanoparticles prepared by the direct precipitation technique. It has been shown that polymers with thiol groups provide much higher adhesive properties than polymers generally considered to be mucoadhesive.

To increase patient compliance and to overcome the undesirable side effects, MET could be entrapped into biodegradable nanoparticles for sustained delivery so that it can inhibit estrogen biosynthesis for a prolonged time by virtue of increased local concentration of the drug at the receptor site. To date, three different TCS derivatives have been synthesized: CS-thioglycolic acid conjugates, CS-cysteine conjugates and chitosan-4- thio-butyl-amidine (CS-TBA) conjugates. These TCS have numerous advantageous features in comparison to unmodified CS, such as significantly improved mucoadhesive and permeation enhancing properties. The strong cohesive properties of TCS make them highly suitable excipients for controlled drug release dosage forms. We have prepared MET-loaded TCS-PMAA nanoparticles by a emulsion solvent evaporation technique to obtain smaller particle size with high entrapment efficiency and a sustained release profile. Particle size, morphology, entrapment efficiency, drug – polymer interaction and *in vitro* release of MET-TCS-PMAA nanoparticles were evaluated. The influence of % of drug (relation to polymer mass) on formulation performance including particle size, entrapment efficiency and *in vitro* release were investigated. The synthesized polymer was characterized by IR spectroscopy. The FT-IR spectrum of TCS-PMAA is shown in Figure 26. In the spectra of TCS –OH and –NH stretch were clearly seen at 3385.292 and 3261.866 cm^{-1}, respectively. The additional presence of amidine I and amidine II bonds are seen at 1703.855 and 1611.087 cm^{-1} corresponding to $\geq NH_2^+$ stretch and $NH \geq NH_2^+$, respectively, being two coupled vibrations. The presence of an additional bond at 1495.3783 and 1449.0847 cm^{-1} can be assigned for N-H bond of the salt $NH_2^+Cl^-$. Other characteristic peaks of CS O-H stretch, C-H stretch and C-O stretch were present at 3400-3600, 2930, and 1009-1171 cm^{-1}, respectively. This confirmed the synthesis of TCS. The spectrum of TCS was well correlated with reports by Matsuda et al. for TCS. The Thiol content of TCS was found at 214±52 μmol/g.

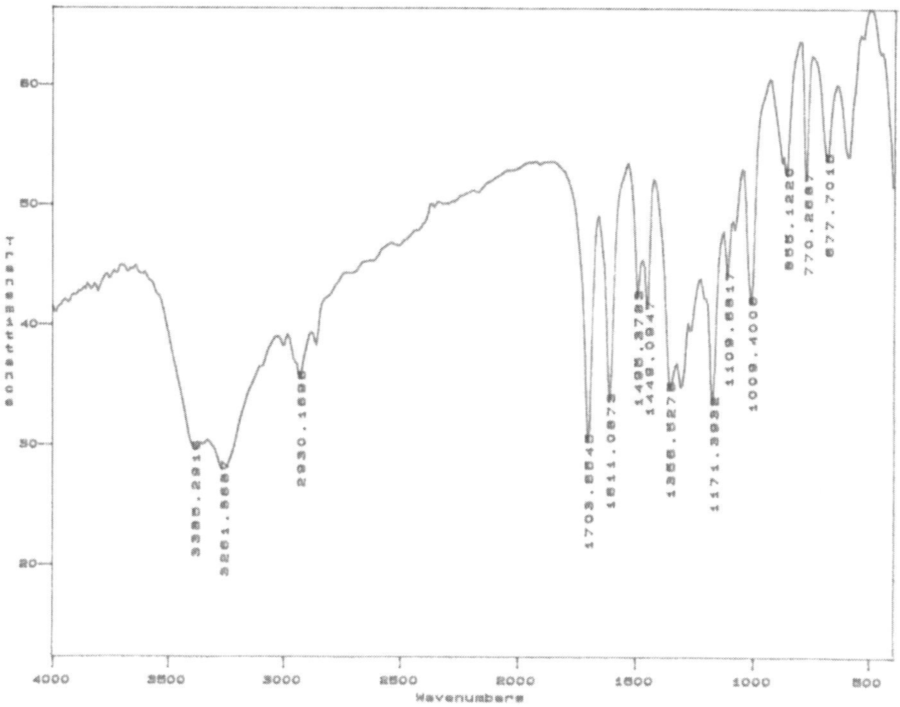

Figure 26. FT-IR spectra of *TCS-PMA* nanoparticles.

Figure 27 shows the percentage swelling of TCS-PMAA nanoparticles at different time intervals. The results revealed that nanoparticles swelled rapidly when immersed in a phosphate buffer (pH 6.8). It is reported that adhesive properties and cohesiveness of mucoadhesive polymers are generally affected by their swelling behavior. Mucoadhesive nanoparticles are anticipated to take up water from the underlying mucosal tissue by absorbing, swelling and capillary effects, leading to considerable stronger adhesion. The percent swelling of nanoparticles was found to follow the rank order 248±18%, 198±15%, 279±26% and 164±15%, respectively. It was observed that TCS-PMAA nanoparticles swelled slowly and produced higher mucoadhesive strength. This is perhaps because slow swelling avoids the formation of over hydrated structure that loses its mucoadhesive properties before reaching the target.

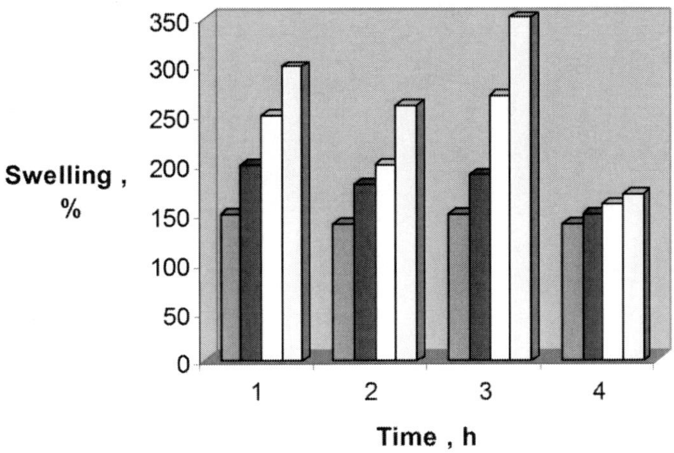

Figure 27. Percentage swelling measurement of *TCS-PMA* nanoparticles Data are presented as mean ± SD (n=3).

Figure 28 shows the release of MET from mucoadhesive nanoparticles. Drug powder was completely released (95.3±4.1%) within 1 hour. The time taken to release 75% of MET from TCS-PMAA nanoparticles was 5.0±0.6, 9.6±1.0, 4.2±0.2 and 5.3±0.5 hours, respectively.

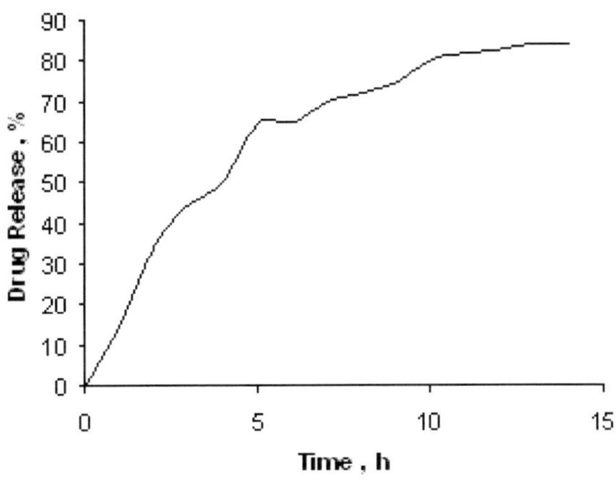

Figure 28. Percentage in vitro drug release of MET from *TCS-PMA* nanoparticles.

The significantly higher time required by the TCS-PMAA nanoparticles to release MET was due to its better stability in acidic medium, which contributed significantly to a lower amount of drug release during the initial 1 hour of dissolution (30.1±1.5% and 21.2±0.6%) drug released from TCS-PMAA nanoparticles may be attributed to the higher solubility of CS in acidic medium. CS is soluble in acidic medium but crosslinking with glutaraldehyde through its amino group stabilized the nanoparticles matrix and provides the sustained release. The significantly lesser drug release from TCS-PMAA nanoparticles is due to the presence of disulphide bonds in the nanoparticle matrix that further stabilized the structure along glutaraldehyde as the cross linking agent.

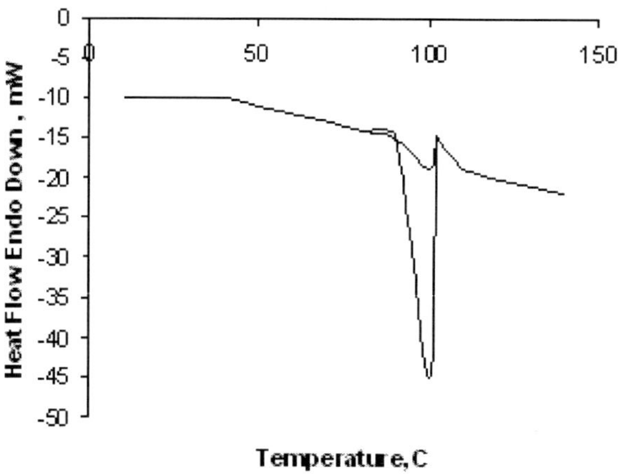

Figure 29. DSC thermograms of (a) *TCS-PMA* nanoparticles (b) MET Drug.

The thermal behavior of a polymer is important in relation to its properties for controlling the release rate in order to have a suitable drug dosage form. The glass transition temperature (Tg) was determined from the DSC thermograms (Figure 29). The higher Tg values probably related to the introduction of cross- links, which would decrease the flexibility of the chains and the ability of the chains to undergo segmental motion, which would increase the Tg values. On the other hand, the introduction of a strongly polar carboxylic acid group can increase the Tg value because of the formation of internal hydrogen bonds between the polymer chains. The morphology of nanoparticles was examined by scanning electron microscopy (SEM, Philips XL-30 E). The nanoparticles were mounted on metal stubs using double – sided tape and coated with a 150 Å layer of gold under vacuum. Stubs were visualized under a scanning electron microscope.

Figure 30 shows the SEM of TCS-PMAA nanoparticles that are synthesized by chemical reaction. This nanoparticle is very sensitive to the temperature that is due to the interaction of the electron and the sample. Scanning electron micrography images were obtained from a diluted solution of the MET particle. The white spots are MET nanoparticles. The SEM image shows the presence of MET spherical particles in polymer matrix, which are evenly distributed throughout the polymer, which is also confirmed from ^{1}H-NMR studies. As observed from SEM photomicrographs, the crystals of MET have a different appearance than recrystallized MET. These nanoparticles do not have clearly defined crystal morphological features in the SEM photomicrographs.

The most important characteristic of any drug is efficiency. This characteristic may often be reduced because of the inability to deliver the drug to the specific cells or tissues. After administration, the drug may pass through different physiologic barriers and /or pathways, decreasing the actual amount of the drug that reaches the site. In the search for an ideal carrier system, the dendrimers may have significant potential. Dendrimers are synthetic macromolecules with well-defined globular structures. The need for advanced materials with improved and new properties for a variety of technological applications has created a demand for both new forms of matter and for polymers that have highly controlled molecular architectures. The established approach to dendritic macromolecules has traditionally involved a divergent process in which growth is started from a polyfunctional core and continued outwards in a stepwise manner that affords larger and larger macromolecules as the process is continued.

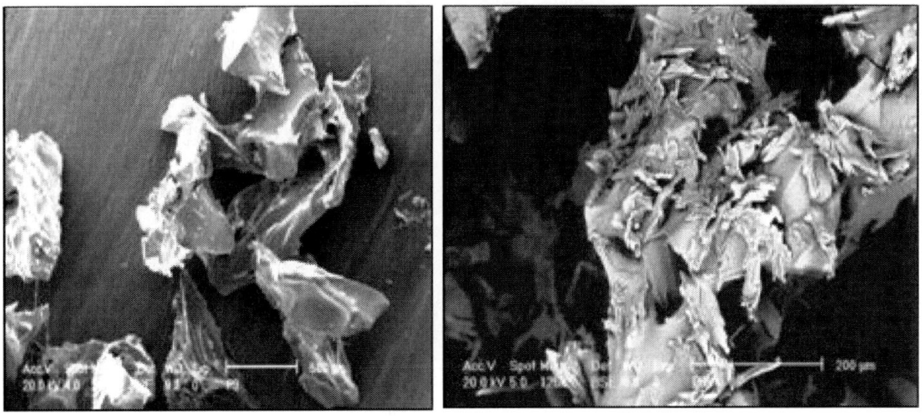

Figure 30. SEM photomicrograph of *TCS-PMA* nanoparticles with *MET* drug.

The fundamental attribute of the convergent approach is that it begins at what will be the periphery of the molecule, proceeding inwards. It is this feature more than any other that allows for unparalleled control over molecular architecture.

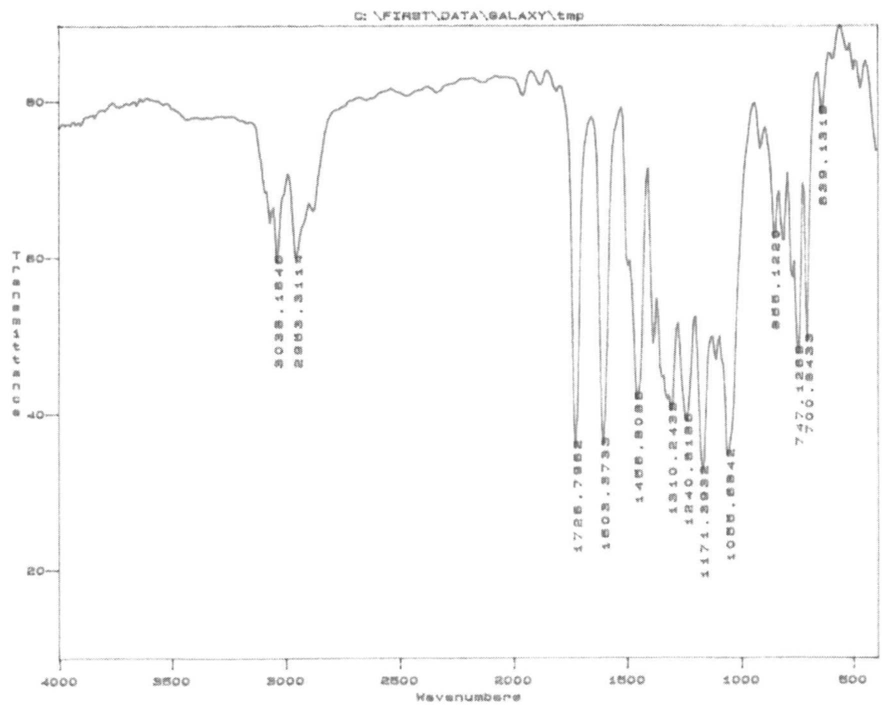

Figure 31. FT-IR spectrum of [G2] –$COOCH_3$ Dendrimer.

We have investigated the potential of dendrimers and hyperbranched polymers as a drug carrier system using poly (2-hydroxy ethyl methacrylate), since the methodologies for evaluation of the cellular activity of dendrimer are well known. We have also explored the interaction of the poly (2-hydroxy ethyl methacrylate), with an aromatic polyetheric dendrimer. The nature of the interaction was characterized by FT-IR and ^{1}H-NMR spectroscopy. Figure 31 shows the FT-IR spectrum of the G2-$COOCH_3$ dendrimer (AB8, heptamer) where the % of transmittance is plotted as a function of the wave number (cm^{-1}). The characteristic FT-IR peaks at 3078, 3053, 2961, 1718 cm^{-1} are due to the presence of =CH bond stretching vibrations , the aliphatic CH bond in protected groups (t-butyldimethylsilyloxy) and carbonyl (C=O) group, respectively.

Figure 32 shows the FT-IR spectrum of the G2-CH_2OH dendrimer (AB8, heptamer). The characteristic FT-IR peaks at 3408,2960,2932,2860,1544 cm^{-1} are due to the presence of the OH phenolic group, the aliphatic and aromatic =CH bond in protected methyl groups and C=C bond of aromatic group, respectively.

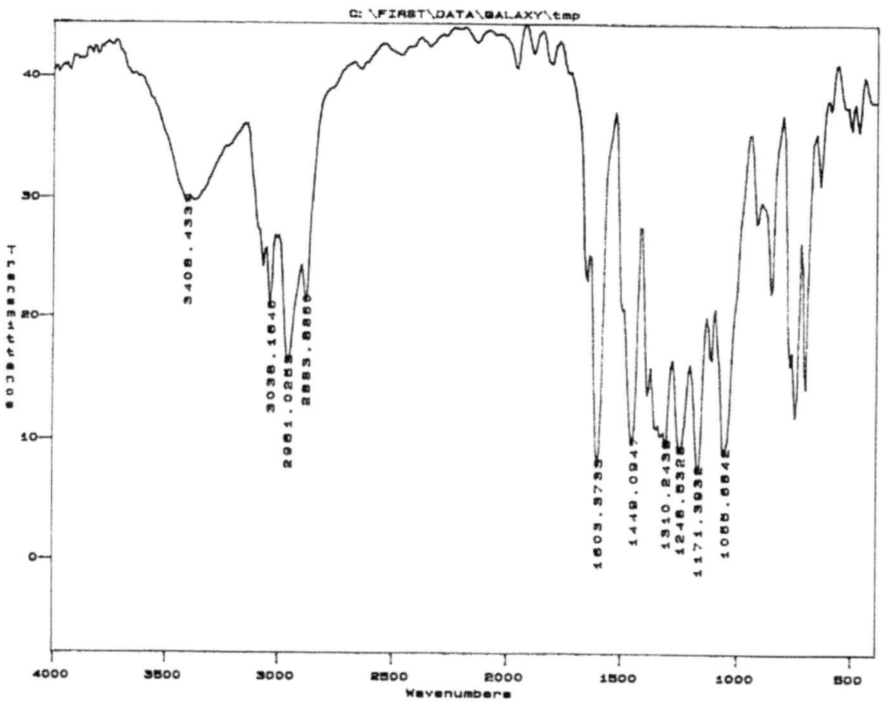

Figure 32. FT-IR spectrum of [G2] – CH_2OH Dendrimer.

Figure 33 also shows the FT-IR spectrum of the G2-CH_2OH dendrimer (AB8, heptamer)/ poly (2-hydroxy ethyl methacrylate) nanocomposite. The characteristic FT-IR peaks at 3408, 2960, 2932, 2860, 1544and 1726cm^{-1} are due to the presence of the OH phenolic group, the aliphatic and aromatic =CH bond in protected methyl groups and C=C bond of aromatic group and carbonyl (C=O) group of 2-hydroxy ethyl methacrylate, respectively.

Figure 34 shows the SEM of aromatic polyether dendrimer/poly (2-hydroxy ethyl methacrylate) nanocomposite with 5-aminosalicylic acid (5-ASA) as a drug model. This nanocomposite is very sensitive to the temperature that is due to the interaction electron and the sample. Scanning electron micrography images were obtained from a diluted solution of the nanocomposite particles. The white particles are 5-ASA nanoparticles. The SEM image shows the presence of 5-aminosalicylic acid (5-ASA) particles in a polyfunctional dendrimeric matrix, which were distributed throughout the nanocomposite, which is also confirmed from ^{1}H-NMR studies.

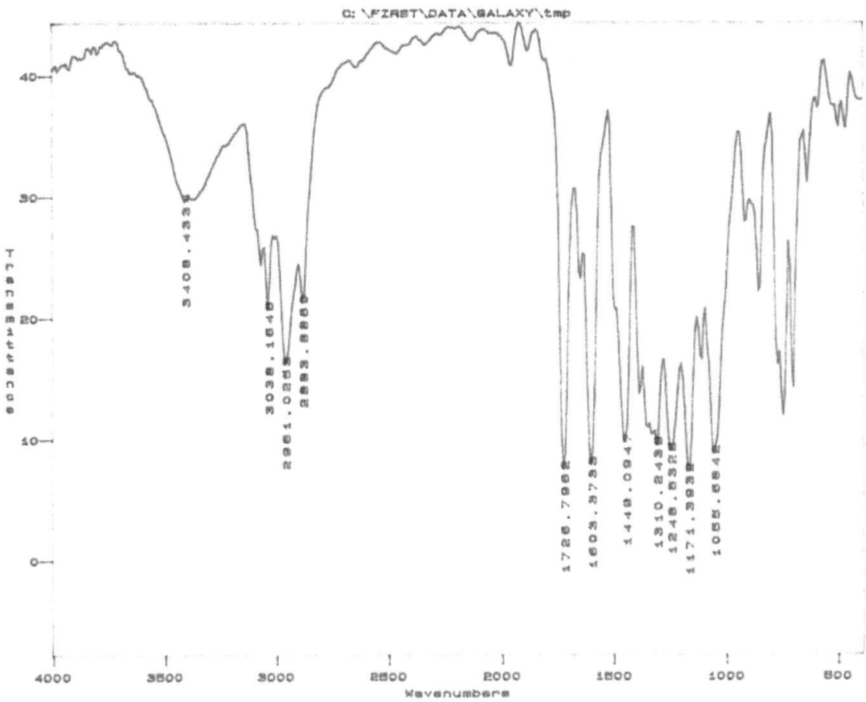

Figure 33. FT-IR spectrum of [G2] – CH_2OH Dendrimer/poly (2-hydroxy ethyl methacrylate)/5-ASA nanocomposite.

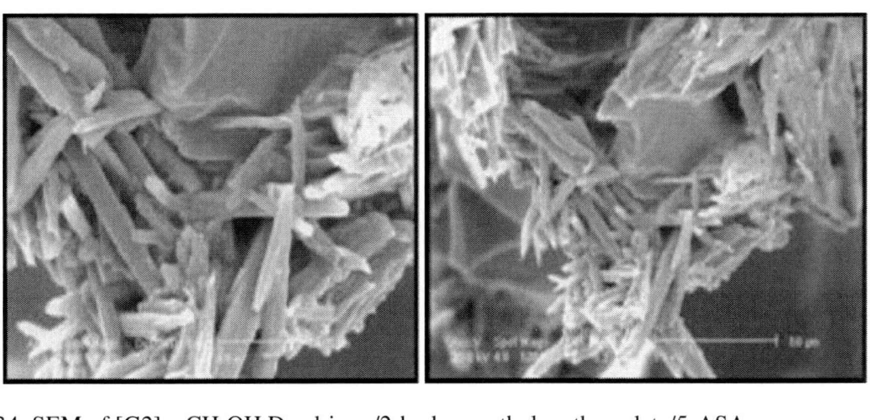

Figure 34. SEM of [G2] – CH_2OH Dendrimer/2-hydroxy ethyl methacrylate/5-ASA nanocomposite.

The ability of the dendrimer to form a complex with a 2-hydroxy ethyl methacrylate monomer depends on the core- surface groups of dendrimer, electrostatic interactions between the dendrimer and the monomer, and the ability of the drug to form a conjugate with the dendrimer through chemical bonding. Therefore, it is possible to manipulate the incorporation process for a given monomer by the appropriate selection of the dendrimer and the surface functionality. One might expect that the 2-hydroxy ethyl methacrylate with the carboxylic group may form a complex with surface OH groups of polyether dendrimer. Figure 35 shows the 400 MHz ^{1}H-NMR spectrum of [G2] – CH_2OH Dendrimer/2-hydroxy ethyl methacrylate conjugate in which three regions can be seen. The silane group 48 and 72 protons are in the regions of 0.37 and 0.85 ppm, respectively. The resonance of CH_2 protons (c group), CH_2, 4 protons (d group), CH_2 8 protons (Heptamer protons), benzylic protons of heptamer(a group) , vinylic CH_2 protons are 4.09, 4.15, 4.27, 4.86, 5.66 ppm, respectively. The resonances for the aromatic protons of the monomer units at the dendrimer occur in the region 6.73 ppm separate resonances are observed in the appropriate ratio for each layer of monomer units. The solvent peak occurs in the region 7.51 ppm. Figure 36 shows the 400MHz ^{1}H-NMR spectrum of [G2] – CH_2OH Dendrimer/poly (2-hydroxy ethyl methacrylate) nanocomposite in which three regions can be seen. The silane group 48 and 72 protons are in the regions of 0.10 and 0.85 ppm, respectively. The resonance of the methyl group, polymer CH_2 protons, CH_2O protons, CH_2 8 protons (Heptamer protons), CH_2 protons of heptamer (a group), are 1.68, 2.27, 4.17, 4.65, 4.79 ppm, respectively.

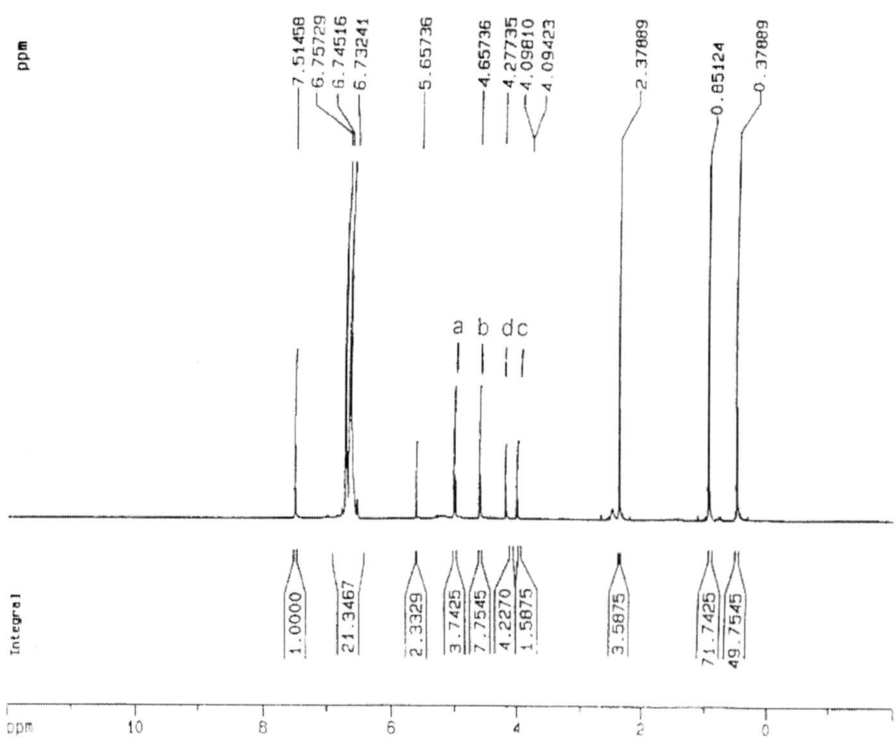

Figure 35. 400MHz ^{1}H-NMR spectrum of [G2] – CH_2OH Dendrimer/2-hydroxy ethyl methacrylate conjugate ($CDCl_3$).

The resonances for the aromatic protons of the monomer units at dendrimer occur in the region 6.28-6.32 ppm. Separate resonances are observed in the appropriate ratio for each layer of monomer units. The DMSO peak occurs in the region 3.34-3.48 ppm.

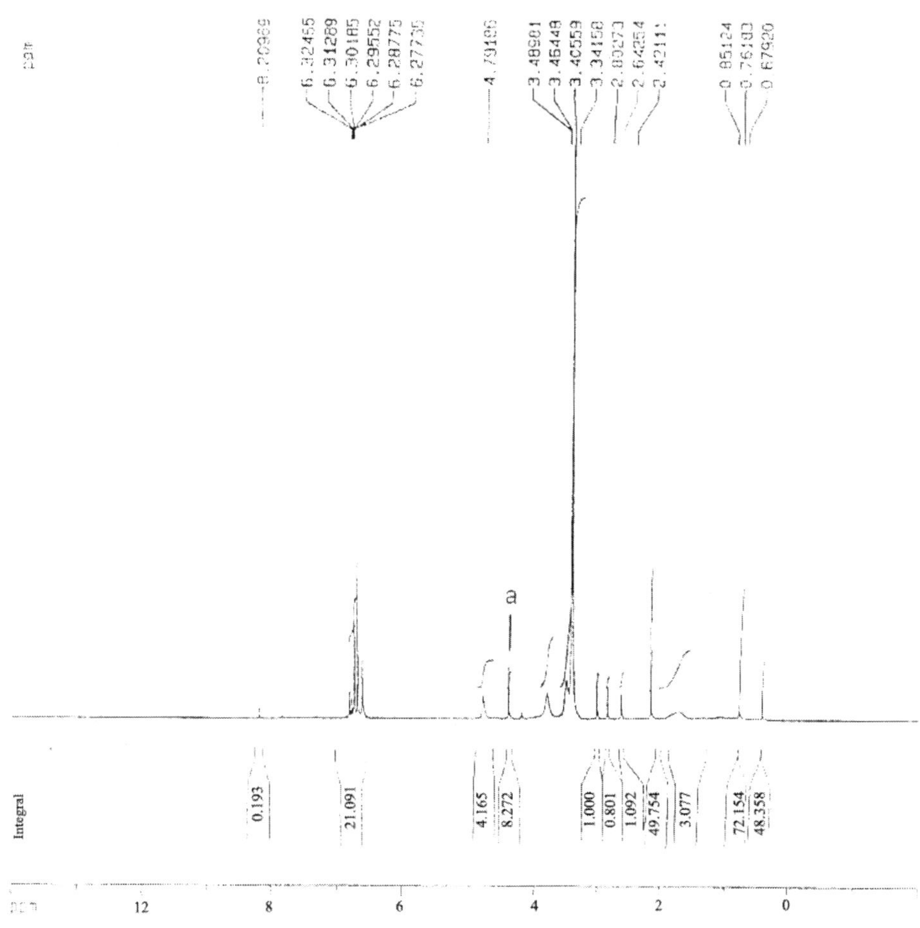

Figure 36. 400MHz ^{1}H-NMR spectrum of [G2] – CH_2OH Dendrimer/poly (2-hydroxy ethyl methacrylate) nanocomposite (DMSO-d6).

The thermal behavior of a polymer is important in relation to its properties for controlling the properties of the nanocomposite. The glass transition temperature (Tg) was determined from the DSC thermograms. The values are given in Table 1. The higher Tg values probably related to the nanocomposite size, which would decrease the flexibility of the chains and the ability of the chains to undergo segmental motion, which would increase the Tg values. On the other hand, the introduction of a strongly phenolic OH group can increase the Tg value because of the formation of internal hydrogen bonds between the polymer chains. Nano and micro polymer bonded drugs (50 mg) were poured into 3 mL of aqueous buffer solution (SGF: pH 1 or SIF: pH 7.4). The mixture was introduced into a cellophane membrane dialysis bag. The bag was closed and transferred to a flask containing 20 mL of the same solution maintained at 37°C. The external solution was continuously stirred, and 3 mL samples were removed at selected intervals. The removed volume was replaced with SGF or SIF. Triplicate samples were used. The sample of hydrolyzate was analyzed and the quantity of 5-ASA was determined using a standard calibration curve obtained under the same conditions. Results of drug content and encapsulation efficiency (EE) demonstrated that drug content increased from 9.24±0.81 mg/100mg to 25.70±0.43mg/100mg with an increase in the amount of drug from 5% to 20% wt/wt. No increase in drug content was observed with the further increase in the amount of the drug, above 15% wt/wt, which could be due to the limited solubility of drug in solvent. This is supported by the presence of drug particles on the surface of the nanocomposite with 20% of drug concentration. In the case of nanocomposite, nearly 80% to 90% of the drug was released in the initial 8 to 10 hours. This situation is not acceptable for drugs that are required to be released locally in the colon. The effect of drug concentration, nanocomposite concentration and cross-linking time on in vitro drug release was observed. In vitro drug release after 10 hours was 84.2%±4.1% in the microsized particles and pH =7.4 having 15% drug, while it was 89.2%±3.78% for nanosized particles with 20% drug. The effect of nanocomposite concentration on the release of drug was found to be limited. It is also observed that the amount of drug released from the nanocomposite decreased with increasing cross-linking time. These properties are probably explained by the promotion of cross-links between nanocomposite chains. Freeze-drying of the samples resulted in larger and more porous nanocomposite and slightly faster disintegration times compared with air-dried nanocomposite. Freeze-drying had the advantage of avoiding drug extraction by immediately freezing and removing the water present within the nanocomposite by sublimation. When air-dried, the nanocomposite was exposed to water for longer periods of time. In the second part of this investigation the cross-linked nanocomposite was used as nanosized particles. The relatively higher drug release after 10 hours reveals a significant increase. This increase in drug release suggests that most of the drug has dissolved and leached out through the nanocomposite. The drug released during 10-hour release rate studies, is because of the presence of nanosized particles. Modified glycol chitosan nanoparticles for cancer therapy have been recently formulated for the administration of docetaxel camptothecin. New modified chitosan nanoparticles are intended as vectors of the antifolic agent methotrexate (MTX). From a therapeutic viewpoint, MTX is substrate of drug efflux transporters located at the apical side of the Blood – Brain – Barrier (BBB) such as p-glycoprotein, multidrug resistance protein and breast cancer protein and thus, the MTX clinical efficacy for the treatment of brain tumors is limited. Therefore, the nanocapsulation of the drug in the polymeric networks of Glycol chitosan network was aimed to mask the drug from the attacks of the aforementioned efflux systems and moreover, to overcome the BBB, the role of polymeric network was investigated. The studies have shown that these nanoparticles had good spherical geometry. It is obvious that the surface of the nanoparticles shrank and a densely crosslinked gel structure was formed. This may explain the greater retardation of drug release from metrics of higher cross linker content. The average drug entrapment was found to be 93.65%±2.15% in the nanoparticles. The results of the drug content and EE demonstrated that drug content increased from 8.32±0.50 mg/100mg to 38.25±0.50 mg/100mg of nanoparticles by increasing the amount of drug from 5% to 15% wt/wt. No significant increase in drug content was observed by further increasing the amount of the drug, i.e., above 15% wt/wt, which could be due to the limited solubility of the drug in DMF and that is endorsed by the presence of drug particles on the surface of the nanoparticles prepared with 15% of drug concentration. The percent EE was increased up to

93.65%±2.47% with increasing polymer concentrations to 4%.Concentration of the cross-linking agent exhibited no significant effect on percent EE. The incorporation of MTX into nanoparticles produced a smooth surface and compact structure.

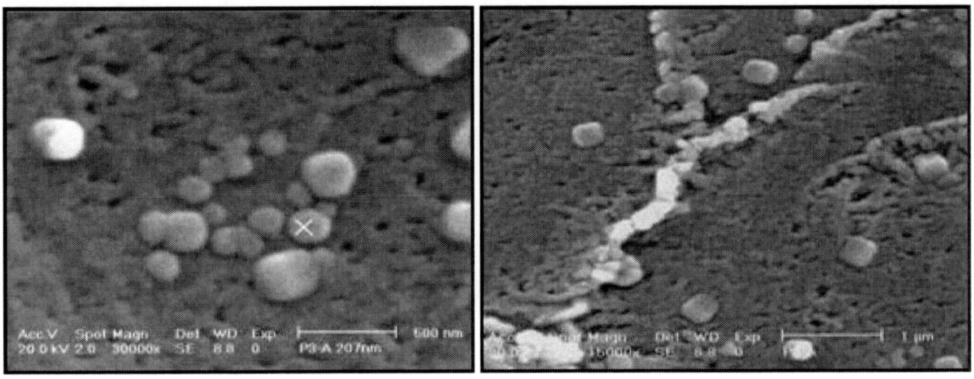

Figure 37. The SEM of Glycol Chitosan hydrogels.

The particle size observed in ESEM is smaller than that measured by the Zetasizer. This is because dried nanoparticles were used in ESEM, whereas particles in the liquid dispersion were analyzed by the Zetasizer. GCS-DS nanoparticles are hydrophilic and would be expected to swell in water, thus producing a large hydrodynamic size when measured by the Zetasizer.

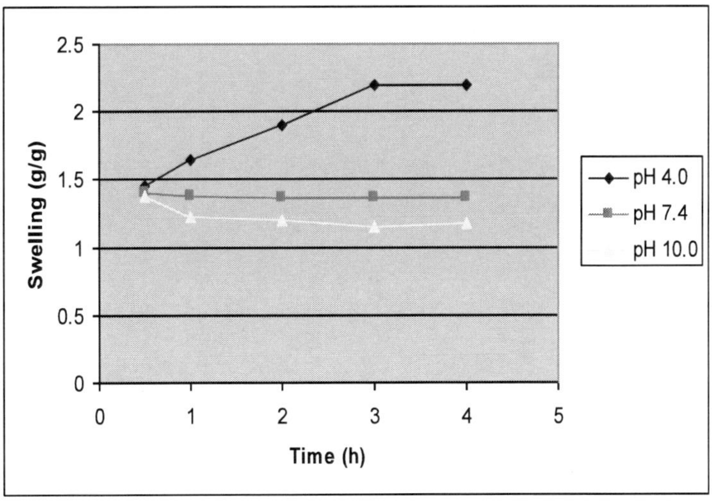

Figure 38. The swelling ratio of Glycol chitosan hydrogels versus time.

MTX – loaded GCS-DS nanoparticles were obtained spontaneously upon the mixing of the DS aqueous solution (0.1 %wt/vol) with the GCS solution (0.1% wt/vol) under magnetic stirring, with MTX dissolved in GCS-DS solution. The incorporation of MTX into the GCS-DS nanoparticles resulted in a sharp increase in the particle size of the nanoparticle dispersion. The significant increases in particle size give a good induction of the incorporation of MTX into GCS-DS nanoparticles. The transport of MTX NPs indicated an enhancement role exerted by the nanoencapsulation in the polymeric matrix with reference to MTX alone used as a control.

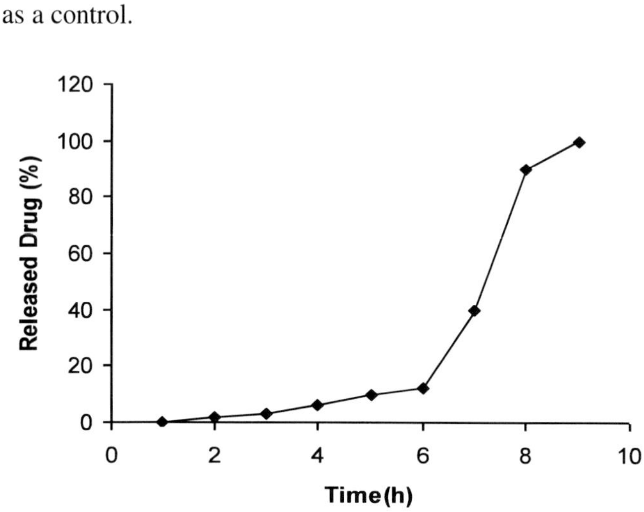

Figure 39. The release Drug from Glycol chitosan hydrogels.

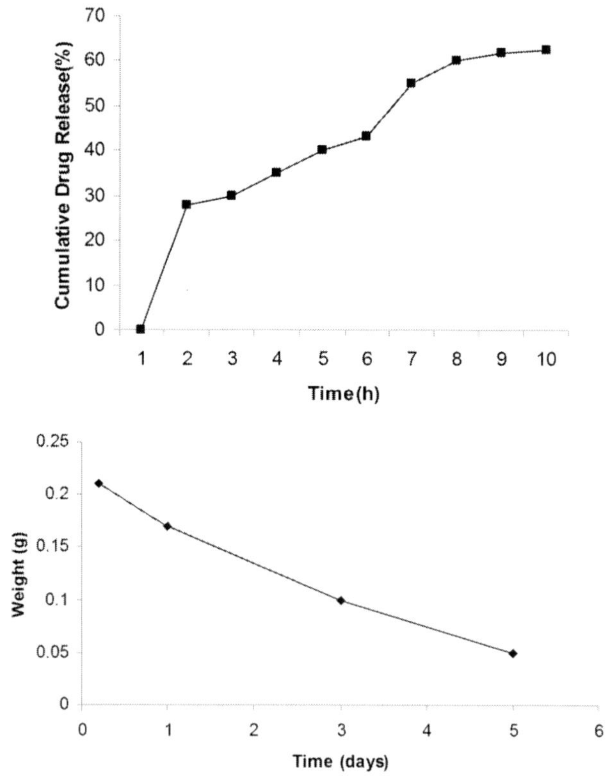

Figure 40. The cumulative drug release of Glycol Chitosan hydrogels.

The effect of MTX concentration, GCS-DS concentration and cross-linking time on in vitro drug release was also observed. In vitro drug release after 5 hours was 93.12%±1.8% in the case of nanoparticles having 15% drug, while it was 94.32%±2.3% for nanoparticles with

20% drug. The effect of GCS-DS on the release of drug was found to be limited. The studies show that the amount of drug released from nanoparticles decreased with increasing cross-linking time. These properties are probably explained by the promotion of cross-links between GCS-DS chains.

Freeze-drying of the samples resulted in larger and more porous nanoparticles compared with air-dried nanoparticles. Freeze-drying had the advantage of avoiding drug extraction by immediately freezing and removing the water present within the nanoparticles and a burst effect during the dissolution study.

The release profiles of MTX-loaded GCS-DS nanoparticles were evaluated in water and a phosphate buffer, which was either in a different ionic strength or with saline, to study the underpinning mechanisms for drug release. The greatest release for MTX –loaded GCS-DS nanoparticles occurred in the release media of a high ionic strength. In contrast, a significantly small portion of MTX was released in water over the release study period. The burst release was observed with nanoparticles, and it may have arisen from desorption of loosely attached MTX from the surface of the matrix polymers [131].

4.7. Synthesis of Architecturally Well-Defined Nanogels via RAFT Polymerization for Potential Bioapplications

As the nanoscale counterpart of hydrogels, nanogels inherit the smart properties of being able to respond to various external stimuli, albeit at a much faster rate, because the response time is proportional to the square of a linear dimension of hydrogels. While some of the well-known properties of nanogels, such as high water content, high surface area and responsive behavior, have been intensively studied and exploited in various applications, new exciting properties distinct to nanogels are continuously being discovered. For example, Lyon and co-workers recently reported self-healing properties of nanogel assemblies in the form of colloidal crystals and films. They also reported the deformable transportation of nanogels through nanosized pores.

These new findings greatly expand the horizon of nanogel properties. By doping nanogel colloidal crystals assembled from one size of nanogels with a certain amount of nanogels of a different size, Iyer and Lyon demonstrated that the assemblies can accommodate the defect nanogels with no observable perturbation of the lattice structure or dynamics, because of the self-healing property of the nanogels due to their softness and compressibility. This is in contrast with hard-sphere colloidal assemblies that require high monodispersity of the particles to form homogeneous crystals. By using a layer-by-layer method, South and Lyon fabricated hydrogels composed of polyelectrolyte nanogels.

They demonstrated that the hydrogel films can be repeatedly healed after damage or mechanical deformation. Upon addition of water, the damage was quickly healed on a time scale of seconds. The excellent self-healing property of hydrogel films assembled from nanogels was possibly due to the migration and reorganization abilities of nanogels. In a recent study of nanogel translocation through pores, Hendrickson and Lyon demonstrated that under pressure differentials relevant to renal filtration, nanogels can easily pass through cylindrical pores whose openings are more than ten-fold smaller than the unperturbed nanogel diameter.

This finding is of particular interest for drug delivery application, which means that by controlling the softness or deformability of the nanogels, nondegradable nanogels having diameters smaller than approximately 80 nm can be used directly for in vivo application because their renal excretion becomes possible due to their high deformability. Of even more interest and also relevance to Lyon's work, DeSimone and co-workers studied the biodistribution and circulation times of hydrogel microparticles of different modulus by controlling the level of cross-linking of the particles. They synthesized the gel particles using a molding technique called PRINT (Particle Replication in Nonwetting Templates) to mimic the size and shape of red blood cells.

Decreasing the modulus, thus increasing the deformability, of these particles altered their biodistribution properties, allowing them to bypass several organs, such as the lung, that entrapped their more rigid counterparts. The modulus of the particles also exhibited a significant influence on the particle circulation times. An 8-fold decrease in hydrogel modulus correlated to a greater than 30-fold increase in the elimination phase half-life for these particles. These studies demonstrate that hydrogel particles are an ideal material platform for use in bioapplications in vivo.

4.7.1. Preparation of Nanogels

One strategy involves the formation of nanogels using preformed polymers whereas the other entails the formation of nanogels via direct polymerization of monomers. No matter which strategy is used, the two important factors that need to be carefully considered are (1) the capability of the strategy to form stable nanogels under synthetic conditions such that the nanogels do not dissociate into polymer chains and (2) the capability of the strategy to endow colloidal stability of the formed nanogels. For practical applications, the latter requirement means that not only macrogelation, flocculation or precipitation should be prevented during the synthesis but also the nanogels should possess excellent colloidal stability without aggregation (at least for the time scale of interest) when used in a complex biological milieux. However, it is not trivial to achieve the required colloidal stability of nanogels in biological milieux because of the complex contents (proteins, cells, salts, reactive compounds etc.) and the dynamic nature of biological fluid. In order to assess the colloidal stability of nanogels in a more bio-related context, it is perhaps more meaningful to test the nanogels in undiluted blood, because many of the claimed antifouling nano-sized materials would just quickly flocculate or precipitate when tested in undiluted blood. The perceived methods of imparting colloidal stability of nanogels or imparting antifouling properties of nanogels in biological fluid, share many common features with those of antifouling surface coatings. Steric, electrostatic repulsion and electrosteric coatings are the typical means to ensure antifouling coatings of nanogels.

Coating nanogels with water soluble and biocompatible polyethylene glycol (PEG) is perhaps the most widely used method to endow antifouling properties of nano-sized materials. However, it should be noted that many PEG alternatives that show similar or even improved properties are also being intensively pursued. Since proteins and cell surfaces are charged, coating nanogels with charged groups usually causes unfavorable interactions through direct electrostatic interactions of oppositely charged nanogels with proteins/cells, or through indirect interactions of nanogels with proteins/cells of the same type of charge via a salt bridge effect. Keeping the charged groups electrically neutral, as in the case of zwitterionic groups, has proven to be an effective strategy in the synthesis of nanosized materials with excellent colloidal stability. Parallel with the development of antifouling coating materials, the development of physical methods to enhance antifouling properties of existing antifouling materials is equally important. The combination of both chemical and physical approaches may finally lead to the development of robust antifouling platforms that could be widely used in antifouling surfaces both at the macroscopic and nanoscopic scales.

4.7.2. Preparation of Nanogels Using Preformed Polymers

Preformed polymers used to prepare nanogels include polymers having only one block and block copolymers having at least two blocks. For polymers with only one block, nanogel synthesis typically involves two steps. The first step is modifying existing polymers with groups that can be subsequently used to form physical or chemical cross-links in the second step. This method is frequently used for water-soluble naturally occurring polymers such as polysaccharides. While for synthetic polymers, it is also possible to incorporate cross-link-forming groups during the polymerization process. To obtain nanogels without macrogelation, flocculation or precipitation, synthesis of nanogels using preformed polymers is typically carried out in dilute aqueous solutions, preferably in combination sometimes with ultrasonication, in organic solvents or with the aid of surfactants in microemulsion-based heterogeneous systems. The formation of physical cross-links in such polymers relies on either self-association or complementary association forces. For nanogels cross-linked by self-association, hydrophobic groups such as cholesterol, hydrophobic polymer chains such as poly (lactic acid), or responsive polymer chains such as poly (N-isopropylacrylamide) (PNIPAM) have been grafted onto water-soluble polymers such as polysaccharides in relatively low densities. The association of these groups or side polymer chains forms multiple hydrophobic domains that only account for a low volume percentage of the nanogels. It is this feature that distinguishes them from core–shell type micelles in which the hydrophobic core accounts for a large volume percentage. For nanogels cross-linked by complementary association, electrostatic interaction, complexation of complementary DNA strands, and host–guest interaction such as cyclodextrin-based inclusion have been employed to prepare stable nanogels.

Preparation of nanogels with chemical cross-links requires a certain degree of population of reactive sites along the polymer chains. Some polymers possess inherent reactive sites that allow convenient cross-linking reactions to occur. For example, polyethyleneimine (PEI) has a high density of amine (primary, secondary and tertiary) functional groups, all of which can be used directly with selected cross-linking reagents to form nanogels. In addition, many natural polymers have various reactive groups such as amine and carboxylic acid that can be used directly for cross-linking reactions. Otherwise, (latent) reactive groups need to be first incorporated into the polymers via modification of existing polymers or incorporated during the polymerization process. Depending on the desired applications, the cross-links can be either permanent or cleavable or a combination of both. Frequently, the cleavable cross-links are able to (reversibly) degrade in response to the application of physical or chemical signals, whereas the permanent crosslinks can be used to maintain the integrity of the nanogels especially when the nanogels are designed for multiple uses. Reactions that are used to form chemical cross-links are usually simple, efficient and high yielding. In addition, when water-soluble polymers are cross-linked by another nanosized entity such as protein, hybrid nanogels are prepared.

4.7.3. Preparation of Nanogels via Monomer Polymerization

While cross-linking preformed polymers, physically or chemically, represents an important strategy for the preparation of nanogels, especially for naturally occurring water-soluble polymers, the preparation of nanogels via direct monomer polymerization combines the two processes of polymerization and formation of nanogels in a one-pot protocol. Thus, in a sense, synthesizing nanogels via monomer polymerization highlights an accelerated strategy, which in some cases involves polymerization-induced self-assembly of the in situ formed polymers. In some other cases, the formation of nanogels via polymerization requires the application of templating methods to realize the desired size and colloidal stability of the nanogels. Given the current intensive pursuit of highly efficient chemical approaches to the preparation of polymers and polymeric nanostructures, such an accelerated strategy of preparing nanogels is playing an increasingly important role in modern polymer chemistry and materials.

The most predominantly used polymerization technique to fabricate nanogels is free-radical polymerization, because of its ease of manipulation, efficiency, tolerance of functionality and adaptability to water-based heterogeneous systems. Recently, other polymerization techniques have begun to emerge in the fabrication of nanogels. For example, ring opening polymerization has been used for the synthesis of polyether-based nanogels. Given the recent development of polymerization techniques, it is not unreasonable to expect that the scope of polymerization techniques that can be used to fabricate nanogels will be significantly expanded. The most well-known example of nanogel synthesis through direct monomer polymerization is the synthesis of thermosensitive nanogels of PNIPAM via surfactant-aided or surfactant-free precipitation polymerization initiated by water-soluble free-radical initiators. In such a process, the monomer, cross-linker (typically methylenebisacrylamide), surfactant (optional), and radical initiator form a homogeneous solution prior to polymerization. Once initiated by thermolysis of the radical initiator, polymerization occurs by the addition of the ionic radical to the monomer, and, thereby, the formed polymers are end-functionalized with a charged ionic group. Since PNIPAM has a lower critical solution temperature (LCST) of 32 1C, when the polymerization temperature is higher than its LCST, the polymers that are formed change their solvation state from being hydrophilic to being hydrophobic, and thus precipitate from the solution to generate nanoparticles, which are stabilized by the charged ionic group derived from the radical initiator and by the surfactant as well if surfactant is used. Inspired by this success, a variety of other responsive (co)polymers have been used for the synthesis of responsive nanogels.

Nanogel synthesis in water via the direct polymerization of monomers is obviously advantageous because this process uses water as the dispersant. However, this process requires the polymer to have a transition in solubility, a property that not every hydrophilic polymer has. A more general process for the synthesis of nanogels of hydrophilic polymers is inverse phase polymerizations, in which the continuous phase is oil and the dispersed phase is water containing hydrophilic monomers. The dispersed aqueous phase is stabilized by a surfactant, which also serves as a template for the formation of nanogels. For some monomers, nanogel synthesis by direct polymerization in an aqueous dispersion system is impossible due to the good water solubility of both the monomer and the corresponding polymer. However, it is possible to design a dispersion polymerization system through the judicious choice of solvents or cosolvents. This point has been demonstrated by Armes group through a series of publications.

4.8. Large Payloads of Gold Nanoparticles into the Polyamine Network Core of Stimuli-Responsive PEGylated Nanogels for Selective and Noninvasive Cancer Photothermal Therapy

A biocompatible photothermal nanomedicine based on a PEGylated nanogel containing gold nanoparticles (GNPs) in a cross-linked network core of stimuli-responsive poly- [2-(N,N-diethylamino)ethyl methacrylate] (PEAMA) gel for cancer photothermal therapy (PTT) was prepared through the reduction of Au(III) ions without any reducing agents. The influence of the reduction conditions, such as pH, temperature, and N/Au ratio (molar ratio of the amino groups in the PEGylated nanogel to the Au (III) ions), on the formation of the GNPs in the stimuli-responsive PEAMA gel core (reducing environment) was also studied. Note that the PEGylated nanogel containing GNPs prepared at pH 6, 60 °C and N/Au ¼ 1 (PEGylated GNG (1)) was found to have the highest GNP-loading capacity with a diameter of about 8 nm, as observed by TEM; viz., about 27 GNPs formed in a single PEAMA gel core. PEGylatedGNG (1) showed a remarkable photothermal efficacy (DT ¼ 7.7 °C) under irradiation with an Ar ion (Ar^+) laser (514.5 nm) at a fluence of 39Wcm^{-2} for 6 min (14 kJ cm^{-2}). Note that PEGylated GNG (1) showed non-cytotoxicity in the absence of irradiation with Ar^+ laser (480 mg mL^{-1}: > 90% cell viability), whereas pronounced cytotoxicity (IC50 ¼ 110 mg mL^{-1}) was observed for PEGylated GNG (1) under irradiation with Ar^+ laser at a fluence of 26 W cm^{-2} for 5 min (7.8 kJ cm^{-2}), because of the heat-generation from the GNPs in the cells, which resulted in selective and noninvasive cancer PTT. Thus, PEGylated GNG (1), which has a high GNP-loading capacity, would be a promising nanomedicine for cancer PTT.

4.9. Synthesis and Characterization of Nanogels of Poly (N-Isopropylacrylamide) by a Combination of Light And Small-Angle X-Ray Scattering

Thermo-responsive crosslinked nanogels of N-isopropylacrylamide (NIPAM) were synthesized by emulsion polymerization and the size was varied using different concentrations of surfactant (sodium dodecyl sulfate, SDS) in the polymerization process. The collapse behavior of the nanogels at the lower critical solution temperature at around 32°C was investigated by dynamic light scattering, combined static light scattering (SLS) and small-angle X-ray scattering (SAXS). The combined data from SLS and SAXS were analyzed by a model for the nanogels which at intermediate temperatures included a central core and a more diffuse outer layer describing pending polymer chains with a low degree of cross linking. In the expanded state, the particles were modeled with a single component with a broad graded surface. In the collapsed state the nanogels were modeled as homogeneous and relatively compact particles. The amount of surfactant used had a profound effect on the final size of the nanogels due to the phenomenon of colloidal stabilization of the emulsion droplets during polymerization. The combination of SLS and SAXS as applied to the nanogels is an attractive method for particle characterization since it spans a very large range of scattering vectors from q = 0.0004 to 0.22 $A^{\circ -1}$.

4.10. Nanogel Surface Coatings for Improved Single-Molecule Imaging Substrates

Surfaces that resist protein adsorption are important for many bioanalytical applications. Bovine serum albumin (BSA) coatings and multi-arm poly (ethylene glycol) (PEG) coatings display low levels of non-specific protein adsorption and have enabled highly quantitative single-molecule (SM) protein studies. Recently, a method was developed for coating a glass with PEG–BSA nanogels, a promising hybrid of these two low-background coatings. We characterized the nanogel coating to determine its suitability for SM protein experiments. SM adsorption counting revealed that nanogel-coated surfaces exhibit lower protein adsorption than covalently coupled BSA surfaces and monolayers of multi-arm PEG, so this surface displays one of the lowest degrees of protein adsorption yet observed. Additionally, the nanogel coating was resistant to DNA adsorption, underscoring the utility of the coating across a variety of SM experiments. The nanogel coating was found to be compatible with surfactants, whereas the BSA coating was not.

Finally, applying the coating to a real-world study, we found that single ligand molecules could be tethered to this surface and detected with high sensitivity and specificity by a digital immunoassay. These results suggest that PEG–BSA nanogel coatings will be highly useful for the SM analysis of proteins [132-135].

References

[1] Peppas NA, Zach Hilt J, Khademhosseini A, Langer R, Hydrogels in biology and medicine: from molecular principles to bionanotechnology, *Adv Mater*, 18, 1345-1360, 2006.

[2] Huang X, Brazel CS, On the importance and mechanisms of burst release in matrix-controlled drug delivery systems, *J. Control Rel*, 73, 121-136, 2001.

[3] Hoffman AS, Hydrogels for biomedical applications, *Adv. Drug Deliv Rev*, 43, 3-12, 2002.

[4] Hoare TR, Kohane DS, Hydrogels in drug delivery: progress and challenges, *Polymer*, 49, 1993- 2007, 2008.

[5] Brannon-Peppas L, Polymers in controlled drug delivery, medical plastic and biomaterials magazine, November 1997, Medical device link, http://www.devicelink.com/mpb/archive/97/11/00 3.html, 10 August 2008.

[6] Peppas NA, Bures P, Leobandung W, Ichikawa H, Hydrogels in pharmaceutical formulations, *Eur. J. Pharm Biopharm*, 50, 27-46, 2000.

[7] Baroli B, Hydrogels for tissue engineering and delivery of tissue-inducing substances, *J. Pharm Sci*, 96, 2197-2223, 2007.

[8] Ganta S, Devalapally H, Shahiwala A, Amiji M, A review of stimuli-responsive nanocarriers for drug and gene delivery, *J. Control Rel*, 126, 187-204, 2008.

[9] He Ch, Kim SW, Lee DS, In situ gelling stimulisensitive block copolymer hydrogels for drug delivery, *J. Control Rel*, 127, 189-207, 2008.

[10] Schmaljohann D, Thermo- and pH-responsive polymers in drug delivery, *Adv Drug Delivery Rev*, 58, 1655-1670, 2006.

[11] Peppas NA, Khare AR, Preparation, structure and diffusional behavior of hydrogels in controlled release, *Adv. Drug Delivery Rev*, 11, 1-35, 1993.

[12] Entezami AA, Massoumi B, Artificial muscles, biosensors and drug delivery systems based on conducting polymers: a review, *Iran Polym J*, 15, 13-30, 2006.

[13] Ghazizadeh Y, Mirzadeh H, Amanpour S, Ahmadi H, Rabbani Sh, Investigation of effectiveness of chitosan hydrogel to stop bleeding and air leakage from lung fistula: an in vivo study, *Iran Polym J.*, 15, 821-828, 2006.

[14] Casadei MA, Pitarresi G, Calabrese R, Paolicelli P, Giammona G, Biodegradable and pH-sensitive hydrogels for potential colon-specific drug delivery: characterization and in vitro release studies, *Biomacromolecules*, 9, 43-49, 2008.

[15] Chiu HC, Wu AT, Lin YF, Synthesis and characterization of acrylic acid-containing dextran hydrogels, *Polymer*, 42, 1471-1479, 2001.

[16] Tavakoli J, Jabbari E, Etrati Khosroshahi M, Boroujerdi M, Swelling characterization of anionic acrylic acid hydrogel in an external electric field, *Iran Polym J*, 15, 891-900, 2006.

[17] Gao D, Xu H, Philbert MA, Kopelman R, Ultrafine hydrogel nanoparticles: synthetic approach and therapeutic application in living cells, *Angew Chem Int Ed*, 46, 2224-2227, 2007.

[18] Lin CC, Metters AT, Hydrogels in controlled release formulations: network design and mathematical modeling, *Adv. Drug Delivery Rev*, 58, 1379-1408, 2006.

[19] Satish CS, Satish KP, Shivakumar HG, Hydrogels as controlled drug delivery systems: synthesis, cross-linking, water and drug transport mechanism, *Ind J Pharm Sci*, 68, 133-40, 2006.

[20] Hennink WE, Van Nostrum CF, Novel cross-linking methods to design hydrogels, *Adv Drug Delivery Rev*, 54, 13-16, 2002.

[21] Wang Zh, Hou X, Mao Zh, Ye R, Mo Y, Finlow DE, Synthesis and characterization of biodegradable poly(lactic acid-*co*-glycine) via direct melt copolymerization, *Iran Polym J*, 17, 791-798, 2008.

[22] Bagheri Sh, Mohammadi-Rovshandeh J, Hassan A, Synthesis and characterization of biodegradable random copolymers of *L*-lactide, glycolide and trimethylene carbonate, *Iran Polym J*, 16, 489-494, 2007.

[23] Zhang Y, Zhu W, Ding J, Preparation of thermosensitive microgels via suspension polymerization using different temperature protocols, *J Biomed Mater Res Part A*, 75, 342-349, 2005.

[24] Baroli B Photopolymerization in drug delivery, tissue engineering and cell encapsulation: issues and potentialities, *J. Chem Tech Biotech*, 81, 491- 499, 2006.

[25] Lopes MA, Felisberti MI, Mechanical behavior and biocompatibility of poly(1-vinyl-2-pyrrolidinone)- gelatin IPN hydrogels, *Biomaterials*, 24, 1279-1284, 2003.

[26] Oh EJ, Kang SW, Kim BS, Jiang G, Cho IH, Hahn SK, Control of the molecular degradation of hyaluronic acid hydrogels for tissue augmentation, *J. Biomed Mater Res Part A*, 86, 685-693, 2008.

[27] Coviello T, Grassi M, Rambone G, Alhaique F, A crosslinked system from Scleroglucan derivative: preparation and characterization, *Biomaterials*, 22, 1899-1909, 2001.

[28] Garcia Y, Collighan R, Griffin M, Pandit A, Assessment of cell viability in a three-dimensional enzymatically cross-linked collagen scaffold, *J. Mater Sci Mater Med*, 18, 1991-2001, 2007.

[29] Ganji F, Abdekhodaie MJ, Ramazany-Sadtabadi A, Gelation time and degradation rate of chitosan as a thermosensitive injectable hydrogel, *J. Sol- Gel Sci Technol*, 42, 47-53, 2007.

[30] Mohamadnia Z, Jamshidi A, Mobedi H, Ahmadi E, Zohuriaan-Mehr MJ, Full natural hydrogel beads for controlled release of acetate and disodium phosphate derivatives of betamethasone, *Iran Polym J.*, 16, 711-718, 2007.

[31] Aalaie J, Vasheghani-Farahani E, Rahmatpour A, Semsarzadeh MA, Effect of montmorillonite on gelation and swelling behavior of sulfonated polyacrylamide nanocomposite hydrogels in electrolyte solutions, *Eur. Polym J.*, 44, 2024-2031, 2008.

[32] Aalaie J, Vasheghani-Farahani E, Semsarzadeh MA, Rahmatpour A, Gelation and swelling behavior of semi-interpenetrating polymer network hydrogels based on polyacrylamide and poly(vinyl alcohol), *J. Macromol Sci, Part B*, 47, 1017-1027, 2008.

[33] Stenekes RJ, Talsma H, Hennink WE, Formation of dextran hydrogels by crystallization, *Biomaterials*, 22, 1891-1898, 2001.

[34] Kimura M, Fukumoto K, Watanabe J, Ishihara K, Hydrogen-bonding-driven spontaneous gelation of water-soluble phospholipid polymers in aqueous medium, *J Biomater Sci. Polym Ed*, 15, 631- 644, 2004.

[35] Oh KS, Han SK, Choi YW, Lee JH, Lee JY, Yuk SH, Hydrogen-bonded polymer gel and its applicationcation as a temperature-sensitive drug delivery system, *Biomaterials*, 25, 2393-2398, 2004.

[36] Cappello J, Crissman JW, Crissman M, Ferrari FA, Textor G, Wallis O, Whitledge JR, Xia Zhou, Burman D, Aukerman L, Stedronsky ER, In-situ self-assembling protein polymer gel systems for administration, delivery, and release of drugs, *J. Control Rel*, 53, 105-117, 1998.

[37] Lowman AM, Smart Pharmaceuticals, www.gatewaycoalition. org/files/NewEH/ htmls/lowman.do c, 7 September 2008.

[38] Flory PJ, Rehner J, Statistical mechanics of crosslinked polymer networks. II. Swelling, *J. Chem. Phys*, 11, 521-526, 1943.

[39] Vasheghani-Farahani E, Vera JH, Cooper DG, Weber ME, Swelling of ionic gels in electrolyte solutions, *Ind. Eng. Chem. Res*, 29, 554-560, 1990.

[40] Grassi M, Grassi G, Mathematical modeling and controlled drug delivery: matrix systems, *Curr. Drug Deliv*, 2, 97-116, 2005.

[41] Marchetti M, Prager S, Cussler EL, Thermodynamic predictions of volume changes in temperature-sensitive gels. 1. Theory, *Macromolecules*, 23, 1760-1765, 1990.

[42] Achilleosa EC, Christodouloub KN, Kevrekidis IG, A transport model for swelling of polyelectrolyte gels in simple and complex geometries, *Comput Theor Polym Sci*, 11, 63-80, 2001.

[43] Oliveira D, Silva AFS, Freitas RFS, Contributions to the thermodynamics of polymer hydrogel systems, *Polymer*, 45, 1287-1293, 2004.

[44] Yang JY, Yan QL, Liu HL, Hu Y, A molecular thermodynamic model for compressible lattice polymers, *Polymer*, 47, 5187-5195, 2006.

[45] Orlov Y, Xu X, Maurer G, An experimental and theoretical investigation on the swelling of *N*-isopropyl acrylamide based ionic hydrogels in aqueous solutions of (sodium chloride or di-sodium hydrogen phosphate), *Fluid Phase Equilib*, 254, 1-10, 2007.

[46] Wallmersperger T, Ballhause D, Kro¨plin B, On the modeling of polyelectrolyte gels, *Macromol Symp*, 254, 306-313, 2007.

[47] Huang y, Jin X, Liu H, Hu Y, Amolecular thermodynamic model for the swelling of thermo-sensitive hydrogels, *Fluid Phase Equilib*, 263, 96-101, 2008.

[48] Park GS, The glassy state and slow process anomalies, In: *Diffusion Polymers*, Crank J, Park GS, (Eds) Academic, London, 1968.

[49] Singh J, Weber ME, Kinetics of one-dimensional gel swelling and collapse for large volume change, *Chem. Eng. Sci*, 51, 4499-4508, 1996.

[50] Li Y, Tanaka T, Kinetics of swelling and shrinking of gels, *J. Chem Phys*, 92, 1365-1371, 1990.

[51] Crank J, Park GS, Diffusion in high polymers, *Trans Faraday Soc*, 47, 1072-1084, 1951.

[52] Joshi S, Astarita G, Diffusion-relaxation coupling in polymers which show two-stage sorption phenomena, *Polymer*, 20, 455-458, 1979.

[53] Mazich KA, Rossi G, Smith CA, Kinetics of solvent diffusion and swelling in a model electrometric system, *Macromolecules*, 25, 6929-6933, 1992.

[54] Singh SK, Fan LT, A generalized model for swelling controlled release systems, *Biotechnol Prog*, 2, 45-156, 1986.

[55] Vasheghani-Farahani E, *Swelling and exclusion behavior of hydrogels*, PhD Thesis, McGill University, June 1990.

[56] Karadag E, Saraydin D, Swelling of superabsorbent acrylamide/sodium acrylate hydrogels prepared using multifunctional crosslinkers, *Turk J. Chem*, 26, 863-875, 2002.

[57] Khorram M, Vasheghani-Farahani E, Golshan- Ebrahimi N, Fast responsive thermosensitive hydrogels as drug delivery systems, *Iran Polym J.*, 12, 315-322, 2003.

[58] Bouquerand PE, Maio S, Normand V, Singleton S, Atkins D, Swelling and erosion affecting flavor release from glassy particles in water, *AIChE*, 50, 3257-3270, 2004.

[59] Vasheghani-Farahani E, Letter to the Editor, *AIChE*, 52, 1970, 2006.

[60] Boral S, Gupta AN, Bohidar HB, Swelling and de-swelling kinetics of gelatin hydrogels in ethanol-water marginal solvent, *Int. J. Biol. Macromol*, 39, 240-249, 2006.

[61] Hosseinkhani H, Hosseinkhani M, Tianc F, Kobayashi H, Tabata Y, Osteogenic differentiation of mesenchymal stem cells in self-assembled peptide-amphiphile nanofibers, *Biomaterials*, 27, 4079-4086, 2006.

[62] Hosseinkhani H, Hosseinkhani M, Tianc F, Kobayashi H, Tabata Y, Ectopic bone formation in collagen sponge self-assembled peptideamphiphile nanofibers hybrid scaffold in a perfusion culture bioreactor, *Biomaterials*, 27, 5089- 5098, 2006.

[63] Hosseinkhani H, Hosseinkhani M, Khademhosseini A, Enhanced angiogenesis through controlled release of basic fibroblast growth factor from peptide amphiphile for tissue regeneration, *Biomaterials*, 27, 5836-5844, 2006.

[64] Langer R, Editorial: tissue engineering: perspectives, challenges, and future directions, *Tissue Eng*, 13, 1-2, 2007.

[65] Hosseinkhani H, Hosseinkhani M, Kobayashi H, Design of tissue-engineered nanoscaffold through self-assembly of peptide amphiphile, *J. Bioact Compat Polym*, 21, 277-296, 2006.

[66] Peppas NA, Hilt JZ, Khademhosseini A, Langer R, Hydrogels in biology and medicine: from molecular principles to bionanotechnology, *Adv. Mater*, 18, 1345-1360, 2006.

[67] Kopeceka J, Hydrogel biomaterials: a smart future?, *Biomaterials*, 28, 5185-5192, 2007.

[68] Hamidi M, Azadi A, Rafiei P, Hydrogel nanoparticles in drug delivery, *Adv. Drug Deliv Rev*, 60 1638-1649, 2008.

[69] Goto T, Morishita M, Kavimandan NJ, Takayama K, Peppas NA, Gastrointestinal transit and mucoadhesive characteristics of complexation hydrogels in rats, *J. Pharm Sci*, 95, 462-469, 2006.

[70] Frokjaer S, Otzen DE, Protein drug stability: a formulation challenge, *Nat. Rev. Drug Discov*, 4, 298-306, 2005.

[71] Esmaeili F, Ghahremani MH, Esmaeili B, Khoshayand MR, Atyabi F, Dinarvand R, PLGA nanoparticles of different surface properties: preparation and evaluation of their body distribution, *Int. J. Pharm*, 349, 249-255, 2008.

[72] Yin X, Hoffman AS, Stayton PS, Poly(*n*-isopropylacrylamide-*co*-propylacrylic acid) copolymers that respond sharply to temperature and pH, *Biomacromolecules*, 7, 1381-1385, 2006.

[73] Ulijn RV, Enzyme-responsive materials: a new class of smart biomaterials, *J. Mater. Chem.*, 16, 2217-2225, 2006.

[74] Angius R, Murgia S, Berti D, Baglioni P, Monduzzi M, Molecular recognition and controlled release in drug delivery systems based on nanostructured lipid surfactants, *J. Phys Condens Matter*, 18, 2203-2220, 2006.

[75] Qiu Y, Park K, Environment-sensitive hydrogels for drug delivery, *Adv. Drug Deliv Rev.*, 53, 321- 339, 2001.

[76] 76. Jeong b, Gutowska A, Lessons from nature: stimuli responsive polymers and their biomedical applications, *Trends Biotechnol*, 20, 305-311, 2002.

[77] 77. Kaneko Y, Nakamura S, Sakai K, Aoyagi T, Kikuchi A, Sakurai Y, Okano T, Rapid deswelling response of poly(*N*-isopropylacrylamide) hydrogels by the formation of water release channels using poly(ethylene oxide) graft chains, *Macromolecules*, 31, 6099-6105, 1998.

[78] Bulmus V, Ding Z, Long CJ, Stayton PS, Hoffman AS, Site-specific polymer–streptavidin bioconjugate for pH controlled binding and triggered release of biotin, *Bioconjugate Chem.*, 11, 78-83, 2000.
[79] Atyabi F, Khodaverdi E, Dinarvand R, Temperature modulated drug permeation through liquid crystal embedded cellulose membranes, *Int J. Pharm*, 339, 213-221, 2007.
[80] Katono H, Maruyama A, Sanui K, Okano T, Sakurai Y, Thermo-responsive swelling and drug release switching of interpenetrating polymer networks composed of poly(acrylamide-*co*-butyl methacrylate) and poly(acrylic acid), *J. Control Rel*, 16, 215-227, 1991.
[81] Aoki T, Kawashima M, Katono H, Sanui K, Ogata N, Okano T, Sakurai Y, Temperature-responsive interpenetrating polymer networks constructed with poly(acrylic acid) and poly(*N,N*-dimethylacrylamide), *Macromolecules*, 27, 947-952, 1994.
[82] Ichikawa H, Fukumori Y, A novel positively thermosensitive controlled-release microcapsule with membrane of nano-sized poly(*N*-isopropylacrylamide) gel dispersed in ethylcellulose matrix, *J. Control Rel.*, 63, 107-119,2000.
[83] Xiao XC, Chu LY, Chen WM, Wang S, Li Y, Positively thermosensitive monodisperse coreshell microspheres, *Adv. Funct Mater*, 13, 847- 852.
[84] Van-Tomme SR, Storm G, Hennink WE, In situ gelling hydrogels for pharmaceutical and biomedical applications, *Int. J. Pharm.*, 355, 1-18, 2008.
[85] Xiong XY, Tam KC, Gan LH, Polymeric nanostructures for drug delivery applications based on pluronic copolymer systems, *J. Nanosci Nanotechnol*, 6, 2638-2650, 2006.
[86] Ghahremankhani AA, Dorkoosh F, Dinarvand R, PLGA-PEG-PLGA tri-block copolymers as in situ gel-forming peptide delivery system: effect of formulation properties on peptide release, *Pharm Develop Technol*, 13, 49-55, 2008.
[87] Lee WC, Li YC, Chu IM, Amphiphilic poly(*D,L*lactic acid)/poly(ethylene glycol)/poly(*D,L*-lactic acid) nanogels for controlled release of hydrophobic drugs, *Macromol Biosci*, 6, 846-854, 2006.
[88] Qiao MX, Chen DW, Ma XC, Liu YJ, Injectable biodegradable temperature-responsive PLGAPEG- PLGA copolymers: synthesis and effect of copolymer composition on the drug release from the copolymer-based hydrogels, *Int. J. Pharm*, 294, 103-112, 2005.
[89] Ruel-Gariepy E, Leclair G, Hildgen P, Gupta A, Leroux J-C, Thermosensitive chitosan-based hydrogel containing liposomes for the delivery of hydrophilic molecules, *J. Control Rel*, 82, 373-383, 2002.
[90] Ganji F, Abdekhodaie MJ, Synthesis and characterization of a new thermoreversible chitosan- PEG diblock copolymer, *Carbohydr Polym*, 74, 435-441, 2008.
[91] Bhattarai N, Matsen FA, Zhang M, PEG-grafted chitosan as an injectable thermosensitive hydrogel, *Macromol Biosci*, 5, 107-111, 2005.
[92] Bhattarai N, Ramay HR, Gunn J, Matsen FA, Zhang M, PEG-grafted chitosan as an injectable thermosensitive hydrogel for sustained protein release, *J. Control Rel*, 103, 609-624, 2005.
[93] Ha DI, Lee SB, Chong MS, Lee YM, Kim SY, Park YH, Preparation of thermo-responsive and injectable hydrogels based on hyaluronic acid and poly(*N*-isopropylacrylamide) and their drug release behaviors, *Macromol Res*, 14, 87-93, 2006.

[94] Bae JW, Go DH, Park KD, Lee SJ, Thermosensitive chitosan as an injectable carrier for local drug delivery, *Macromol Res*, 14, 461- 465, 2006.

[95] BrØndsted, H.; Kopeček, J.; (1990);" Hydrogels for site-specific oral delivery"; *Proc. Int. Symp. Control. Rel. Bioact. Mater*, 17, 128-129.

[96] Giammona,G.; Pitarresi,G.;et al.; (1999);" New biodegradable hydrogels based on an acryloylated polyaspartamide crosslinked by gamma irradiation"*; J. Biomed. Sci. Polym. Ed*, 10, 969-987.

[97] Krogars, K.; Heinamaki, J.; et al.; (2000);"Extrusion-spheronization of pH-sensitive polymeric matrix pellets for possible colonic drug delivery"; *Int J Pharm*, 199, 187-194.

[98] Chiu, H. C.; Hsiue, G. H.; et al.;(1999);" Synthesis and characterization of pH-sensitive dextran hydrogels as a potential colon-specific drug delivery system"; *J. Biomater.Sci. Polym.* Ed, 10, 591-608.

[99] Jabbari,E.; Nozari,S.;(2000);" Swelling behavior of acrylic acid hydrogels prepared by γ- radiation crosslinking of polyacrylic acid in aqueous solution*"; Eur. Polym. J*; 36, 2685-2692.

[100] Fanta G.F. and Doane, W. N.; (1986);" Grafted Starches. In Modified Starches: Properties and Uses"; Wurzburg, O.B., Ed., CRC, Boca Raton (Florida), 149-178.

[101] Furda I.; (1983);"Aminopolysaccharides-their potential as dietary fiber.In: Furda I, ed."; *Unconventional Sources of Dietary Fiber, Physiological and In vitro Functional Properties*, Washington, DC: American Chemical Society;:105-122.

[102] Saboktakin M. R.; Maharramov, A.et al.;(2007);" Synthesis and characterization of aromatic polyether dendrimer / Mesalamine (5-ASA) nanocomposite as drug carrier system"; *Journal of American Science*, 3(4), 45.

[103] Peppas, N. A.; (1987);" hydrogels in medicine and pharmacy"; CRC Press: Boca Raton, FL.

[104] Honghua Xu.; Tiejing Li.; (2005);" The analysis of boundary functions of CMS reaction factors"; *J. Nature and Science*, 3(2),25-28.

[105] Jabbari, E.; Nozari, S.; (2000); Swelling behavior of acrylic acid hydrogels prepared by γ- radiation crosslinking of polyacrylic acid in aqueous solution, *Eur. Polym. J*, 36, 2685-2692.

[106] Ratner, B. D.; (1989); Comprehensive Polymer Science–The Synthesis, Characterisation, Reactions & Applications of Polymers"; Aggarwal, S. K., ed., vol 7, Pergamon Press, Oxford, pp. 201-247.

[107] Thierry B.; Winnik, F. M.; et al.; (2003);" A New $Y_3Al_5O_{12}$ Phase Produced by Liquid-Feed Flame Spray"; *J. Am. Chem. Soc.*,125,7494.

[108] Mahfouz. A.; Hamm B.; et al.; (1997);*"* Contrast agents for MR imaging of the liver: clinical overview"*; Eur. Radiol*, 7,507.

[109] Schmitz. S.A.; Winterhalter S.; et al.; (2000);*"* Superparamagnetic iron oxide nanoparticles functionalized polymers"; *Invest. Radiol*; 35,460.

[110] Bloembergen N.; P. S. Pershan P. S. ; (1967);" Model Catalysis of Ammonia Synthesis ad Iron-Water Interfaces - A Sum Frequency Generation Vibrational Spectroscopic Study of Solid-Gas Interfaces and Anion Photoelectron Spectroscopic Study of Selected Anion clusters"; *Physical review*,128(2),606.

[111] Saboktakin M. R.; Maharramov A.; et al.; (2008); "Poly(amidoamine) (PAMAM) /CMS Dendritic nanocomposite for controlled drug delivery", *Journal of American Science*, 4(1), 48.

[112] Davis S. S. "Assessment of gastrointestinal transit and drug absorption", In: Prescott LF, Nimmo WS, eds. *Novel Drug Delivery and Its Therapeutic Application*, Chichester, UK:Wiley;1990:89-101.

[113] Chourasia M. K., Jain S., K., "Pharmaceutical approaches to colon targeted drug delivery systems", *J. Pharm Pharm Sci.*, 2003;6:33-66.

[114] Davaran S., Hanaee J., Khosravi A.,"Release of 5-aminosalicylic acid from acrylic type polymeric prodrugs designed for colon-specific drug delivery", *J. of Control Release*, 1999; 58:279-287.

[115] Schacht E., Gevaert A., Kenawy E., R., et al.;" Polymers for colon specific drug delivery", *J. of Control Release*, 1996;58:327-338.

[116] Chung K., T.,Stevens S., E., Cerniglia C., E.;"The reduction of azo dyes by the intestinal microflora"; *Crit. Rev. Microbiol.*;1992;18:175-190.

[117] Yamaoka T., Makita Y., Sasatani H., Kim S. I., Kimura Y.;" Linear type azo-containing polyurethane as drug-coating material for colon-specific delivery: its properties degradation behavior and utilization for drug formulation"; *J. of Control Release*, 2000;66:187-197.

[118] Shantha K. L., Ravichandran P., Rao K. P.;" Azo polymeric hydrogels for colon targeted drug delivery"; *Biomaterials.*;1995;16:1313-1318.

[119] Van den Mooter G., Samyn C., Kinget R.; "Azo polymers for colon-spicific drug delivery"; *Int. J. Pharm.*; 1992;87:37-46.

[120] Ghandehari H., Kopeckova P., Kopecek J.; "In vitro degradation of pH sensitive hydrogels containing aromatic azo bonds"; *Biomaterials*;1997;18:861-872.

[121] Kakoulides E. P., Smart J. D.,Tsibouklis J.; "Azo crosslinked poly(acrylic acid) for colonic delivery and adhesion specificity synthesis and characterization"; *J. of Control Release*; 1998;52"291-300.

[122] Furda I., "Aminopolysaccharides-their potential as dietary fiber.In: Furda I, ed.; *Unconventional Sources of Dietary Fiber, Physiological and In vitro Functional Properties*, Washington, DC: American Chemical Society; 1983:105-122.

[123] Ormrod D.J., Holmes C.C., Miller T.E., "Dietary Chitosan inhibits hypercholesterolaemia and atherogenesis in the apolipoprotein E-deficient mouse model of atherosclerosis"; *Atherosclerosis*; 1998:138;329-334.

[124] Yamamoto A., Tozaki H., Okada N., Fujita T.; "Colon specific delivery of peptide drugs and anti-inflammatory drugs using chitosan capsules"; *STP Pharma Science*; 2000;10:23-43.

[125] Agnihotri S. A., Mallikarjuna N. N., Aminabhavi T. M., "Recent advances on chitosan-based micro- and nanoparticles in drug delivery"; *J. Control Release*; 2004;100:5-28.

[126] Calvo P., Remunan-Lopez C., Vila-Jato J. L., Alonso M. J.; "Novel hydrophilic chitosan-polyethylene oxide nanoparticles as protein carriers"; *J. Appl. Polym. Sci.*; 1997; 63:125-132.

[127] De S., Robinson D.; "Polymer realationships during preparation of chitosan-alginate and poly-l-lisyne-alginate nanospheres"; *J. of Control Release*; 2003;89:101-112.

[128] Chen Y., Mohanraj V., Parkin J.; "Chitosan-dextransulfate nanoparticles for delivery of an anti-angiogenesis peptide"; *Int. J. Pept. Res. Ther.*; 2003;10:621-629.

[129] Ma Z.,Yeoh H. H., Lim L. Y.; "Formulation pH modulates the interaction of insulin with chitosan nanoparticles"; *J. of Pharm Sci.*;2002;91:1396-1404.
[130] Tiyaboonchai W., Woiszwillo J., Sims R. C., Middaugh C. R.; "Insulin containing polyethylenimine-dextran sulfate nanoparticles"; *J. of Pharm. Sci.*;2003;255:139-151.
[131] Pan Y., Li j., Zhao H., et al.; "Bioadhesive polysaccharide in protein delivery system: chitosan nanoparticles improve the intestinal absorption of insulin in vivo"; *Int. J. of Pharm.*; 2002;24:139-147.
[132] Zesheng An, Qian Qiu and Guangyao Liu; "Synthesis of architecturally well-defined nanogels via RAFT polymerization for potential bioapplications"; *Chem. Commun.*, 2011, 47, 12424-12440.
[133] Prateek K. Jha, Jos W. Zwanikken, François A. Detcheverry, Juan J. de Pablo and Monica Olvera de la Cruz; "Study of volume phase transitions in polymeric nanogels by theoretically informed coarse-grained simulations"; *Soft Matter*, 2011,7, 5965-5975.
[134] Takahito Nakamura, Atsushi Tamura, Hiroki Murotani, Motoi Oishi, Yuta Jinji, Kiyoto Matsuishi and Yukio Nagasaki; "Large payloads of gold nanoparticles into the polyamine network core of stimuli-responsive PEGylated nanogels for selective and noninvasive cancer photothermal therapy"; *Nanoscale*, 2010,2, 739-746.
[135] Lee A. Tessler, Casey D. Donahoe, Daniel J. Garcia, Young-Shin Jun, Donald L. Elbert and Robi D. Mitra; "Nanogel surface coatings for improved single-molecule imaging substrates"; Published online before print February 16, 2011, doi: 10.1098/rsif.2010.0669.

Chapter 5

MAGNETIC NANOPARTICLES FOR CANCER THERAPY

5.1. BIOMEDICAL APPLICATIONS OF MAGNETIC NANOPARTICLES

Nanotechnology is associated with the handling of matter in a nanometric scale, that is, one meter contains one billion nanometers. In this scale, the atoms reveal unique features, and they may present tolerance to temperature, colors, chemical reactivity, electrical conductivity, or even show an extraordinary intensity power. These characteristics explain the industry interest for nanomaterials that are already manufactured to be employed in cosmetics, paints, linings, clothes, catalyzers or to provide additional strength to materials. In the biomedical field, nanotechnology (or nanobiotechnology) will produce important developments in diagnosis, treatment and bioengineering. Basic issues still need to be solved, however. Magnetic nanoparticles shall cross the endothelial barrier and accumulate specifically in the target cells without causing any damage to the normal cells. These features can be achieved through the coating of particles with a biologically active material, the control of their physical parameters, such as particle size, magnetic susceptibility of the solution and knowledge of its behavior in the body. In addition to providing biocompatibility to the magnetic fluids, the coating of magnetic nanoparticles with specific ions, nucleotides, olygonucleotides, peptides, vitamins, antibiotics, analog substrates or antibodies allow numerous biomedical applications. On the other hand, instead of existing as a magnetic fluid, the magnetic nanoparticles may be encapsulated in lysosomes, which are vesicles that, similar to cell structures, present appropriate features for applications in cell therapy. Overall, the magnetic nanoparticles contained in a magnetic fluid or incorporated into lysosomes have received special attention for being able to be directed or located in a specific target by external magnetic fields. This location in a preferential site by gradients of magnetic fields suggested that magneto liposomes and magnetic fluids, among other magnetic systems, become effective carriers of drugs with site specificity for the controlled release of chemotherapy agents. Scientific data report that the biomedical applications of magnetic systems exceed the activity of drug carriers, and they can also be used as contrast agents in nuclear magnetic resonance imaging, magnetic separation of cells or varied biological molecules, in markers for target cells and the treatment of cancer.

5.2. Use of Nuclear Magnetic Resonance in the Study of Tumors

Nuclear magnetic resonance has gained great importance compared with other imaging techniques. Based on the different behavior of protons in different tissues, the technique provides an image of anatomical structures, which can be better enhanced with the use of contrast agents. Magnetic nanoparticles, especially those smaller than 10 nm, represent an alternative class of contrast agents for nuclear magnetic resonance with advantages from a physical point of view because they further enhance the behavior of protons from different tissues. A special case of nuclear magnetic resonance imaging is obtained through the association of monoclonal antibodies to the magnetic nanoparticles. The viability of this strategy is backed by recent studies showing the ability to visualize specific cells through magnetic resonance, by labeling with monoclonal antibodies conjugated with commercial microbeads (MACS) composed of superparamagnetic nanoparticles. A group of researchers identified the activity of an endothelial tumor through the use of nanoparticles associated with liposomes showing that the angiogenic process can be detected by the magnetic resonance images and that multimode nanoparticles provide the perception of the exact mechanism of tumor accumulation.

5.3. Site-Directed Administration of Drugs in Tumor Target Cells

Drug carrying by nanoparticulate systems is considered a valuable technique to optimize controlled release of drugs. Miniaturization of a drug carrying system to a nanometric scale allows good stability, absorption and excellent quantitative tissue transference and, hence, the expected pharmacodynamic activity. Additionally, the side effects and foreign body reactions may be simultaneously prevented with local and systemic tolerance. These goals may be reached when the correct dose of the drug is released specifically in target tissues or cells without overloading the body with massive doses. This is especially true in cases of drugs with severe side effects such as chemotherapeutic agents for cancer. Several nanoparticulate systems are currently being used and, among them, the ones based on magnetic nanoparticles play an important role for their property to be conducted and retained in a specific body region by means of a gradient of an external magnetic field. With the purpose of increasing specificity, the MNP (magnetic nanoparticle)-drug conjugate can be associated with another molecule that is able to recognize and to bind specifically to the target site. Such molecules can be antibodies, proteins, lectins, and hormones, among others. Therefore, a site-directed administration of the MNP drug conjugate associated with the specific monoclonal antibody would reach the target tumor cells.

5.4. Magnetohyperthermia

Hyperthermia is a therapeutic procedure employed to cause increased temperature in a certain region in the body affected by a neoplasm with the purpose of causing the lysis of cancer cells (Figure 1). Its action is based on the fact that temperatures as high as 41°C-42°C (106°F- 107°F) are able to directly destroy the tumor cells since they are less resistant to sudden increases in temperature than normal surrounding cells. The temperature increase required by hyperthermia can be achieved, among other methods, by the use of magnetic nanoparticles.

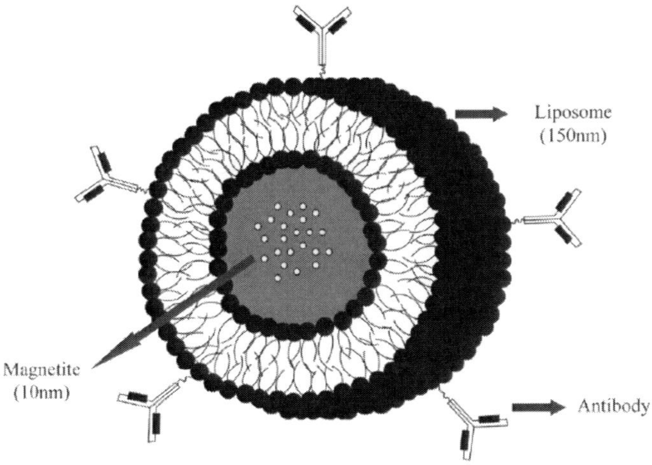

When undergoing the action of an external magnetic field with alternate frequency, the magnetic nanoparticles are heated. The use of magnetic nanoparticles (magnetic monodomains) is preferable to microparticles (magnetic multidomains) because magnetic nanoparticles respond more efficiently to external fields thereby absorbing more energy from them. In the process of magnetohyperthermia, also known as magneto thermocytolysis, the biocompatible magnetic nanoparticles (MNP) are associated with monoclonal antibodies (MAb) specific for proteins of tumor cell membranes. When administered, the MNP-MAb conjugate enhances the contrast of magnetic resonance images thus allowing earlier and more efficient localization of metastases. After absorption by the cancer cells, magnetic nanoparticles are submitted to a magnetic field which results in local increased temperature and subsequent lysis of the tumor cell. Alternatively, the magnetic nanoparticles may be attracted and retained in the tumor area by means of the gradient of an external magnetic field or also directly injected in the tumor. Regardless of the mode of conduction of the magnetic nanoparticles to the tumor, it is viable to localize the heat to the tumor tissue, thus minimizing the damages to the surrounding normal tissues. This makes magnetohyperthermia a promising technique for the treatment of several cancers. Studies related to the use of magnetic nanoparticles as mediators in the heating and lysis of tumor cells have been carried out with magnetoliposomes or magnetic fluids presenting several coatings. Magnetohyperthermia has been used in conjunction with other cancer treatment modalities, especially to improve the efficacy of antineoplastic drugs without increasing their cytotoxic potential, and also being favorable to increase the clinical response to radiation therapy, immune therapies and gene therapy. In the presence of these important clinical applications, the granular materials that make up the nanoparticles should present low levels of toxicity, as well as a high saturation moment that allows the minimization of the required doses.

Magnetite (Fe_3O_4) seems to be a promising candidate, since it presents a high Curie temperature (CT), a high saturation moment (90-98 emu/g or 450-500 emu/cm^3) and low rates of toxicity in preclinical tests. Nanoscale magnetism has stimulated great interest due to its importance in mapping the scaling limits of magnetic information storage technology and understanding spin dependent transport phenomena. Recent progress in the production of nearly monodisperse magnetic nanoparticles from metallic Fe, Co and Ni, to iron oxides, MFe_2O_4 and FePt and CoPt intermatellic compounds provides various systems suitable for nanomagnetic studies. An assembly of monodisperse magnetic nanoparticles with controlled interparticle spacing will allow detailed studies on magnetization, anisotropy, as well as magnetization reversal processes and interparticle interactions of the particles with different sizes and surface properties. An interesting magnetic nanoparticle system is that of core/shell structured nanoparticles in which the magnetic core is coated with a layer of a nonmagnetic, antiferromagnetic, or ferro/ferri-magnetic shell. A nonmagnetic coating is used routinely for magnetic core stabilization and surface functionalization for biomedical applications. An antiferromagnetic coating over a ferromagnetic core leads to exchange bias (a shift of the hysteresis loop along the field axis), and improvements in the thermal stability of the core. Compared with these two different types of core/shell systems, a bimagnetic core/shell system, where both core and shell are strongly magnetic (ferro- or ferri-magnetic) is less often studied yet is more interesting due to its potential in electromagnetic and permanent magnetic applications. In such a system, the intimate contact between the core and the shell leads to effective exchange coupling and therefore cooperative magnetic switching, facilitating the fabrication of nanostructured magnetic materials with tunable properties.

Here are the magnetic properties of a bimagnetic core/ shell nanoparticle system with ferromagnetic FePt core and thickness-tunables 0.5–3 nm d, ferrimagnetic MFe_2O_4 sM=Fe,Co d shell. A single-phase-like smooth variation of magnetization with field, indicating that the core and the shell are effectively exchange coupled and magnetization of both core and shell reverses cooperatively. As a result, the magnetic properties, such as magnetization and coercivity, of these core/shell nanoparticles can be readily controlled by tuning the core/shell geometrical parameters and chemical compositions.

Monodisperse 3.5 nm FePt nanoparticles were synthesized by the thermal decomposition of Fe-CO and polyol reduction of Pt simultaneously as reported. The iron oxide coating was achieved via mixing and heating the FePt nanoparticle seeds with Fe / polyol, or Co/Fe / polyol precursors as published elsewhere. The structure of the nanoparticle assemblies was examined by transmission electron microscopy (TEM), electron diffraction, and x-ray diffraction (XRD). The magnetic properties were measured by a superconducting quantum interference device magnetometer. The cores appear darker and the shells lighter in the images due to the large difference in electron penetration efficiency on FePt and oxides. The FePt core has a diameter of 3.5 nm, the shell thicknesses are about 2 nm in both images. The shell thicknesses observed are quite uniform throughout the particles, with a standard deviation of about 10%. For FePt/Fe_3O_4 nanoparticles, the main phase of the shell is composed Fe_3O_4 as it is obtained from the same chemistry used to make Fe_3O_4 particles, and its structure is confirmed by both XRD and high resolution TEM (HRTEM). Results reveal the crystalline shell with the distance between two lattice fringes matching with the lattice spacing of Fe_3O_4. The HRTEM also shows a partially coherent interface between the FePt core and the Fe_3O_4 shell, indicating possible epitaxial growth of the shell over the core during the coating process.

Both FePt and Fe_3O_4 nanoparticles are ferromagnetic at 10 K. The coercivity for 3.5 nm FePt nanoparticles is 5.5 kOe, while that for 4 nm Fe_3O_4 is only 200 Oe. The large coercivity from disordered fcc FePt nanoparticles likely originates from a uniaxial surface anisotropy. The FePt/Fe_3O_4 core/shell nanoparticle is therefore a two-phase system consisting magnetically of a hard sFePtd and a soft sFe_3O_4d phase.

Despite consisting of both hard and soft phases, the hysteresis loop shows a single-phase-like behavior, with the magnetization changing with the applied field smoothly. The coercivity is determined to be 2.3 kOe, a value in between that of FePt and Fe_3O_4. This indicates that the intimate contact between the FePt core and Fe_3O_4 shell leads to an effective interphase exchange coupling, which results in cooperative magnetization switching of the two phases. Earlier theoretical studies suggest that for hard and soft phases to reverse cooperatively in a hard–soft composite system, the critical dimension of the soft phase should be less than twice the domain wall width of the hard phase. Using the measured 10 K of the FePt nanoparticles as an approximation for the effective uniaxial anisotropy field, the effective anisotropy constant *Ku* for the FePt nanoparticles is calculated to be approximately 53106 erg/cm^3. 310−6 erg/cm^3, we can estimate it to be about 10 nm. The $Fe_3$04 shell in this study is less than 3 nm and well within the limit of 20 nm. Hence, the switching of the hard and soft phase should indeed occur coherently, leading to a smooth magnetization transition.

The coercivity of the core/shell particles is observed to decrease with increasing shell thickness. The coercivity of a hard–soft exchange-coupled system with collinear easy axis is:

$$H_c = 2\frac{K_H f_H + K_S f_S}{M_H f_H + M_S f_S}, \tag{1}$$

where *K* is the anisotropy constant, *M* the saturation magnetization, *f* the volume fraction, and the subscripts *H* and *S* denote the hard and soft phases, respectively. Equation (1) is only strictly valid for uniaxial anisotropy, with the easy axes of the hard and soft phases being collinear. Nevertheless, since *KH* of 53106 erg/cm^3 is about 20 times larger than that of Fe_3O_4, we can ignore the term *KSfS* in Eq. (1) for the thickness range we studied without introducing significant error. Based on this, Eq. (1) can be transformed into where *hc* is the coercivity of the core/shell particles normalized by the coercivity *HcH* of the hard phase *shc=Hc /HcH*d. If the above-presented analysis is correct, *hc* should decrease monotonically with increasing volume fraction of the soft phase, following Eq. (2). We can see that they match each other reasonably well. This indicates that the coercivity of the FePt/Fe_3O_4 nanoparticles depends only on the volume ratio of core/shell, not on the actual size or thickness of the core and the shell.

$$h_c \approx \frac{1}{1 + \frac{M_S}{M_H}\frac{f_S}{1 - f_S}}, \tag{2}$$

The situation of FePt/$CoFe_2O_4$ nanoparticles is different from that of FePt/Fe_3O_4. $CoFe_2O_4$ has much larger magnetocrystalline anisotropy than Fe3O4, and exhibits higher coercivity at low temperatures. The results show that at 10 K, the 8 nm CoFe2O4 nanoparticle assembly has an *Hc* of 12 kOe and the 18 nm one has *Hc* of 21 kOe. Compared to FePt/Fe_3O_4, the hard–soft phases in FePt/Fe_2CoO_4 core/shell system are therefore reversed, with FePt being magnetically softer and CoFe2O4 harder. Figure 3(c) shows the 10 K hysteresis loop of FePt/Fe_2CoO_4 with 2 nm shell. The *Hc* increases from 5.5 kOe for FePt to 8 kOe, as expected for such an exchange-coupled system. Since the anisotropy of the hard and soft phase in this system is rather close, Eq. (2) cannot be used to describe such systems. In conclusion, a class of bimagnetic core/shell nanoparticles can be readily synthesized via solution phase chemistry.

The magnetic properties of these core/shell nanoparticles can be tailored by controlling the core/shell dimensions, and by tuning the material parameters of both core and shell. Such systems may show interesting nanomagnetism emerging from the exchange-coupling between the core and the shell, and may yield finely tailored materials for various nanomagnetic applications sensors with appropriate magnetical shielding suppressing environmental magnetic interferences for in-vivo quantification of MNP distributions in small animals up to rabbit size, which avoids the need for a bulky and expensive mu-metal shielding. An integrated superconducting Niobium shield allows the operation of the system in a conventional laboratory environment. Up to 18 liquid sensors are each connected to a pick-up coil of variable size arranged regularly around the circumference of the sample that are used for spatially resolved MRX. The results of the physical characterization of the device focus on the behavior of the superconducting shield. The shielding turned out to be better than 105 for radial and 107 for axial external magnetic fields. Furthermore, the residual magnetic fields in the sensor area frozen in during the transition of the shield into its superconducting state were found to be smaller than 100 nT. This together with the robustness of the device against mechanical vibrations results in a total system noise level of only 2 to 3 fT Hz-1/2. In addition to MRX measurements on MNP assemblies, the magnetocardiogram of a rabbit detected by the system demonstrates its excellent performance in a conventional laboratory environment.

A novel way of Magnetic Drug Targeting is the application of tiny aerosol droplets containing magnetic nanoparticles to specific regions in the lung of a specimen guided by an external magnetic field. Thus, higher doses of drugs can be delivered to a cancerous lung region without increasing side effects. Recently, the method was transferred from a mouse model to a pig lung model since it is more similar to a human lung. The development of this procedure requires detailed knowledge about the bio-distribution of the nanoparticles in the different parts of the lung dependent on the applied magnetic field. Magnetorelaxometry (MRX) is an effective and reliable technique for the quantitative detection of magnetic nanoparticles in biological systems. After switching off a magnetizing field, the relaxating magnetic field distribution caused by the magnetic nanoparticles in the whole lung is recorded as a function of time for 60 seconds.

By coating specific bio-molecules onto magnetic nanoparticles, magnetic nanoparticles become bio-functionalized. The synthesis of bio-functionalized magnetic nanoparticles coated with various kinds of bio-probes, depend on bio-applications. Using magnetic nanoparticles bio-functionalized with anti-vascular cell adhesion molecule-1, which served as an indicator for the lesions prone to vulnerable atherosclerotic plaque formation, the enhancement in the MRI contrast of the vulnerable atherosclerotic plaque is demonstrated. In addition, by coating anti-CD34 onto magnetic nanoparticles, magnetic nanoparticles are able to associate with stem cells. The results show that stem cells bound with anti-CD34 bio-functionalized magnetic nanoparticles can be extracted from umbilical cord blood through magnetic separation. Furthermore, the magnetoreduction assays on proteins using magnetic nanoparticles as markers are explored. These results reveal potential in bio-medical applications such as magnetic resonance imaging (MRI), cell sorting, and immunoassay using bio-functionalized magnetic nanoparticles. Ingested ferromagnetic particles in the micron size range are novel tracers for the study of cytoskeleton mediated cell functions (phagocytosis, intracellular transport, integrity). The target cells for these studies are alveolar macrophages, which are defense cells in the lung. Ingested particles are incorporated into phagosomes, and the transport of such magnetic phagosomes is detected by cytomagnetometry. Ferromagnetic particles can also be inhaled and used to measure macrophage functions in vivo in the human lung. The purpose of this study was to investigate the uptake and intracellular transport of ferromagnetic nanoparticles in macrophages. The fate of inhaled nanoparticles in living cells and in the lung is unclear. Nanoparticles are not only phagocytized by macrophages, but they can penetrate into epithelial cells and may enter circulation and secondary organs.

The relaxation characteristics of ferromagnetic $CoFe_2O_4$-nanoparticles were studied in (1) dry form, and (2) after incubation with J774A.1 macrophages. The particles were aligned in a strong magnetic field pulse (100mT) and relaxation (loss of alignment) was recorded for 5 minutes using fluxgate sensors. Relaxation studies of fixed particles (Brownian motion is inhibited) showed that about 25% of the remanent magnetization followed the Neel relaxation.

The relaxation further accelerated and after 24 hours of incubation with macrophages, relaxation was fastest and complete, indicating intracellular particle movement and transport (Brownian-like relaxation). Electron microscopy shows aggregates of particles within phagosomes and intracellular vacuoles. Only a few particles were located within the cytoplasm and no particles were found within the nucleus. The preliminary data show that relaxation studies of ferromagnetic nanoparticles can discriminate between incorporated and free particles in cells and in the human lung. Using magnetic nanoparticles for drug delivery and for hyperthermia therapy has significant advantages, such as the ability of targeting a specific lesion, and reduction in the quantity of drugs used. Therefore, this technology has attracted great interest in medical diagnosis and treatment. In these applications, it is important to monitor the location and distribution of nanoparticles in the regions of interest within the human body. A novel technique, called magnetorelaxometry (MRX), has been introduced to image nanoparticles by measuring their remanent magnetization after switching off a magnetizing field. However, it requires magnetization of the particles and the magnetic signals are observable only within a few seconds. In our previous work, we have demonstrated that the magnetic signals from certain nanoparticles can be detected without pre-magnetization by Magnetoencephalography (MEG). The results reveal that the enhanced magnetic field of nanoparticles by magnetization can remain detectable hours after magnetization. These results are encouraging, since using our method the magnetization procedure may either be eliminated or performed hours before measurement. In order to study this method for in vivo imaging, the properties of several types of nanoparticles were studied and experiments were performed on animals. The liver, spleen, and blood of rodents were extracted after the injection of nanoparticles. These results promise new applications of our method in imaging in vivo substances labeled by magnetic nanoparticles.

The evoked magnetic field has been measured from the human spinal cord with an array of vector magnetometers. The results propose to apply the sLORETA spatial filter technique to reconstruct current density distributions associated with the spinal-cord evoked magnetic field, thus providing a novel imaging diagnosis tool for orthopedic surgeons. The results also show that both conduction velocity and intensity of the volume current are decreased at the location of the incomplete conduction block. These results demonstrate that the spinal cord imaging from its evoked magnetic field can be an effective tool for diagnosing spinal cord injuries. The results show that the reconstructed results from the normal-direction-only measurement suffer from artifacts that are overlapped onto the reconstructed volume current. These artifacts may mislead the interpretation of the imaging results. By contrast, the results from the vector measurement are free from these artifacts, and demonstrate the advantage of the vector measurement for developing a spinal-cord imaging system. Results from our computer simulations and animal experiments are reported as a separate paper in this conference.

The results describe the application of the proposed method to the spinal-cord evoked data obtained from patients, and show its effectiveness in actual clinical environments. The patient's X-ray image was acquired with the spinal-cord evoked magnetic field. The reconstructed current density distributions are overlaid onto the patient X-ray image. The X-ray image is also used for compensating for the false intensity change of reconstructed current caused by the depth non-uniformity of the spinal cord of a patient with respect to the sensor array. In the results of these experiments, both the quadrupole current conducting along the spinal cord and the volume current flowing into it can be reconstructed, and the conduction velocity and the intensity of the volume current are decreased at the location of the conduction block of spinal-cord in a patient. This location of the conduction block was estimated by using the subject MRI.

The colonic motility is poorly understood when compared to the superior GI (gastrointestinal) tract. Animal models employing radiation free and low cost methodologies can be used to evaluate the effects of several stimuli and drugs on colonic motility. The results were to employ the EMG (Electromyography) and ACB (AC Biosusceptometry) techniques simultaneously to demonstrating stimuli-induced motility changes in dogs. We use a quartz stamp, fabricated using electron-beam lithography, which has 100nm diameter pillars at 300nm pitch in a square array. Inexpensive stamps have also been produced using self-assembled polymer spheres which cover 4" Si wafers (1 inch¼2.54 cm) and result in a 100-fold increase in the number of nanoparticles per imprint (Figure S2 in the supporting information).

The imprinted templates are used as substrates for the deposition of thin multilayer magnetic films with precise thickness control. A diagram of the substrate, release layer, resist bilayer, and metallization is shown in Figure 1c. The resist layers and overlying metal film are removed using sonication in solvents, and the nanoparticles are released using liquid etches and surfactants to stabilize them in solution. A yield of over 85% is achieved, as indicated by inductively coupled plasma–optical emission spectroscopy and magnetic measurements.

A top view scanning electron microscopy (SEM) image of patterned metal multilayer nanoparticles is shown in Figure 1d. SEM and transmission electron microscopy (TEM) images of released nanoparticles indicate that nanoparticle sizes and shapes are very uniform. The magnetic properties of substrate-bound SAF nanoparticles, as well as released nanoparticles in aqueous solution, were measured by alternating gradient magnetometry (AGM) at room temperature.

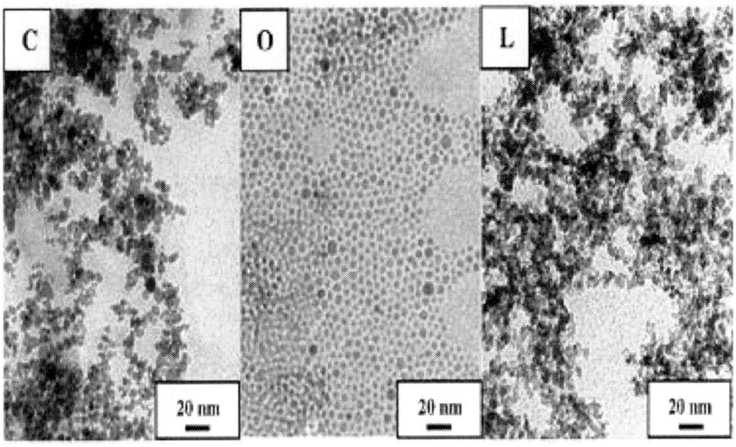

Figure 1. TEM images of magnetic nanoparticles prepared by cooperation (sample C) and by decomposition of a metal-organic precursor, either in solution (sample O) or by laser pyrolysis of an aerosol (sample L).

Figure 2a shows hysteresis loops (normalized by CoFe mass) of substrate-bound 120nm diameter SAF nanoparticles composed of stacks of Ta (5 nm)/Ru (2)/CoFe(t)/Ru(2.5)/ CoFe(t)/Ru(2)/Ta(5) for magnetic bilayer thicknesses of t¼3nm (trace A), 6nm (trace B), and 12nm (trace C). The remanence and coercivity of these nanoparticles are nearly zero, dramatically reduced from those of single layer CoFe nanoparticles of the same size (Figure S3 in the supporting information). The functional dependence of the magnetization M is M (H)¼MS(H/HS) until M attains a constant value MS when H reaches the saturation field HS. MS and HS are both proportional to the magnetic bilayer thickness because interlayer magnetic repulsion increases linearly with t, as expected from considerations of demagnetizing fields. Deviations from linearity at small fields are associated with as-deposited anisotropies which cause a non-zero ''spin flop'' field. Equally sized SAF nanoparticles using only this thickness dependent magnetostatic repulsion have identical moments at low fields, but their moments differ at high fields (after the onset of saturation). The magnetic saturation fields can be further tailored by employing a special spacer between the magnetic layers that produces strong magnetic interfacial exchange coupling, which has been widely used in magnetic information storage technologies. This subtle quantum phenomenon depends very strongly on the nonmagnetic spacer material, often ruthenium, and its thickness. The effects on hysteresis loops can be included by considering the interfacial exchange coupling as producing an effective magnetic field which adds to, or subtracts from, the magnetostatic demagnetizing field.

This effect is quite pronounced for thin (<1 nm) ruthenium spacers which provide strong antiferromagnetic coupling and thus increase the saturation field. Figure 2b shows the expected increase in saturation fields obtained by changing the Ru spacers from 2.5nm (trace A) to 0.6nm (trace B) while keeping the two CoFe layers at a thickness of 6nm each. The saturation field can be further increased by increasing the number of interfaces and magnetic layers while keeping the sum of magnetic layer thickness constant (trace C in Fig. 2b, where 0.6nm Ru spacers separate 3nm CoFe layers). SAF nanoparticles exploiting variable interfacial exchange interactions can thus be made to have identical saturation moments at high fields, while their moments differ at low fields (before the onset of saturation).

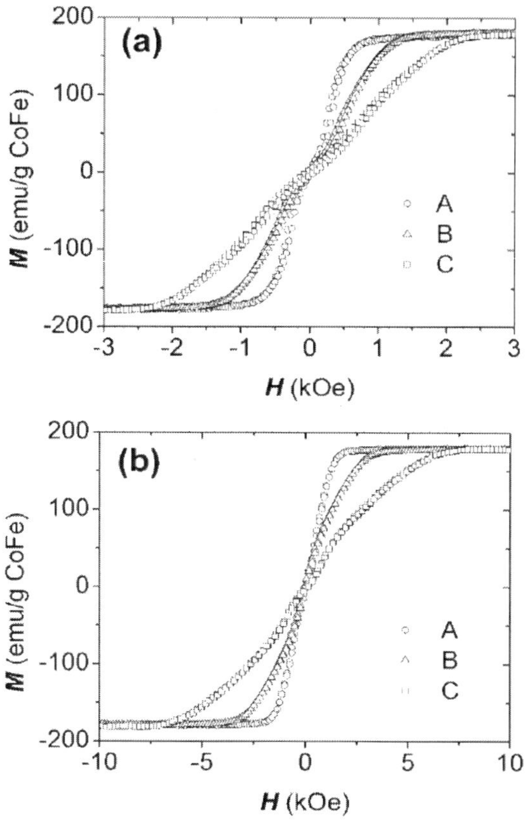

Figure 2. Tailoring magnetic properties with magnetostatic interfacial coupling.

The hysteresis loops of SAF nanoparticles released in aqueous solution are measured to be similar to those of substrate-bound nanoparticles. Individual 100nm metallic nanoparticles strongly scatter light and thus can be detected in solution by conventional optical microscopy with an immersion objective lens. In a thick optical cell, individual particles diffuse through the 4 mm thick object focal plane in a fraction of a second and are visible only intermittently. Continuous optical tracking of individual nanoparticles is accomplished by confining diluted solutions in glass cells with a depth of <4 mm to limit vertical diffusion. At high magnification, Brownian motion is apparent and allows determination of the diffusion constant as D"6 mm2 s#1, confirming the 100nm diameter. The digitized coordinates of two particles, subjected to a magnetic field of ca. 1 kOe and a field gradient of ca. 1 kOe cm#1, are shown in Figure 3a for 100 charge-coupled device (CCD) frames. Plotting these coordinates as a function of time (Figure 3b) yields a magnetically induced velocity of ca. 3 mm s#1, consistent with estimates from the Stokes–Einstein equation. The particle-to-particle variation in magnetic drift velocity is negligible, confirming the monodispersity of the SAF nanoparticles. The saturation magnetization of SAF nanoparticles with 12nm magnetic layer thickness (trace C in Figure 2a) is measured to be ca. 850 emu cm#3, which is lower than that of Co90Fe10 (1600 emu cm#3) due to incorporation of nonmagnetic layers, but still 2.5 times that of iron oxide nanoparticles (ca. 340 emu cm#3). Because the magnetically induced velocity is proportional to the magnetization and the square of the nanoparticle radius, these SAF nanoparticles will drift >100 times faster than 10nm diameter iron oxide nanoparticles under similar magnetophoretic conditions. The NIL-based fabrication of the SAF nanoparticles not only provides the desired tunability of the magnetic properties, but also allows customized incorporation of materials with unique properties to achieve multimodality.

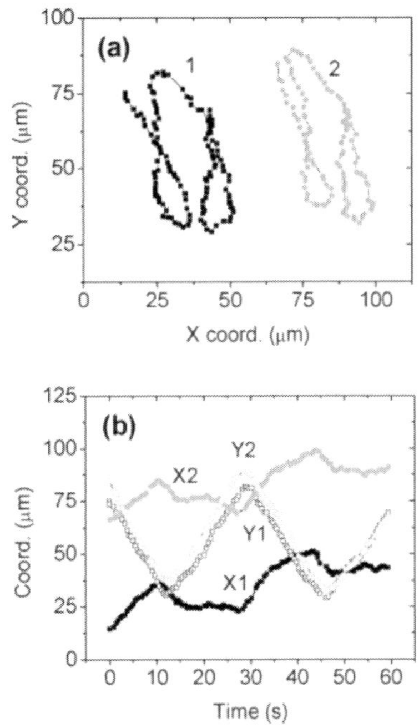

Figure 3. Optical tracking of SAF nanoparticles in solution.

For example, an optional Au layer can be deposited on the top or bottom of the tantalum layer during the fabrication process, resulting in nanoparticles with a localized surface plasmon band (Figure 4a). This band, positioned at 730 nm, falls within the range of surface Plasmon wavelengths reported for Au nanodisks at similar sizes. Fluorescent nanoparticles are obtained via surface attachment of dyes. Figure 4b shows the paths of AlexaFluor 594 labeled SAF nanoparticles integrated over the course of a rotation of a magnetic field gradient, demonstrating magnetic and fluorescent multifunctionality.

These SAF nanoparticles have magnetic moments hundreds of times larger than those of 10nm diameter iron oxide nanoparticles, and yet have very small remanence. As a result, they have great advantages for magnetic detection of nanoparticles because the signal per nanoparticle is usually proportional to the magnetic moment per nanoparticle. Their adjustable saturation fields are also valuable for multiplex magnetic labeling in biodetection, which is akin to multicolor fluorescent labeling. Another potential application of these nanoparticles is multiplexed magnetic sorting, wherein distinct magnetic nanoparticles, which have different magnetic moments (at one or several magnetic fields) and distinct affinity targets, confer different magnetophoretic velocities to their respective analytes.

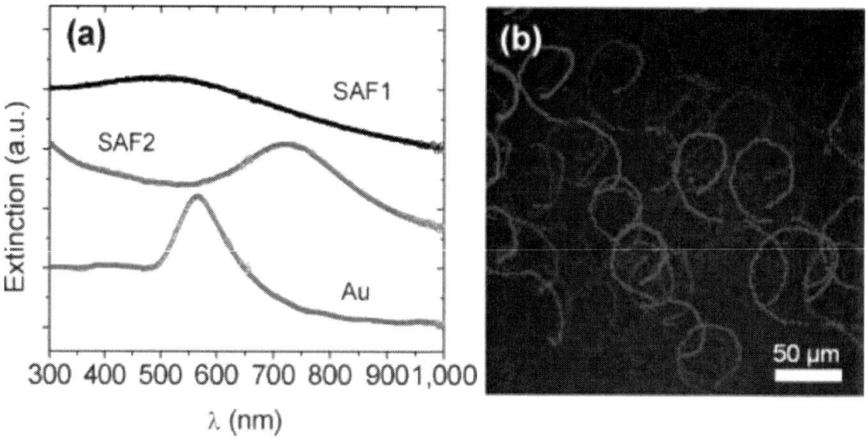

Figure 4. Multiplex functionality of SAF nanoparticles.

We have witnessed breathtaking biomedical progress in recent years, including the widespread application of superparamagnetic nanoparticles, by their virtue of zero remanence and negligible interactions in the absence of magnetic fields, as capture probes for magnetic separation and purification, and biosensor labels. We envision that the synthetic antiferromagnetic nanoparticles may offer new possibilities by contributing added advantages of high moment, manipulability, multiplex labeling, and multifunctionality.

Cytokine adsorption/release on uniform magnetic nanoparticles for localized drug delivery cancer immunotherapy is at present an active field of biomedical research given its relevance in the design of new anti-tumor devices. Treatments based on activating an anti-tumor immune response would not only allow removal of the primary tumor, but could also provide sufficient protection to eliminate neoplastic cells that spread to form distal metastases. There is considerable recent pre-clinical and clinical evidence showing that manipulation of the innate and adaptive immune responses to neoplastic cells can control development of tumors, leading to their elimination. This has led to the concept that immunotherapy might be a central element in the development of anti-cancer therapies, in combination with surgery as well as chemo- and radiotherapy. Studies of the anti-tumor immune response indicate that the T helper type 1 (Th1) lymphocyte-mediated response is much more effective than that mediated by Th2 lymphocytes. Due to their increased cytotoxic activity, Th1 cells have a greater capacity than Th2 cells to eradicate tumors in murine models. Type Th1 responses are characterized by the production of interferon gamma (IFN-γ), interleukin-2 (IL-2) and IL-12, whereas IL-4, IL-5 and IL-10 production is typical of Th2 responses. Administration of specific cytokines that promote expansion and/or activation of one type of lymphocyte or another may thus be useful in cancer immunotherapy. Many studies point to IFN-γ as the most effective cytokine in tumor elimination. Injection of IFN-γ-blocking monoclonal antibodies into experimental animals suppresses their ability to reject transplanted tumors. Compared to normal mice, 129/SvEv mice are deficient in proteins involved in IFN-γ receptor signaling development of primary tumors more frequently, which grow more rapidly at lower carcinogen doses. The results were similar for IFN-γ-deficient C57BL/6 mice. In models of genetically induced tumor formation (p53−/− mice), lack of IFN-γ or of its signaling pathway proteins further enhances tumor development. IFN-γ acts at two levels; first, it promotes the differentiation and expansion of tumor-specific Th1 CD4+ T cells and cytolytic T cells (CTL), and also activates macrophages. Secondly, IFN-γ acts directly on tumor cells, increasing their immunogenicity.

Local cytokine administration in the vicinity of neoplastic cells improves the immune response to solid tumors. The most commonly used method to increase local cytokine concentration is tumor cell transduction with cDNA encoding a cytokine of interest, although this approach poses serious problems for clinical application. Synthetic polymers and microspheres generated with gelatin and chondroitin sulfate have also been tested as controlled cytokine release systems, although to be effective, the microspheres must reach the tumor at the appropriate concentration. Nanotechnology provides valuable new tools that would allow controlled local release of cytokines. One of the most promising methods for delivery of drugs to their points of action is by binding them to magnetic particles that are then targeted using an external magnetic field. The first trial of magnetism to attract a drug to the site of action used intravenously injected epirubicin-coated magnetic particles. After injection, a magnetic field was applied as close to the tumor as possible for 45 minutes, resulting in efficient drug administration to the site. To enable a more extensive study of the IFN-γ effect on tumor progression by modulating the Th1 immune response in the tumor microenvironment, it was necessary to optimize a magnetic nanoparticle based system for IFN-γ delivery. Three different methods were used to produce uniform magnetic nanoparticles (polydispersity degree b20%) with different crystallinity and surface characteristics. One was a conventional method of coprecipitation of Fe salts in water, widely used for commercial products and for previous studies of biomolecule immobilization. Another was a new method based on the decomposition of an organic precursor in organic media, which assures the formation of uniform, highly crystalline particles. Finally, we used the laser pyrolysis technique, a unique method for producing dry powders in a single step starting from a gaseous precursor. We analyzed the adsorption/ release of murine interferon gamma (IFN-γ) onto these iron oxide nanoparticles with various coatings. Particle size and crystal structure, as well as the coating, affect the magnetic properties of the material and are therefore predicted to have an effect on its adsorption/release capacity. Magnetic nanoparticles were prepared by coprecipitation and by decomposition of a metal-organic precursor, either in solution or from an aerosol. These methods were optimized to obtain uniform particles, with a polydispersity degree b20% and thus within the monodispersed range.The particles prepared by decomposition and laser pyrolysis had an average diameter of 4.5 and 10 nm, respectively, and standard deviations in the monodispersed range (b0.20). Iron oxide nanoparticles synthesized by coprecipitation in the presence of dextran (sample C) had a 10 nm diameter and a standard deviation of 35%.

The presence of dextran reduces and regulates particle growth, but for sample C, the particle size distribution (a consequence of the mechanism of particle formation) was still wider than that for the decomposition synthesis methods (samples O, L). We characterized the nature of the iron oxide nanoparticles and their surface coating by X-ray diffraction and IR spectroscopy.

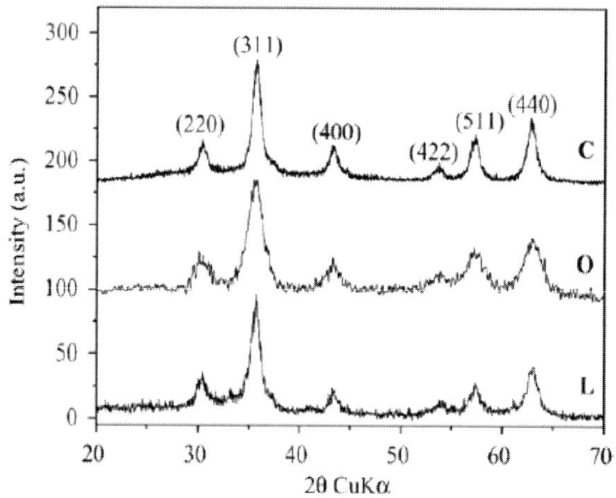

Figure 5. X-ray diffraction patterns for magnetite nanoparticles by different methods.

All diffraction peaks corresponded to spinel structure similar to magnetite or maghemite, and the crystal size calculated from the peak broadening (sample C=10 nm, sample O=5 nm, sample L=9 nm) concurred with mean TEM size.

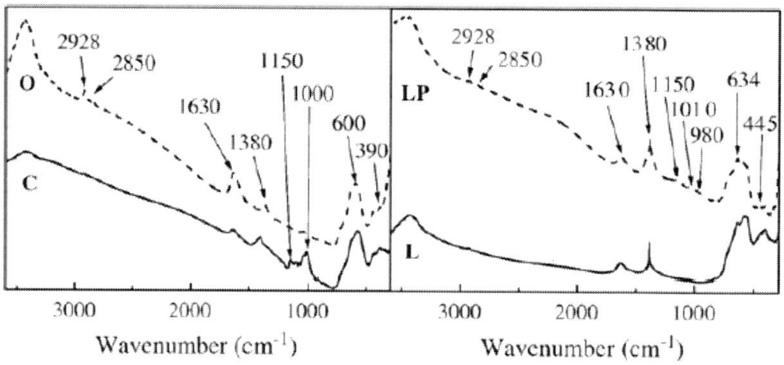

Figure 6. Infrared spectra for magnetite nanoparticles coated with distinct carboxyl molecules.

In the low frequency IR area, two strong absorption bands were assigned at 390 and 600 cm−1 to the Fe–O stretching of iron oxide. The shape and position of these bands in samples C and O suggest that the majority of iron oxide is magnetite, although some oxidation has taken place and maghemite is also present. The appearance of new shoulders at higher frequency in this IR region indicated that laser pyrolysis samples L and LP consist mainly of maghemite rather than magnetite. This is due to the oxidant conditions of the recrystallization process.

In the intermediate IR region, we identified surface coating molecules as well as synthesis-derived impurities, i.e., the sharp band at 1380 cm^{-1} due to nitrate impurities remaining from acid recrystallization (samples L, LP). From 1150 cm^{-1} to 1000 cm^{-1}, multiple strong absorption bands appear in the C and LP spectra due to the C–O single bond stretching modes of the dextran and phosphonoacetic acid, respectively. Sample LP showed broad absorption bands in this region, with bands at 1050 and 1010 cm^{-1} corresponding to the split triply degenerate υ3 (P–O–) vibration and at 980 cm, corresponding to the υ1 (P–O). This assignment is commonly associated with a monodentate bond between particle and phosphonate. The small differences in band intensity and position between this sample and those previously reported are due mainly to pH and to phosphonate concentration. In the high energy range, the two absorption bands near 3000 cm^{-1} were due to C–H stretching (anti-symmetric and symmetric). These bands were partially masked for samples C and LP, because of the intense O–H stretching modes of dextran hydroxyl groups and water molecules. At 1630 cm^{-1}, we observed a weak absorption band in all spectra due to adsorbed water. This peak was more intense in sample O because of the overlapping of the DMSA C=O stretching mode.

Thermogravimetric analyses of the powder samples were used to quantitate the coating. In the case of the phosphate-coated sample (LP), phosphate is not lost after heating up to 1000 °C. Analysis of Fe content was, therefore, carried out by elemental analysis by ICP. The weight percentage of coating molecules adhered to the nanoparticle surface increased from 14% for the dextran-coated sample C to 24% for the DMSA-coated sample O and 64% for the carboxydextran coated sample (LCD). This LCDNONC coating percentage trend did not correlate as predicted with the increase in particle size (ONCNLCD), indicating that surface nature and negative charge are the principal factors that affect the amount of coating adsorbed on the nanoparticle surface. All negatively charged samples had similar hydrodynamic sizes (~100 nm), which is essential for comparison of the adsorption/ release process between IFN-γ and the coated nanoparticles. The amount of IFN-γ bound might otherwise vary as a function of aggregate size, interfering with sample comparison. The hydrodynamic size values were stable for months and sedimentation was not observed. Nanoparticle dispersion stability depends on charge and surface chemistry, which give rise to both steric and electrostatic repulsion. Conventionally, a zeta potential higher than ±30mV would be considered stable, i.e., the dispersion will resist aggregation.

In the case of DMSA-, phosphonoacetic- and citric acid-coated samples (O, LP and LC), the ionizable carboxylate groups on their surfaces govern the pH dependent evolution of the Z-potential, whose high values resemble those reported in similar cases. A pure polymer-coated sample, such as dextran sample C, shows lower electrostatic stabilization (lower Z-potential), due to the lack of ionizable surface groups; the steric effect helps to maintain a stable dispersion, independently of the pH. For the carboxydextran-coated sample (LCD), the charged polymer on the surface provides both stability factors. The electrostatic effect is reflected in the Z-potential values, higher than those for sample C and similar to samples LC and LP. The pH dependence of the Z-potential thus showed intermediate variation between sample C and samples LC and LP. Finally, the uncoated pyrolysis sample (L) shows the typical Z-potential values reported for bare iron oxide nanoparticles, which is λ_0 at pH 7.

5.5. Tunable Magnetic Arrangement of Iron Oxide Nanoparticles *In Situ* Synthesized on the Solid Substrate from Diblock Copolymer Micelles

The synthesis of magnetic nanoparticles is an interesting and important area of research due to the many potential applications such as high-density recording media. In solution-phase synthesis, magnetic nanoparticles were efficiently produced in a controlled way on their size and dispersity. However, effective deposition of magnetic nanoparticles in an ordered pattern on the solid substrate via solvent evaporation cannot be easily achieved particularly over an extended area.

To fabricate magnetic nanopatterns, a top-down lithographical approach with electron-beam, interference, or X-ray lithography can be employed although it is not cost effective. Alternatively, bottom-up self-assembling techniques can be engaged to produce an ordered pattern of magnetic nanoparticles without separating their synthesis and placement on the substrate. For example, a self-assembled nanodomain of diblock copolymers was utilized as a lithographic mask in the etching process or a nanotemplate in the deposition process to generate periodic nanopatterns transferred from the nanostructure of copolymers. Magnetic patterns of cobalt nanodots or nanowires were successfully fabricated from ordered nanodomains of diblock copolymers in thin films. For the lithographical mask application of diblock copolymers in a reactive ion-etching process, nanodomains of one block had a stronger etching resistance than those of the other block so that the nanopattern of diblock copolymers was transferred to the cobalt layer. An additional ashing process was generally necessary to remove residues of block copolymers.

In the process of generating a perpendicular arrangement of magnetic cobalt nanowires, cylindrical nanodomains of diblock copolymers aligned perpendicularly to the substrate by electric fields were selectively removed, and then cobalt was deposited in the nanotemplate of the empty cylindrical domains. In these fabrication processes, an etching/deposition process of magnetically active components and a removing process of diblock copolymers were separately required to produce a magnetic pattern. Ordered patterns of nanoparticles on the solid substrate can also be fabricated from a single-layered film of diblock copolymer micelles in a hexagonal arrangement. For example, diblock copolymer micelles containing $HAuCl_4$, a precursor of gold nanoparticles, were coated on various substrates to form a micellar monolayer in a hexagonal order. An ordered pattern of gold nanoparticles was then fabricated in situ on the substrate by plasma treatment without deteriorating the original order of micelles. During the plasma treatment, the copolymers were also completely removed so that a pure pattern of nanoparticles was fabricated in a controlled manner by a single procedure after coating of diblock copolymer micelles on the substrate without an additional removing process of copolymers. In addition, the approach with diblock copolymer micelles would not require an annealing process after coating because nanometer- sized micelles consisting of a soluble corona and an insoluble core already formed in a selective solvent for one of the blocks. In contrast, annealing should be performed to induce nanodomain formation in thin films of diblock copolymers themselves. Thus, synthesis of a pattern of nanoparticles on the substrate with diblock copolymer micelles can be process efficient. However, it was very limited to fabricate patterns of electrically, optically, or magnetically active nanoparticles by employing monolayers of diblock copolymer micelles. We demonstrate instant synthesis of magnetic patterns of iron oxide nanoparticles on silicon wafers by a single step of plasma treatment on a monolayer of diblock copolymer micelles. Moreover, we were able to generate ferrimagnetic or superparamagnetic patterns with taking control of the size of iron oxide nanoparticles simply by changing the amount of precursors in the micelle. Size-controlled synthesis and deposition in an ordered pattern of magnetic nanoparticles were simultaneously achieved.

In toluene, a selective solvent for the PSblock, spherical micelles of PS-PVP diblock copolymers consisting of a PS corona and a PVP core were formed at the concentration studied, which was well above the critical micelle concentration. A variety of metal salts, precursors of nanoparticles, can be selectively coordinated to the PVP core. Ferric chloride ($FeCl_3$) as a precursor of magnetic nanoparticles was selectively loaded into the PVP core.

From the PS-PVP micellar solution with $FeCl_3$, a single layer of the micelles in a short-range hexagonal order was fabricated simply by spin coating and was confirmed by dark PVP cores, due to selectively loaded $FeCl_3$, in the TEM images shown in our previous report. To synthesize iron oxide nanoparticles in hexagonal order while maintaining the two-dimensional micellar arrangement, a monolayer film of PS-PVP micelles containing $FeCl_3$ was treated with oxygen plasma, resulting in the formation of iron oxide nanoparticles and the removal of organic PS-PVP micelles. As shown in Figure 1a, a pattern of iron oxide nanoparticles (18.0 nm in average diameter) was fabricated with the preservation of a short range hexagonal order of micelles. The molar ratio of $FeCl_3$ to vinyl pyridine was 0.5 in the case of Figure 7. To verify the crystalline structure of iron oxide nanoparticles, the X-ray diffraction experiment was performed on a pattern of nanoparticles synthesized on the silicon wafer. However, the amount of nanoparticles on the substrate was not enough to obtain an adequate signal-to-noise ratio.

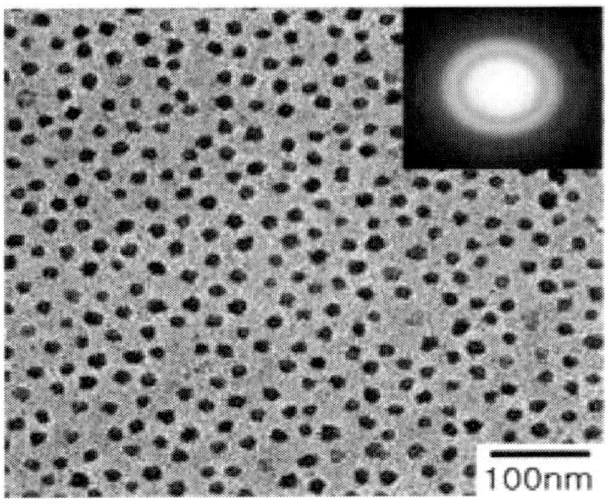

Figure 7. TEM image of a self-assembled pattern of iron oxide nanoparticles in hexagonal order.

Alternatively, the selected-area electron diffraction pattern shown in the inset of Figure 7 verified the iron oxide structure as ç-Fe_2O_3. Nonspecific orientation of crystal planes of nanoparticles to the substrate, polycrystalline nanoparticles, or both would result in a concentric diffraction pattern.

The Fe (2p3/2) photoelectron peak at 711.2 eV shown in the inset of Figure 2b was also in agreement with the value reported for ç-Fe_2O_3 in the literature. In addition, XPS results in

Figure 8 confirmed the complete removal of PS-PVP copolymers by the disappearance of the carbon photoelectron peak at 284.5 eV after oxygen plasma treatment.

We synthesized a pattern of pure iron oxide nanoparticles in short-range hexagonal order without polymer residues. In atomic force microscopy (AFM) and FE-SEM images of iron oxide nanoparticles on silicon wafers (Figure 9), each iron oxide nanoparticles appeared as a hemisphere.

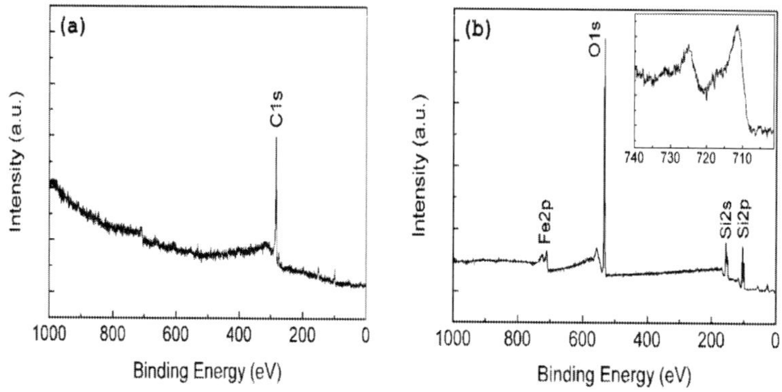

Figure 8. X-ray photoelectron spectra: (a) monolayer film of PS-PVP micelles containing $FeCl_3$; (b) a pattern of iron oxide nanoparticles on silicon wafer after O_2 plasma treatment.

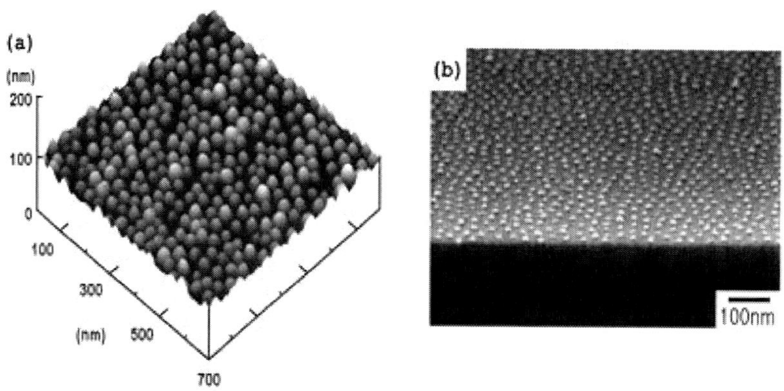

Figure 9. Hexagonal pattern of iron oxide nanoparticles on silicon wafers.

During the formation of iron oxide nanoparticles, the hemispherical morphology was induced presumably due to the chemical similarity between iron oxide and native silicon oxide on the top of silicon wafers, instead of spherical nanoparticles, which were observed in the case of gold nanoparticles synthesized in a similar condition. The average height was 10.0 nm with a 0.5 molar ratio of $FeCl_3$ to vinyl pyridine. Since the oxygen plasma process was executed at room temperature, iron oxide nanoparticles were annealed to improve the crystallinity of each nanoparticle.

After the annealing of iron oxide nanoparticles, selected-area electron diffraction patterns became sharp and clear. However, the shape of iron oxide nanoparticles directly synthesized on the silicon nitride membrane of TEM grids became irregular. If adhesion of the oxide nanoparticles on the nitride membrane were poor, it would be a cause for the irregular shape of nanoparticles. However, the shape of oxide nanoparticles fabricated on the silicon wafer was not changed significantly when FE-SEM images of Figures 9b and 10b before and after annealing were compared. The average diameter and height of iron oxide nanoparticles were slightly reduced after annealing for all cases of iron oxide nanoparticles synthesized with different molar ratios of $FeCl_3$ to vinyl pyridine.

By the use of the same procedures of coating, plasma treatment, and annealing, we effectively altered the diameter of iron oxide nanoparticles with the preservation of a hexagonal order on the silicon substrate only by changing the amount of $FeCl_3$ per micelle. As shown in Figure 10, the average diameters of iron oxide nanoparticles synthesized with 1.0, 0.5, and 0.2 molar ratios of $FeCl_3$ to vinyl pyridine were 25.0, 18.0, and 6.0 nm, respectively. The height of hemispherical iron oxide nanoparticles was also decreased accordingly as shown in the tilt-view images. However, the spacing of a pattern of iron oxide nanoparticles was almost same with the average value of 40.0 nm in all three cases from the fast Fourier transformation (FFT) patterns (the insets of Figure 10), although the short-range order of nanoparticles blurred the hexagonal FFT patterns.

The area density of iron oxide nanoparticles on the substrate was ca. 3.7×10^{11} nanoparticles/ in for all cases. Since each iron oxide nanoparticle was synthesized in each core part of hexagonally ordered PS-PVP micelles, the spacing of patterns of iron oxide nanoparticles was the same as that of the original pattern of micelles regardless of the particle size.

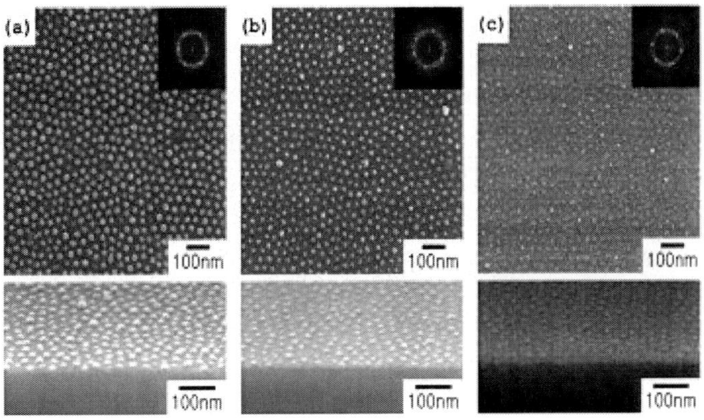

Figure 10. FE-SEM images of hexagonal pattern of size-controlled iron oxide nanoparticles in plain tilt views.

We effectively fabricated patterns of size-controlled iron oxide nanoparticles with the maintenance of their inter distance on the solid substrate, which cannot be created easily with presynthesized magnetic nanoparticles. Since different sizes of magnetic iron oxide nanoparticles in the same order were synthesized, their size-dependent magnetic properties were measured at room temperature as shown in Figure 11.

Magnetization was normalized by the total weight of the sample including the substrate. A hexagonal arrangement of iron oxide nanoparticles of 25.0 nm in average diameter, corresponding to Figure 10a, was ferrimagnetic with the coercivity of 50 Oe (Figure 11a), whereas that of the nanoparticles of 6.0 nm in average diameter, corresponding to Figure 4c, was superparamagnetic without hysteresis (Figure 11b). Since the area density of iron oxide nanoparticles on the substrate and the volume of each iron oxide nanoparticle can be estimated from the images shown in Figure 10, saturation magnetizations can be normalized by the total weight of iron oxide nanoparticles themselves without the substrate. The density of the nanoparticles was assumed to be equal to that of iron oxide in bulk. The estimated values of the saturation magnetizations (ca. 100 emu/g for 25.0 nm diameter and ca. 30 emu/g for 6.0 nm diameters) were similar to those reported in the literature, although the estimated values were not exact due to the limitation of calculation.

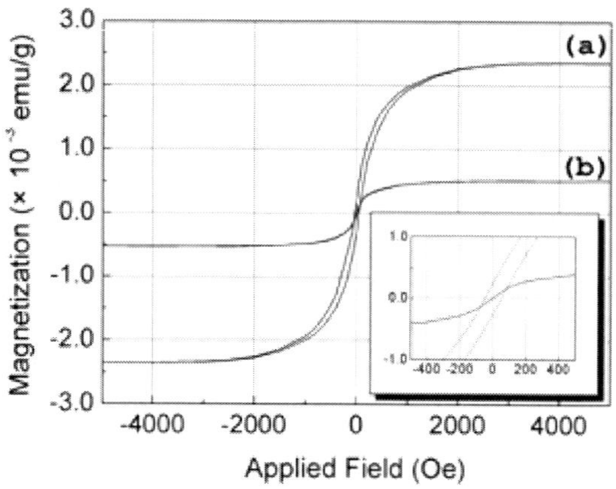

Figure 11. Magnetization curves at 300 K for the annealed patterns of size-controlled iron oxide nanoparticles.

For the superparamagnetic pattern, temperature-dependent magnetizations at 200 Oe for both the zero-field cooled (ZFC) and the field-cooled (FC) cases were investigated (Figure 12). The ZFC magnetization passed through a maximum at 90 K, which was associated with the blocking temperature. Although iron oxide nanoparticles were uniform, they could be polycrystalline. It would be responsible for a broad peak of the ZFC curve. Thus, we were able to select a ferrimagnetic or a superparamagnetic pattern of iron oxide nanoparticles in situ synthesized on the substrate by adjusting their diameter with a dose of the precursor per micelle.

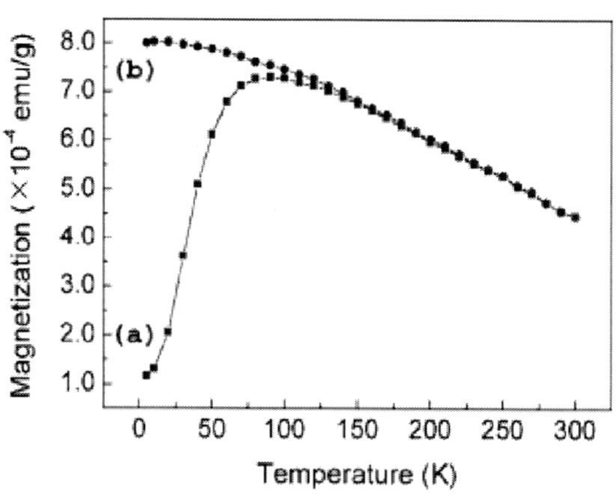

Figure 12. Temperature dependence of magnetization for the annealed pattern of iron oxide nanoparticles of 6.0 nm in average diameter at the applied field of 200 Oe.

5.6. Pullulan Acetate Coated Magnetite Nanoparticles for Hyper-Thermia

Magnetic nanoparticles with tailored surface chemistry have recently attracted a great deal of attention by virtue of their potential biomedical applications including magnetic resonance imaging (MRI) contrast enhancement, hyperthermia, immunoassay, and drug delivery. Hyperthermia involves applying an alternating magnetic field of suitable frequency that produces heat dissipation through the oscillation of the internal magnetic moment of superparamagnetic particles. Hyperthermia using magnetic nanoparticles either alone, or in association with classical chemo- or radiotherapies, has been recognized as a useful therapeutic modality in the treatment of malignant tumors because the neoplastic cells are more thermosensitive than normal cells. For hyperthermic applications, nanomagnetic particles should have high magnetization values in order to generate high thermal energies and size smaller than 50 nm with a narrow particle size distribution. In addition, hyperthermic thermoseeds require a surface coating of the core magnetic particles, which has to be not only nontoxic and biocompatible but also give targetable particles with particle localization in a specific area. Because of their hydrophobic surfaces and large surface area to volume ratio, *in vivo* use of nano-magnetic particles trends to result in agglomeration and very rapid clearance from the circulation. In addition, a common problem of uncoated magnetic nanoparticles is their chemical instability with respect to oxidation in air. In the case of iron oxides, the final compositions are very often intermediate between magnetite (Fe_3O_4) and maghemite (α-Fe_2O_3), due to the oxidation of the particles during the synthesis. Surface modification or coating of iron oxide nanoparticles with biocompatible polymers is often indispensable for *in vivo* biomedical applications in addition to prevent oxidation of the particles. Pullulan is a flexible water soluble, neutral linear polysaccharide consisting of α-1,6 linked maltotriose residues (Figure 13(a)) which is used extensively as a food additive, and has therefore been shown to be safe for human use. However, recently pullulan has also been investigated for its biomedical applications in areas such as targeted drug and gene delivery and diagnostic imaging using quantum dots, for the following reasons: (1) it is non-toxic, (2) its usefulness as a plasma expander, (3) its non-immunogenic properties, and (4) its non-antigenic properties. The pullulan modified magnetic nanoparticles reported previously in the literature were prepared by adding an aqueous solution of pullulan to magnetite particles dispersed in deoxygenated water.

In this method, the individual polymeric chains of pullulan need to be cross-linked with glutaraldehyde. Pullulan modified magnetic nanoparticles can afford reduced cytotoxicity and enhanced cellular uptake. However, the remaining glutaraldehyde may have an adverse effect on the human body. In this study, we describe a novel method to coat magnetic nanoparticles with pullulan acetate (PA) (Figure 13(b)), which is an amphipathic derivative of pullulan. This method does not require a cross-linking agent, and moreover is a rapid and easy process with a variety of ways to control the particle size.

The properties of the PA-coated magnetic nanoparticles (PAMNs), such as their morphology, particle size, and distribution, composition and magnetic properties were investigated by transmission electron microscopy (TEM), dynamic light scattering (DLS), thermogravimetric analysis (TGA), and vibrating sample magnetometry (VSM).

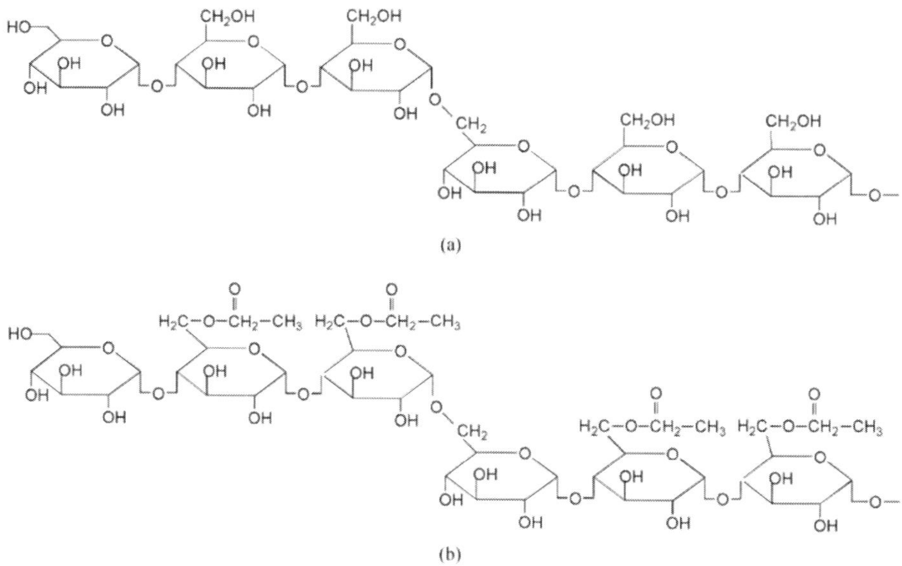

Figure 13. Chemical structure of pullulan (a) and PA (b).

Heating *in vitro* in an alternating magnetic field was carried out. The presence of the coating of PA on the magnetic nanoparticles was confirmed by Fourier transform infrared spectroscopy (FTIR). *In vitro* cytotoxicity of PAMNs was evaluated using L929 cells as model cells. Cellular uptake of nanoparticles, using KB cells as model cells, was visualized by TEM. The *in vitro* hyperthermic effect of PAMNs was examined using KB cells as model tumor cells. These results obtained will be useful for further application of the novel magnetic material in tumor thermotherapy.

A novel approach to the synthesis of FePt magnetic nanoparticles chemically synthesized magnetic nanoparticles has drawn great attention because the magnetic particles have reduced dimensions down to a few nanometers with well controlled size, size distribution and shape. Monodisperse magnetic nanoparticles have potential applications in high performance nanocomposite magnets, high-density magnetic storage media, magnetic resonance imaging contrast agents, biological sensors and directed drug therapy. The polyol reduction methodology used by Fievet et al. and modified by Sun et al. has been developed by many researchers to provide a wide variety of nanoparticle systems for investigations in the mentioned applications. The polyol process requires at its basis, a diol as a reducing agent, metal salt precursors, surfactants and a solvent. The surfactants are added to prevent particles from agglomeration and to inhibit oxidation. In certain cases, surfactants can also induce geometric changes to the particles. In a typical polyol synthesis route, FePt particles are synthesized via decomposition of iron pentacarbonyl, $Fe(CO)_5$ and reduction of ($Pt(acac)_2$, in presence of 1,2-hexadecanediol (HDD). Oleic acid and oleyl amine are used as the surfactants in a proper solvent. It is still a research focus to develop simple, efficient and cost effective methods to synthesize FePt nanoparticles with better control of morphological, structural and magnetic properties of the particles. In our previous studies, modified synthesis recipes have been used to produce FePt nanoparticles with alternative morphological and magnetic properties. By using iron acetylacetonate ($Fe(acac)_3$) to replace iron pentacarbonyl $Fe(CO)_5$, we obtained monodisperse FePt nanoparticles with a particle size of 2 nm. It was also found that the FePt nanoparticles can be synthesized without using any polyol reducing agent in octyl ether. Figure 14 shows a TEM image of the as-synthesized FePt nanoparticles. It can be seen that the average particle size is 2 nm, which is typical for FePt nanoparticles reported earlier for recipes using Fe $(acac)_3$. It is clear that the particles have spherical shape and narrow size distribution.

Figure 15 shows XRD patterns of as-synthesized FePt nanoparticles. From the position of the peaks, it is clear that the as-synthesized FePt nanoparticles have a chemically disordered fcc structure. An average particle diameter of 1.6 nm was calculated from the peak width of the XRD pattern using the Scherer formula, which is consistent with the diameter calculated by statistical analysis of TEM images, and is similar to the particle size reported. Energy dispersive x-ray analysis of the formed FePt particles show the initial feed ratio of Fe and Pt metal precursors are maintained in the fcc FePt nanoparticles. The structural characterizations above clearly show that fcc FePt nanoparticles with a composition of equal atomic Fe:Pt ratio were synthesized by co-reduction of $Fe(acac)_3$ and Pt(acac) in the presence of 1, 2-hecadecanediol in octyl ether.

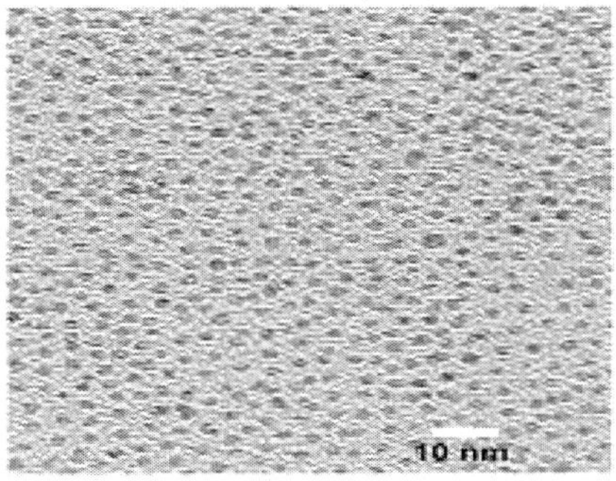

Figure 14. TEM image of the as-synthesized FePt particles.

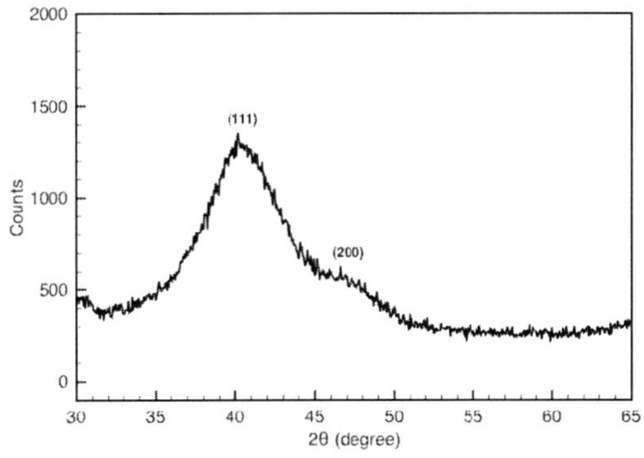

Figure 15. Powder XRD scan of the as- synthesized FePt particles using HDD as the surfactant and reductant.

More interestingly, the HDD also played a role as an effective surfactant. Based on the structure and functionality, using HDD as a surfactant for dispersing the formed nanoparticles is not unreasonable, while also providing the reductive species needed to form the fcc FePt nanoparticles.

In addition, it is possible that the oxidized HDD molecules are also used as capping ligands to protect the particles from oxidation and facilitate dispersion of particles in nonpolar solvents, unlike traditional methods where capping ligands such as oleic acid and oleylamine are used.

The chemically disordered fcc FePt crystal has a low magnetocrystalline anisotropy, and hence is superparamagnetic at room temperature. To obtain high uniaxial magnetocrystalline anisotropy, which is required for many applications, the as-synthesized FePt nanoparticles typically have to be annealed at high temperature to transform to the chemically ordered L10 tetragonal FePt phase. Figure 16 shows the XRD patterns for the samples annealed in nitrogen and Ar +H2 (7%) at 700o C for 1 h, compared with the XRD pattern for the L10 FePt sample produced.

It can be clearly seen from this figure that both the peak position and intensity for the annealed samples prepared by our new method match perfectly with those of the sample made from the standard method, indicating that the Ll0 structure is indeed formed after thermal annealing. This double confirms that the particle composition is in the right range for the formation of the equal atomic L10 phase.

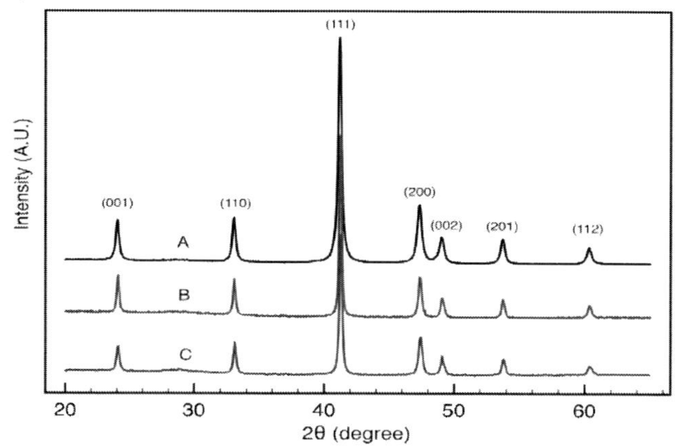

Figure 16. XRD scans of the samples annealed at 700°C.

In addition, the samples annealed in nitrogen do not show any oxide peaks in their XRD patterns which indicates that the HDD effectively protected the fcc FePt nanoparticles from oxidation when exposed to the open atmosphere. Figure 17 shows the coercivity change with annealing temperature of the annealed particles obtained from our new method, compared with the coercivity values of particles obtained by the method using oleic acid and oleylamine as surfactant.

Particles were annealed on Si substrate under Ar + H2 (7%) for 1 hour. It was observed that for the traditional FePt sample maximum coercivity of 1.8 T was obtained at 700°C and a sharp decrease in coercivity was observed above 700°C. On the other hand, for the sample prepared by our new method, coercivity continued to increase above 700°C, and a maximum coercivity of 2.3 T was obtained at 800° C.

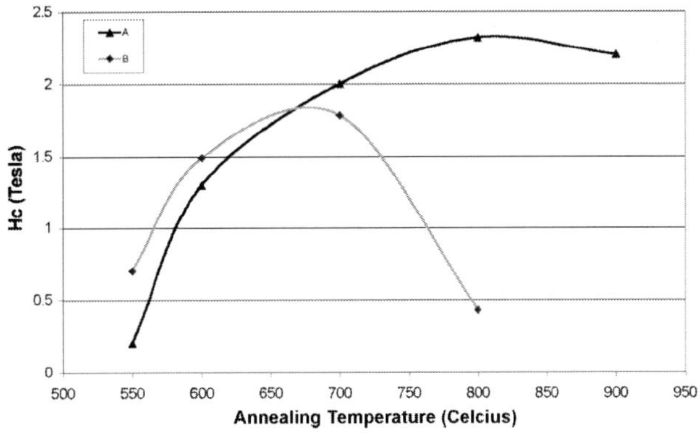

Figure 17. Annealing temperature dependence of coercivity of annealed samples prepared by the current method.

It is interesting to find that there is an approximately 28% increase in coercivity of the particles obtained by our new method as compared to the particles prepared. This result shows that the magnetic hardening of the FePt nanoparticles prepared by our new method can be maintained up to 800°C.

Even when particles were annealed at 900° C, only a slight decrease in coercivity of the FePt nanoparticles was observed. The exact nature of the observed coercivity enhancement by this method is not completely clear. However, it is possible that the nanoparticles derived from this new method did not sinter substantially during annealing, relative to the particles from the traditional method. This degree of isolation would be possible if the HDD is preferentially adsorbed to the particle surface during the synthesis, protecting the particles during annealing from sintering.

The smaller particle size and high degree of chemical ordering within the particles would then result in higher coercivity at the higher annealing temperatures. This magnetic hardening behavior may also be related to the enhanced atomic diffusion during the heat treatment that expedited the L10 phase formation.

5.7. Biocompatible Magnetic Fluid Nanoparticles Internalized in Vegetal Tissue

The biological interest in the magnetic fluid effect in living organisms represents an important application field of magnetic nanoparticles, mainly for biotechnological use. Stable concentrated suspensions of magnetic nanoparticles in either organic or inorganic solvents are called magnetic fluids. Magnetic fluids are suspensions of superparamagnetic particles – small single domain particles – that respond to a magnetic field but completely demagnetize when the field is removed. Colloidal stability is assured by coating the particle with a shell of nonmagnetic molecular surfactants which prevent close approach of the nanomagnetic cores thereby reducing the possibility of aggregation via Van der Waals or magnetic attractions. For medical purposes, the ferrophase also needs to have a small physical diameter in order to pass through the cell biomembranes and to reach various target tissues.

The interest in the study of the biological effects induced by magnetic fluid presence in culture medium upon vegetal organisms and microorganisms as well as upon animals has increased. Special attention was paid to genetic effects of magnetic fluid nanoparticles that are found to lead to chromosomal aberrations in young vegetal plants which may be related to the putative use in plant biotechnology. A relatively small number of experimental studies is dedicated to the influence of magnetic fluids on the photosynthesis process revealing the stimulatory effect of magnetic fluids on the chlorophylls content, which has evidenced some stimulatory effects on the plant growth. This stimulatory effect may be explained on the basis of iron importance in the vegetal organisms.

Lobreaux et al. have reported that iron treatment of *Zea mays* induced ferritin protein accumulation in roots and leaves over a period of 3 days. Recently investigations have shown that phytoferritin occurs in plant cells as crystalline magnetite (Fe_3O_4), maghemite (α-Fe_2O_3), and hematite (γ-Fe_2O_3). The biological interest in the magnetic fluid effect in living organisms represents an important application field of magnetic nanoparticles, mainly for biotechnological use. Some quantitative observations have been made regarding the influence of magnetic nanoparticles coated with perchloric acid on the growth of *Zea mays* plants in early ontogenetic stages.

5.8. Synthesis and Characterization of Fe_3O_4 Magnetic Nanofluid

Ferrofluids (FF) or magnetic nanofluids are stable magnetic colloids systems composed of nanoparticles of single-magnetic domain with a mean diameter of around 10 nm, dispersed in organic or inorganic carrier fluids (aqueous or nonaqueous). Recently, FF or magnetic nanofluids have been the subject of much interest because of their unusual optical, electronic, and magnetic properties, which can be changed by applying an external magnetic field. The special feature of FF is the combination of normal liquid behavior with superparamagnetic properties. They are said to be superparamagnetic, meaning that they are attracted by a magnetic field but retain no residual magnetism after the field is removed. Due to their small size and superparamagnetic behavior, magnetic ferrofluids can be viewed as a new class of magnetic materials with potential technological applications in biomedicine and nanotechnology. They are now promising materials for cancer diagnosis and therapy.

There are various methods to prepare the magnetite nanoparticles of nanometer size range, due to their immense technological applications. The surfacted and ionic aqueous Fe_3O_4 magnetic nanofluids have been synthesized and characterized. Magnetic properties of nanoparticles in ferrofluids and their mean crystallite size in powder were investigated through a vibrating sample magnetometer (VSM) and X-ray diffraction (XRD) by using the Scherrer approximation, respectively.

Scanning probe microscopy (SPM) was used in two ways: atomic force microscopy (AFM) to determine the morphology of conglomerates of nanoparticles and to measure their diameter; and magnetic force microscope (MFM) to visualize the magnetic domains of Fe_3O_4 nanoparticles. The IR spectra were used to confirm the formation of bonds, related to metal in the octahedric and tetrahedric sites, allowing the identification of the ferrites spinel structure. The precipitated fine particles were characterized by XRD for structural determination and estimation of crystallite size. The X-ray powder diffraction pattern of the sample was recorded on a *Bruker D8 Advance* diffractometer using CuK (1.5406 Å) radiation at room temperature in the range of 10 to 80° in the 2Θ scale, with a scanning speed of 0.02°/s and a step time of 3 s. Generally, XRD can be used to characterize the crystallinity of nanoparticles, as well as the average nanoparticle diameter. All the peaks of XRD patterns were analyzed and indexed using the ICDD data base, comparing with magnetite standards. The lattice constant *a* was found to be 8.310 Å, which was compared with the lattice parameter for the magnetite of 8.39 Å. Finally, the analysis of the diffraction pattern showed the formation in the sample of a cubic spinel structure, due to the strongest reflection that proceeds from the (311) plane (Figure 18), characteristic of such a phase. The peaks indexed as planes (220), (311), (400), (422), (511) and (440) corresponded to a cubic unit cell, characteristic of a cubic spinel structure. Therefore, it was confirmed that the crystalline structure of obtained magnetite nanoparticles, agreed with the structure of an inverse spinel type oxide. Crystallite size measurements were determined from the full-width at half maximum (FWHM) of the strongest reflection of the (311) peak, using the Scherrer approximation, which assumes the small crystallite size to be the cause of line broadening.

Here, DN is the crystallite mean size, k is a shape function for which a value of 0.9 is used, λ is the wavelength of the radiation, the full width at half maximum (FWHM) in radians in the 2Θ scale, and the Bragg angle. The crystal size calculated were (9.64 ± 0.13) nm.

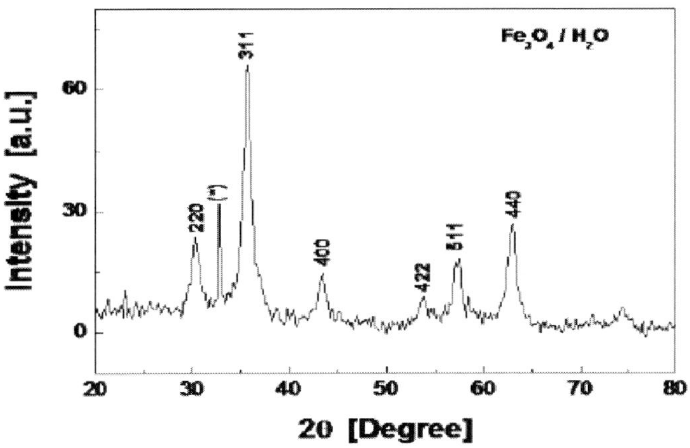

Figure 18. X-ray diffraction pattern of Fe_3O_4 water based magnetic nanofluid.

Due to the broad diffraction pattern lines, it can be said that particles have a size of about nanometers; therefore, the nanofluids can be conveniently prepared by making use of particles in this size range. Magnetic characterization of the nanoparticles was done using a vibrating sample magnetometer (VSM) in the physical property measurements system (PPMS) of *Quantum Design*, at room temperature, with a magnetic field in the range of -5000 to 5000 Oe, where parameters such as saturation magnetization (Ms), and coercive field (Hc) were evaluated. The behavior of nanofluids is mainly determined by their magnetic properties.

Ferrofluids usually show a superparamagnetic behavior, since each of the particles can be treated as a thermally agitated permanent magnet in the carrier liquid. In the presence of a magnetic field H, the magnetic moment (μ) of the particles will try to align with the magnetic field direction, leading to a macroscopic magnetization of the liquid. The magnetization M of the liquid behaves like the magnetization of a paramagnetic system. Furthermore, these types of materials often show hysteresis loops *(M–H* curves), explained by an irreversibility in the magnetization process related to the pinning of magnetic domain walls at impurities or grain boundaries within the material. Another possible cause of such a behavior, is the magnetic anisotropy of the crystalline lattice. The shape of these loops is determined, in part, by particle size. Even at smaller sizes (around tenths of nanometer or less), superparamagnetic behavior can be seen, where the magnetic moment of the particle as a whole is free to fluctuate in response to thermal energy, while the individual atomic moments maintain their ordered state relative to each other. This leads to the low anhysteretic behavior show in the *M–H* curve for the magnetite sample based in water or kerosene like carrier liquid, respectively (Fig. 19 and 20). The saturation magnetization (MS) obtained and the coercive field (Hc) at room temperature was 0.067 emu/cm3 and 16.7 Oe for the *Fe3O4* sample water based magnetic nanofluid, and 0.335 emu / cm3 and 8.6 Oe, for the Fe_3O_4 sample kerosene based, respectively. When magnetic nanoparticles are suspended in kerosene, and coated with oleic acid as a surfactant agent, the difference in the magnetic properties of particles can be attributed to a weak structuration favored by the combined action of interfacial (Van der Waals or acid–base), magnetic attractions between the particles immersed in the carrier liquid and the steric repulsion produced between particles, which avoids the formation of conglomerates. Therefore, they can be considered like single– domain nanoparticles showing a superparamagnetic behavior. Finally, the recorded hysteresis loops shows that the particles are superparamagnetic at room temperature, which is characteristic of a soft ferromagnetic material such as magnetite. AFM investigation provides qualitative and quantitative information on the diameter and height of the scanned particles.

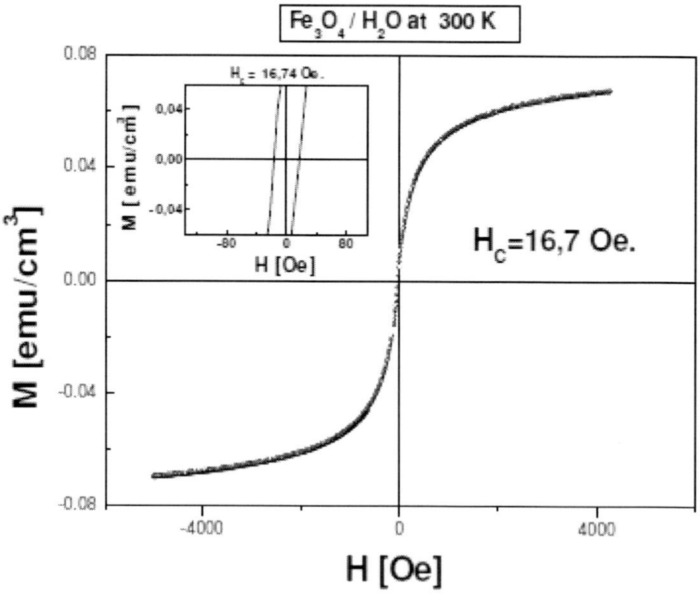

Figure 19. M vs. H hysteresis loop of Fe_3O_4 water based magnetic nanofluid.

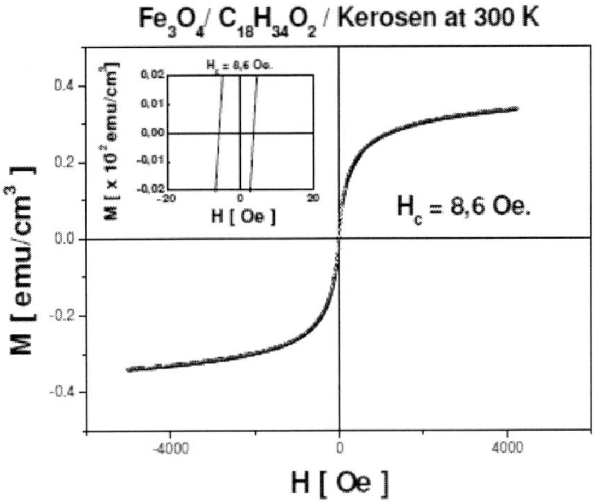

Figure 20. M vs. H hysteresis loop of Fe_3O_4/$C_{18}H_{34}O_2$/kerosene based magnetic nanofluid.

The equipment used to visualize the morphology of nanoparticles was an *Asylum Research MFP-3D* Atomic Force Microscope in both AC non-contact mode in air and magnetic mode (MFM) operating at room temperature. An *Asylum Research* ASYMFM cantilever and tip were used. This cantilever was made up of silicon, coated with a 50nm CoCr layer.

The cantilever had the following characteristics: coercivity in the range of 300-450 Oe, spring constant 1-2 N/m, resonant frequency of 55 to 90 KHz and a tip radius less than 20 nm. AFM measurements have been carried out on dried samples deposited on mica and hard disk substrate (useful immobilizing individual particles). They have been repeated on different sites of the deposited sample, prepared in the same conditions of room temperature and ambient atmosphere. The analysis of magnetic particles consisted in the observation of the surface of Fe_3O_4 nanoparticles. The mean diameter from the height, measured by AFM is larger than the mean size of the nanoparticles measured by XRD. This can be explained because the analysis is statistical and some 'particles' are, in fact, conglomerates, which may result in apparent particles with relatively larger height. The existence of such agglomerates implies that an improvement in the preparation of the samples is required, because the attractive electric and magnetic forces are not totally balanced by the steric repulsion conferred by the oleic acid coating molecules.

The agglomeration of Fe_3O_4 nanoparticles is evident in the topography image, Figure 21a, obtained by means of AFM. Figure 21b, represents the best curve of the particle size histogram using the lognormal distribution function. It shows that the most probable particle diameter of conglomerates was of DN = (15.3 ± 0.2) nm, with a coefficient of variation of 1.3%.

Figure 21c shows a 2D MFM image, where we can see the magnetic domains of Fe_3O_4 nanoparticles conglomerates with a random direction. The samples were magnetized by an external field oriented vertically to the mica substrate plane prior to the measurement. In this case, the interpretation of magnetic domains shows that a dark zone represents a repulsive force of domains, while a bright zone is an attractive force.

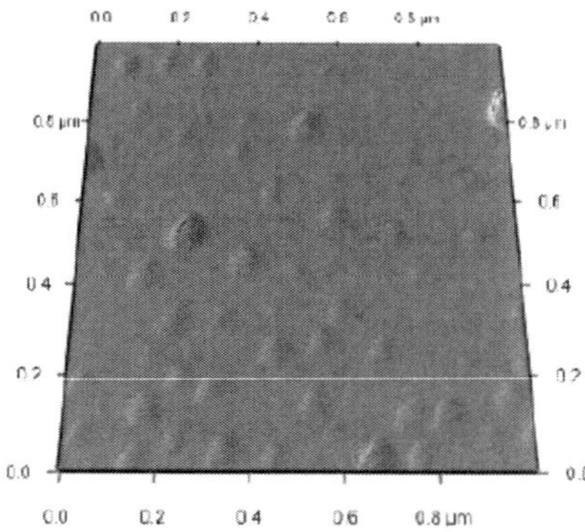

a)

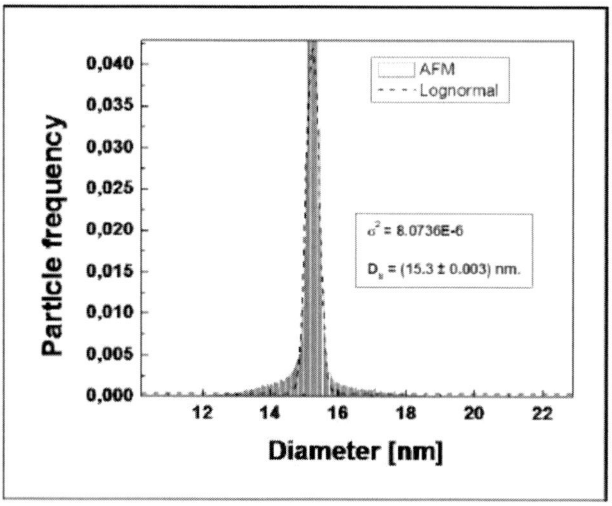

b)

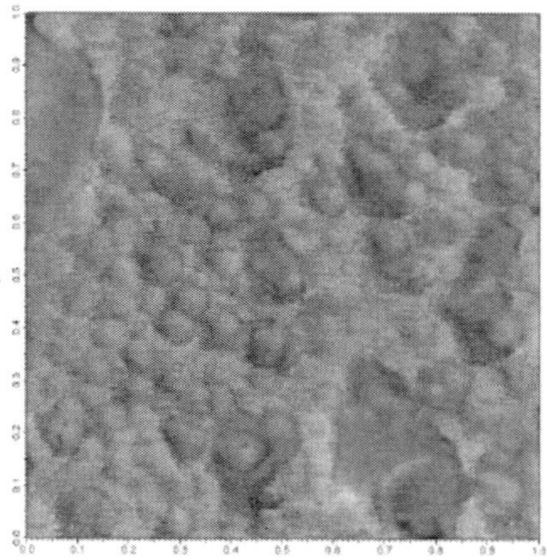

c)

Figure 21. AFM topography images of conglomerates of Fe_3O_4 nanoparticles.

At the present time, studies of the structure and properties of ferromagnetic materials draw the great attention. The most studied and widely used in different investigations are iron oxide nanoparticles–magnetite γ-Fe_2O_3 and magnetic Fe_3O_4. Metal nanoparticles on the basis of iron are prospective material as information carriers, medicine transportation system etc. A valuable property of the iron is its ability to reduce and oxidize, to form complex compounds with different biochemical properties and to participate in electron transmission reactions. Elementary or metallic iron Fe^0 oxidizes to Fe^{3+} with the presence of air oxygen and humidity, forming insoluble Fe_2O_3. Bivalence iron FeO is an intermediate product of iron and ferrites production. It is a constituent of ceramics and heat resistant enamels. Iron oxide Fe_3O_4–tetroxide of three iron (magnetite mineral) ($FeO \cdot Fe_2O_3$ or Fe^{II} $(Fe^{III}O_2)_2$) when heated at air oxidizes to Fe_2O_3. It is ferromagnetic, has high electro conductivity and dissolves in acids with formation of Fe^{+2} and Fe^{+3} salts.

The iron-ligands bonds these complexes are coordination bonds. Donor-atoms of ligands give their unbonding electron pairs to the central atom of metal. The bond length of Fe^{III}-O, defined in various complexes is 2.10Å at S=5/2 and 1.99Å at S=1/2; bond length of Fe^{III}–N is 2.32Å at S=5/2 and 1.98Å at S=1/2, bond length of Fe^{III}–C is 2,31Å at S=5/2 and 1.98Å at S=1/2. There are many investigations in the literature devoted to descriptions of super magnetic iron oxide nanoparticles used in clinical and diagnostic medicine due to their satisfactory biocompatibility as a contrasting agent in magnetic-resonance analysis. Loaded by thousands of molecules of medicine nanoparticles transport medicine to the exact address of affected tissues i.e., they are universal medicine carriers. At the same time, neither medicine nor other components of the complex have an influence on the magnetic properties of iron oxide nanoparticles.

Although modern nano medicine has had impressive success in the creation of a new class of medications, there is currently some gap in terms of full theoretical models describing the electron structure of the complex compounds of iron oxide nanoparticles with medical molecules.

Recent work has studied the spatial and electron structure of iron (II) and iron (III) oxides, complex compounds of Fe_3O_4 with bi-glucose and CREKA pentapeptide, composed of amino acid residues Cys1, Arg2, Glu3, Lys4 and Ala5, by semiempirical molecular mechanic and quantum chemistry methods, parameterized for transition metal atoms. This complex is used for malignant tumorous diseases therapy with to the objective of delivering medicines to affected tissue cells. Calculations have been carried by means of application computer programs package HyperChem 8.03. The next stage of this research describes magnetic particles that not only home to tumors, but also amplify their own homing. The system is based on a magnetic amino dextran – CREKA peptide conjugate that can be used as anti-tumor molecules. Iron oxide nanoparticles are coated with this amino dextran – CREKA peptide conjugate, thereby producing new binding sites for more particles. These results show that the particle-bound peptide retains its binding toward clotted plasma proteins. The detailed study of the spatial and electron structure of iron (II) and iron (III) oxides and Fe_3O_4, preceded the investigation of the complex compound of Fe_3O_4 ($FeO \cdot Fe_2O_3$) with d-glucose and peptide CREKA. It has been used with the resonance model of Fe_3O_4 structure, including Fe=O and Fe_2O_3. That is why there are results of comparable analysis of electron structures FeO, Fe_2O_3 and Fe_3O_4, optimization of valence geometry has been carried out by the semiempirical method of molecular mechanics MM+. Figure 22a shows the calculation models of Fe_2O_3 and Fe_3O_4, and Figure 22b-22d show generalized results of geometrical parameters calculations and calculations of values of partial charges on the atoms. The values of the contributions of various types of electron energy in total energy after optimization of the electron structure of FeO, Fe_2O_3 and Fe_3O_4, calculated semiempirical method of quantum chemistry PM3, are generalized in the Table 1. In accordance with study results of the optimized structure Fe_2O_3 the single bond length of Fe-O is 1.77 Å, meanwhile double bond length of Fe=O is 1.48Å.

The resonance structure of Fe_3O_4 and its low energetic state corresponds to the structure where bond length of Fe-O is 3.9 Å and bond length Fe=O is 1.90 Å. The valence angle of Fe-O-Fe equal 76.4° provides the maximum favorable balance of electrostatic interactions of opposite charged atoms in Fe_2O_3. The angle of $Fe_3O_4$is 109.5°, i.e., a little bit more than expected due to the Fe=O presence. Such increasing of valence angles is characteristic for the O--Fe=O bond. If in Fe_2O_3 the value of this angle is 148.8^0, in Fe_3O_4 it makes ~180^0. It means that iron oxide FeO·Fe_2O_3 has a strictly tetrahedral structure. Despite the attribute difference in total and electron energies of iron oxides, the repulsion energy of nucleases in FeO·Fe_2O_3 is higher than in FeO and Fe_2O_3 58365.9 kcal/mol and 13367.60 kcal/mol, respectively. Consequently, the nucleases repulsion energy in FeO·Fe_2O_3 as whole is 5014.23 kcal/mol which is higher than sum of repulsion energy of separated FeO and Fe_2O_3. The comparison of the bonding energy parameter (Table 1) let us conclude that the difference of these values for Fe_2O_3 and Fe_3O_4 makes 92.69 kcal/mol, i.e. the formation of Fe_2O_3 is preferable to the formation of FeO·Fe_2O_3. One of the important molecules characteristics, that is crucial for their behavior in various force fields is dipole momentum, which has an additive property. Optimized structures FeO, Fe_2O_3 and Fe_3O_4 are characterizing with dipole momentum, equal to 1.61, 0.93 and 2.63, respectively (Table 1).

The obtained results indicate that the character of molecular interactions of Fe_3O_4 with medium because of its high polarity severely differs from Fe_2O_3. Comparison of partial charges on investigated compounds atoms (Figure 22b) shows that formation of Fe_3O_4 leads to redistribution of electron density.

As the result of redistribution of electron density from the p-orbital of oxygen atom of the central atom to the oxygen atom of FeO, observed the decreasing of negative charge value on oxygen atoms (~ 0.233), and iron atoms get a formal positive charge ~ 0.230 (Figure 22,b). For comparison of the reaction ability of iron oxides have been analyzed the values of molecular orbitals energies and carried out comparative analysis of energies of highest occupied molecular orbital HOMO and lowest unoccupied LUMO molecular orbital of Fe_2O_3 and Fe_3O_4 (Figure 23).

In accordance with the approximation of the valence electrons, total number of electrons, considering at PM3 calculations is equal 34 and 48 correspondingly. At the ground state, electrons occupied 17 and 24 molecular orbitals. Ionization energies corresponding to negative values of energy HOMO Fe_2O_3 and Fe_3O_4 is equal to 10.31 and 12.69 eV. The energy difference of the first unoccupied orbital and HOMO is interpreted as the first energy of disturbance and is equal to 6.27 eV for Fe_2O_3 and 8.89 eV for Fe_3O_4. The less difference there is, the more the electron affinity of the system.

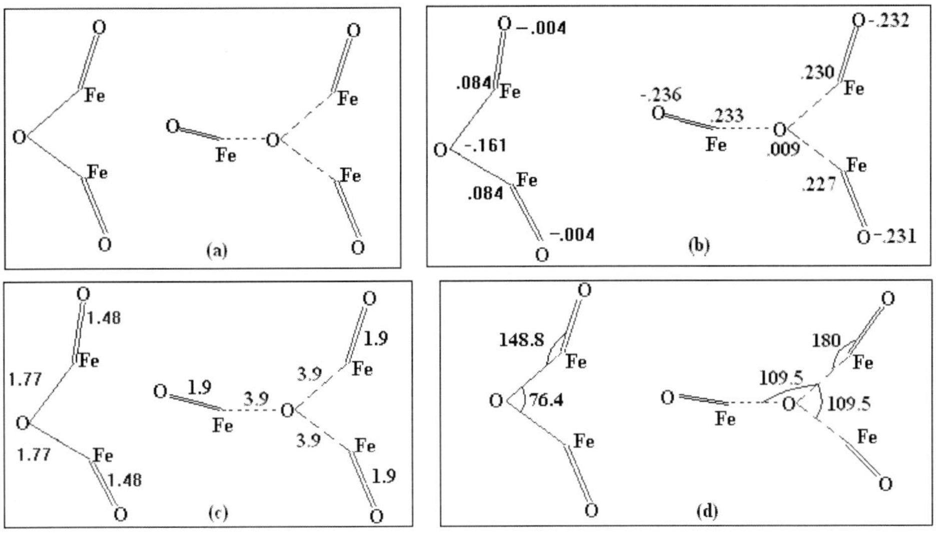

Figure 22. Optimized structures of Fe_2O_3 and Fe_3O_4 on MM+ and PM3 methods data: (a) calculation model; (b) partial charges on atoms in electron charges units; (c) valence bond lengths, Å; (d) valence angles, degrees.

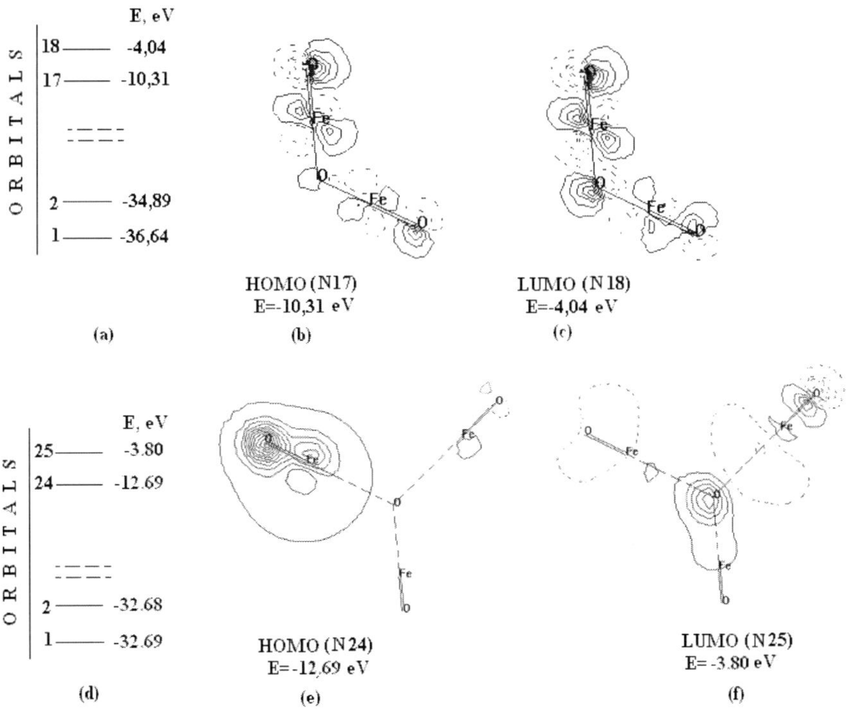

Figure 23. Energetic diagrams and graphic images HOMO and LUMO of Fe_2O_3 (a-c) and Fe_3O_4 (d-e).

Table 1. Energy parameters (kcal/mol) and dipole momentum of iron oxide and Fe_3O_4 complexes with di-glucose and peptide CREKA on PM3 method data

Compound (complex)	Total energy	Electron energy	Nuclease repulsion energy	Binding energy	Dipole momentum, D
FeO	-18686.68	-27040.04	8353.37	-13.95	1.61
Fe_2O_3	-44266.46	-97618.13	53351.67	-248.50	0.93
Fe_3O_4	-62846.49	-129565.76	66719.27	-155.81	2.63
Fe_3O_4 +di-glucose	-164526.47	-1258765.58	1094239.10	-4972.72	4.49
Fe_3O_4 +di-glucose+ CREKA-1	-334034.81	-4790443.65	4456408.85	-12498.93	47.45
Fe_3O_4 +di-glucose+ CREKA-2	-334141.39	-4769214.43	4435073.04	-12605.51	38.86
Fe_3O_4 +di-glucose+ CREKA-3	-334163.16	-4447260.03	4113096.87	-12627.28	32.84

Table 2. Parameters, characterized the complex structure of Fe_3O_4 with di-glucose by MM+ methods

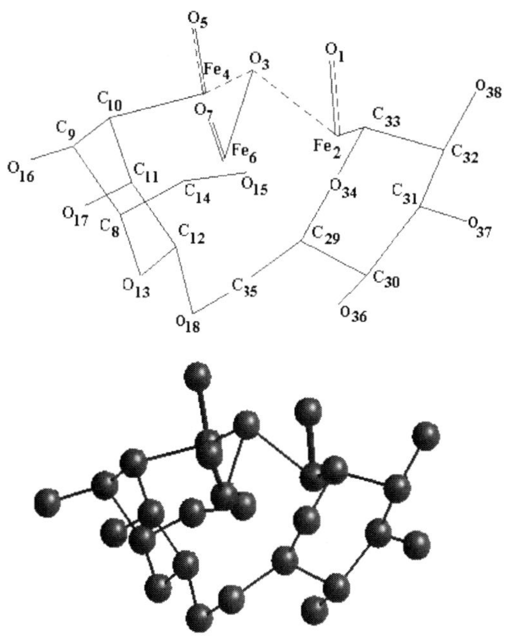

Spatial models of Fe_3O_4 with bi-glucose

Bond	Length, Å
C_{33}-Fe_2	2.00
Fe_2=O_1	1.73
Fe_2-O_3	1.89
Fe_6 -O_3	1.91
Fe_6=O_7	1.73
Fe_4 -O_3	1.89
Fe_4-C_{10}	2.00
Fe_4=O_5	1.73
C_{10}-C_{11}	1.53
Valence angle	Value, degree
C_{33}-Fe_2-O_1	118.78
O_1-Fe_2-O_3	118.83
Fe_2-O_3-Fe_6	103.31
Fe_2-O_3-Fe_4	104.80
O_3-Fe_6=O_7	179.59
Fe_6-O_3-Fe_4	105.14
O_3-Fe_4=O_5	117.76
O_3-Fe_4-C_{10}	120.76
O_5=Fe_4-C_{10}	119.03
Torsion angle	Value, degree
O_{32}-C_{33}-Fe_2 =O_3	-152.41
C_{33}-Fe_2-O_3-Fe_6	104.78
O_1=Fe_2-O_3-Fe_6	-52.84
O_1=Fe_2-O_3-Fe_4	-162.73
Fe_2-O_3-Fe_4=O_5	-68.57
Fe_2-O_3-Fe_6=O_7	-170.99
O_1=Fe_4-C_{10}-C_{11}	-123.36
O_7=Fe_6-O_3-Fe_4	-61.36
Fe_6-O_3-Fe_4=O_5	-177.12
O_3-Fe_4-C_{10}-C_{11}	74.81

On the basis of the results obtained and the generalization of studies, a model of the complex has been built including Fe_3O_4, bi-glucose and CREKA. First, the spatial structure of complex Fe_3O_4 with bi-glucose by semiempirical and molecular mechanic MM+ methods was studied. The results are generalized in Tables 1-3. As follows from these calculations (Table 2), bond length of Fe-C (Fe_2-C_{33} and Fe_4-C_{10}) are 2.0Å, bonds Fe=O (Fe_2=O_1, Fe_6=O_7 and Fe_4=O_5) are equal 1.73Å, and Fe-O are 1.89Å & 1.91 Å.

From a comparison of the results obtained, it follows that the bond length of Fe=O in complex with bi-glucose decreases from 1.9 up to 1.73Å, Fe-O from 3.9Å up to 1.9Å compared with Fe_3O_4. Valence angles characterizing bonding part of Fe_3O_4 with bi-glucose change at following intervals: angles O_3-Fe_2=O_1 and O_3-Fe_4=O_5 decrease up to 118.83 and 117.76°, correspondingly, meanwhile O_3-Fe_6=O_7 save linearity of bond (~180°) (Table 2).

The angle of Fe_2-O_3-Fe_4 decreases to 104.8°. As a result, the complex formation with bi-glucose deforms the structure of Fe_3O_4. The result of the deformation is the redistribution of partial charges on the complex atom (Table 3). It follows from the calculations that the negative charge on the oxygen atom of iron oxide decreases O_1 (Δ=0,167), O_5 (Δ=0,164) and O_7 (Δ=0,110) and increases the electron density of the iron atoms Fe_2, Fe_4 and Fe_6. As a result, partial charges on these atoms have negative values. The changes of partial charges on oxygen atoms of bi-glucose are negligible. The complex formation of Fe_3O_4 with bi-glucose is accompanied by changes of geometry as well as electron and charge densities on Fe and O atoms. The next stage of calculation studied the structure of complex Fe_3O_4 with bi-glucose and CREKA.

Table 3. Charges (electron charge unit) on atoms Fe3O4 with di-glucose before and after optimization on PM3 method data

atom	Charge		Atom	charge	
	before	after		before	after
Fe_2^*	0.227	-0.059	O_{13}	-0.279	-0.264
Fe_4	0.230	-0.060	O_{15}	-0.310	-0.343
Fe_6	0.233	-0.098	O_{16}	-0.306	-0.314
O_1	-0.231	-0.064	O_{17}	-0.317	-0.299
O_5	-0.232	-0.068	O_{18}	-0.290	-0.238
O_3	0.009	-0.002	C_{29}	0.024	0.038
O_7	-0.236	-0.126	C_{30}	0.051	0.051
C_8	0.020	0.058	C_{31}	0.011	0.035
C_9	0.055	0.030	C_{32}	0.018	0.020
C_{10}	0.008	0.055	C_{33}	0.193	0.133
C_{11}	0.018	-0.067	O_{34}	-0.284	-0.219
O_{38}	-0.301	-0.308	C_{35}	0.065	0.048
C_{12}	0.170	0.082	O_{36}	-0.314	-0.309
C_{14}	0.066	0.051	O_{37}	-0.317	-0.318

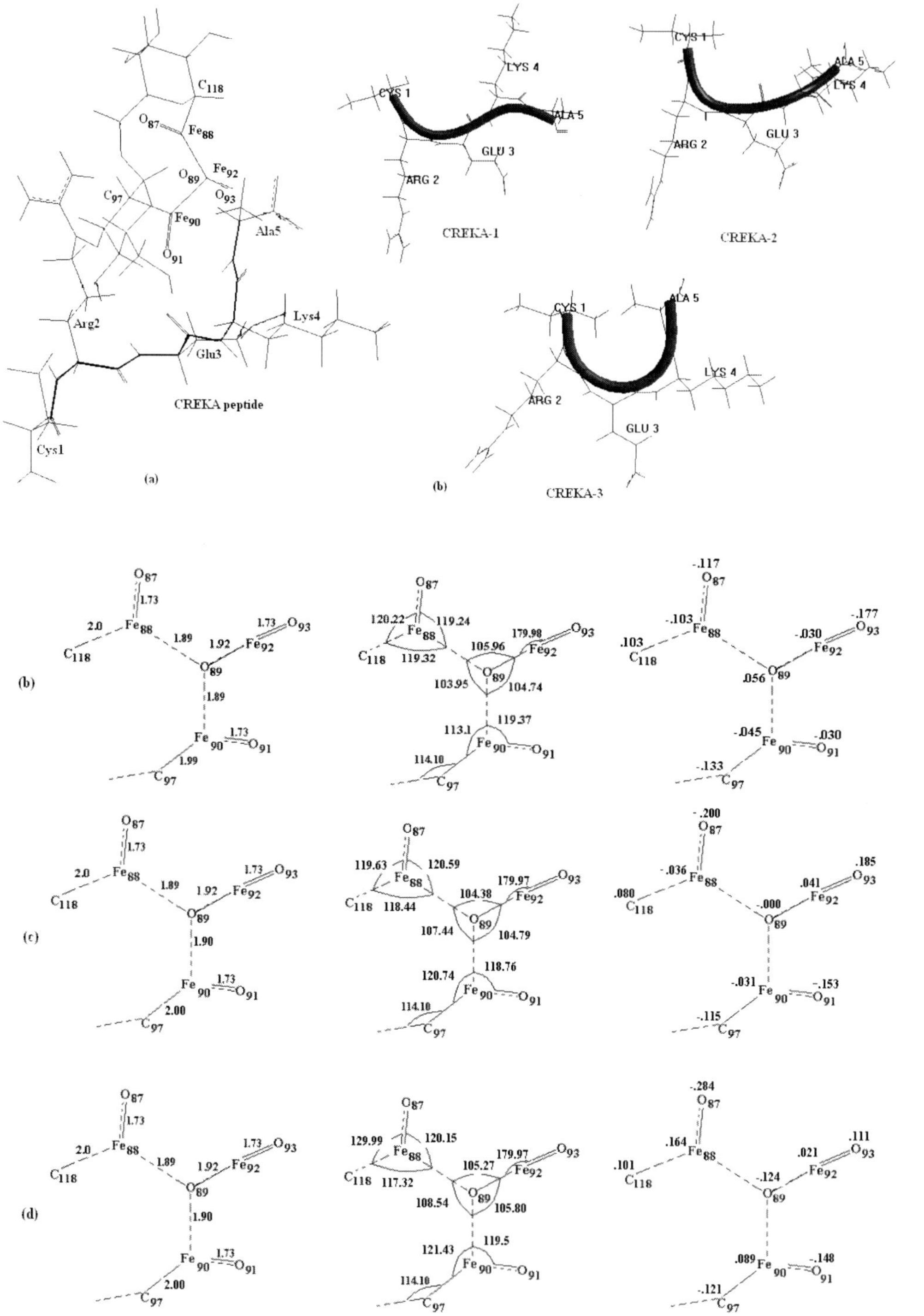

Figure 24. Spatial model of complex composed of Fe_3O_4, di-glucose and CREKA (a); conformations CREKA molecule (b); bond length (Å), valence angles (degree) and partial charges on atoms in various conformations of CREKA (b-d).

CREKA is a peptide composed of five amino acid residues Cys1, Arg2, Glu3, Lys4 and Ala5, which are used for cancer treatment with nitrogen oxide and dextran (glucose polymer) as medication for addressing delivery to tumorous cells. Calculations were used for three low energy conformations of CREKA. They were calculated and described by the theoretical conformation analysis method. They are known as CREKA-1, CREKA-2 and CREKA-3 and differ by dihedral angle values in the main and side chains of amino acid residues (Figure 24). The calculations results are presented in Figure 24 and Tables 1-3. As shown in Table 1, complex formation with bi-glucose accompanied with decreasing of total electron energy up to 101 679.98 kcal/mol, and addition of CREKA-1 in staggered conformation leads to decreasing the system energy to 233354.83 kcal/mol. Dipole momentum of complex rapidly increases 47.45D (in comparison with 4.49 for complex Fe_3O_4 with bi-glucose), which is indicative of high reaction ability of the complex, containing many charged groups. It should be noted that the optimization of complex geometry by MM+ method does not bring changes in the geometrical structure of bonding area $Fe_3O_4$with bi-glucose (Figure 24b-24d). Valence angle values and bond length do not change significantly. It is interesting to compare the calculations of partial charges on complex atoms that are binding with peptide. As shown in the calculations, the essential redistribution of electron density is observed on Fe_{88} and O_{87} atoms. The shifting of electron density from these atoms leads to a decrease of the negative charge on Fe_{88} and O_{87}, up to Δ=0.068 and Δ=0.11, respectively. Analysis of the results obtained (Table 1) reveals that the energy preference of the complex containing CREKA-3 in conformation with approached amino acid residues Cys1 and Ala59 (Figure 3b). In accordance with the distance between C^{α}-atoms of Cys1 and Ala5 in this conformation makes 6.1 Å, meanwhile in CREKA-1 and CREKA-2 this value is equal to 11.1 and 9.5 Å. Despite the energy difference of these conformations, it does not exceed 0.9 kcal/mol, in complexes of Fe_3O_4 with di-glucose. This difference makes 128.35 kcal/mol. This complex is characterized by a compact structure in which Fe_3O_4 has a tetrahedral structure (Figure 24,d). An important choice of dextran as the molecular backbone was based on practicality and availability.

Therefore, active amino dextran molecules have been synthesized in a 2-step process (Figure 25).

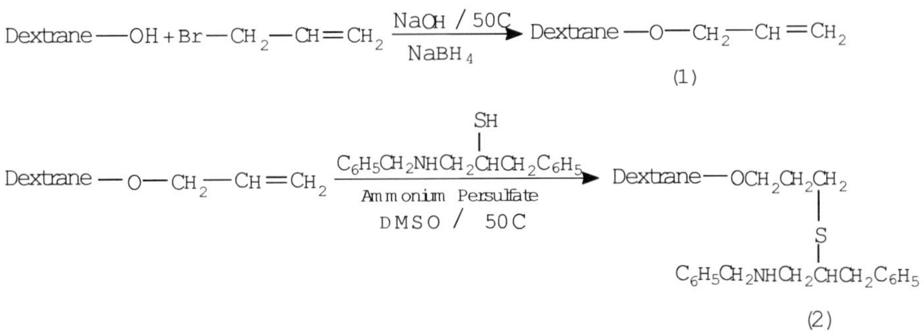

Figure 25. Covalent attachment of amino groups to dextran hydroxyl groups is 2-step process which prevents dextran cross-linking. DMSO = dimethyl sulfoxide.

Figure 26 shows the SEM of amino dextran – coated iron oxide nanoparticles that are synthesized by chemical reaction.

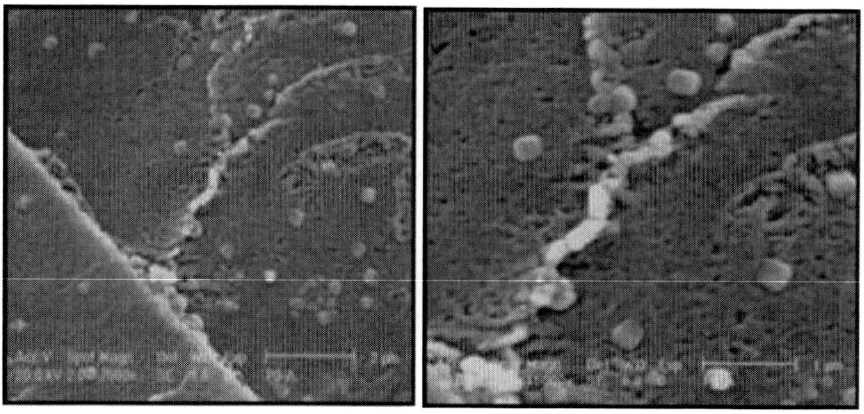

Figure 26. SEM of amino dextran – coated iron oxide SPIO nanoparticles.

These nanoparticles are very sensitive to temperature. Scanning electron micrography images were obtained from a diluted solution of the nanocomposite particles. The ability of the amino dextran – coated iron oxide nanoparticles to form a complex with CREKA peptide depends on the nanoparticles electrostatic interactions between the nanoparticles and the peptide. Therefore, it is possible to manipulate the incorporation process for CREKA peptide by the appropriate selection of the nanoparticles and the surface functionality.

Figure 27 shows X-ray the diffraction pattern of amino dextran – coated iron oxide nanoparticles. Diffraction of amino dextran – coated iron oxide nanoparticles have a strong peak at about 2Θ = 26.46°, which is a characteristic peak of amino-dextran –coated iron oxide nanoparticles. Studies on XRD patterns of nanoparticles are scarce in the literature.

Figure 28 shows the FT-IR spectrum of amino dextran – coated iron oxide nanoparticles, where the % of transmittance is plotted as a function of wave number (cm^{-1}). The wide peak around 3411 cm^{-1} is attributed to the O-H stretching vibrations of amino dextran. The peaks at 1523 and 1714cm^{-1} are attributed to the COO^- unsymmetrical and symmetrical stretching vibration respectively.

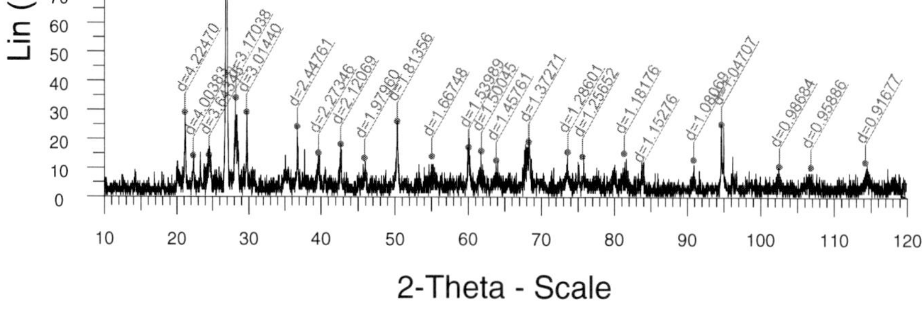

Figure 27. XRD of amino dextran – coated iron oxide SPIO nanoparticles.

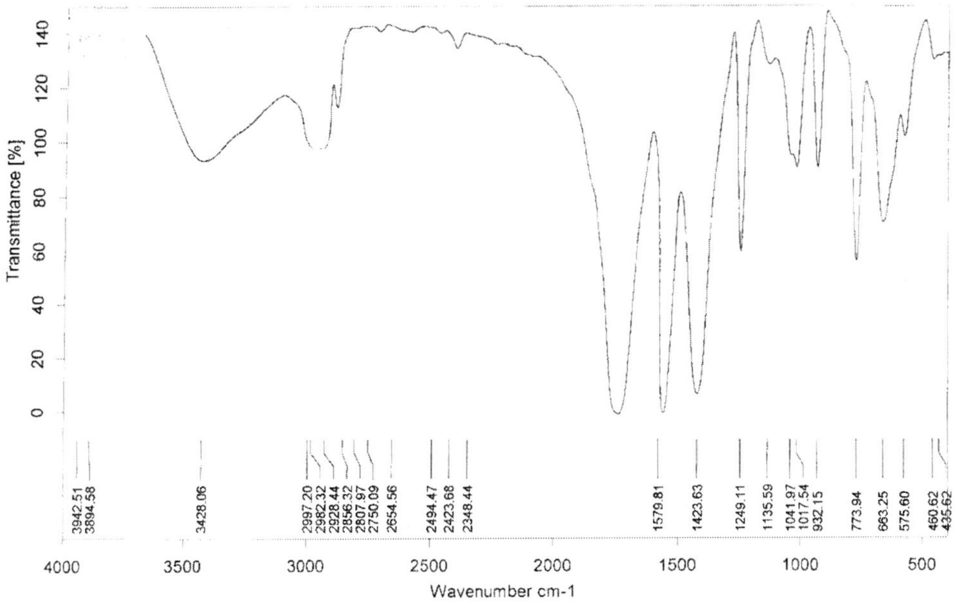

Figure 28. FT-IR Spectrum of amino dextran – coated iron oxide SPIO nanoparticles.

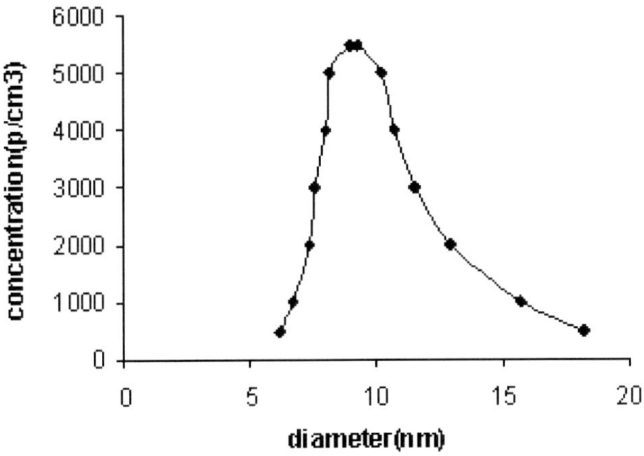

Figure 29. The particle size distribution of amino dextran – coated iron oxide SPIO /CREKA peptide conjugate.

The mean diameter of each conjugate was measured by dynamic light scattering (UPA-150; Honeywell-Microtrac, Clearwater, FL). Each conjugate was assayed for 10 minutes at a concentration > 5mg/mL of 0.9% saline. The refractor index of each sample was assayed (Fisher Scientific, Santa Clara, CA) and did not deviate from 0.9% saline. Latex particle standards of 3 different sizes gave weight-averaged diameters that were within 5% of their mean diameters (19±1.5, 102±3 and 993±21 nm), which were calibrated by photon correlation spectroscopy or electron microscopy. The analyzer software did not assume a Gaussian size distribution. Mean molecular diameters with SDs were calculated from volume distribution data (Figure 29). The average number of amino groups per dextran was measured in the following manner. The lyophilized dextran conjugate was dissolved in saline and the amine concentration was measured by the trinitrobenzene sulfonate assay using hexylamine as a standard. The glucose concentration of the same sample was measured by the sulfuric acid method. The amino density was calculated by dividing the amine concentration by the glucose concentration by multiplication by the average number of glucose units per dextran.

REFERENCES

[1] G. A. Held and G. Grinstein, *Appl. Phys. Lett.* 79, 1501 (2001).
[2] C. T. Black, C. B. Murray, R. L. Sandstrom, and S. Sun, *Science* 290, 1131 (2000).
[3] F. P. Victor, K. M. Krishnan, and A. P. Alivisatos, *Science* 291, 2115 (2001).
[4] T. Hyeon, *Chem. Commun.* (Cambridge) 2003, 927.
[5] S. Sun, H. Zeng, D. B. Robinson, S. Raoux, P. M. Rice, S. X. Wang, and G. X. Li, *J. Am. Chem. Soc.* 126, 273 (2004).
[6] S. Sun, C. B. Murray, D. Weller, L. Folks, and A. Moser, *Science* 287, 1989 (2000).
[7] C. C. Berry and A. S.G. Curtis, *J. Phys.* D 36, R198 (2003).
[8] W. H. Meiklejohn and C. P. Bean, *Phys. Rev.* 105, 904 (1957).
[9] V. Skumryev, S. Stoyanov, Y. Zhang, G. Hadjipanayis, D. Givord, and J. Nogues, *Nature* (London) 423, 850 (2003).
[10] E. E. Carpenter, S. Calvin, R. M. Stroud, and V. G. Harris, *Chem. Mater.* 15, 3245 (2003).
[11] H. Zeng, J. Li, J. P. Liu, Z. L. Wang, and S. Sun, *Nature* (London) 420, 395 (2002).
[12] H. Zeng, J. Li, Z. L. Wang, J. P. Liu, and S. Sun, *Nano Lett.* 4, 187 (2004).
[13] F. Bødker, S. Mørup, and S. Linderoth, *Phys. Rev. Lett.* 72, 282 (1994).
[14] Y. Labaye, O. Crisan, L. Berger, J. M. Greneche, and J. M.D. Coey, *J. Appl. Phys.* 91, 8715 (2002).
[15] R. Skomski and J. M.D. Coey, *Phys. Rev.* B 48, 15812 (1993).
[16] R. Skomski and J. M.D. Coey, in *Permanent Magnetism* (IOP, Bristol, 1999).
[17] Z. J. Guo, J. S. Jiang, J. E. Pearson, S. D. Bader, and J. P. Liu, *Appl. Phys.Lett.* 81, 2029 (2002).
[18] V. F. Puntes, K. M. Krishnan, A. P. Alivisatos, *Science* 2001, 291, 2115.
[19] C. Desvaux, C. Amiens, P. Fejes, P. Renaud, M. Respaud, P. Lecante,.E. Snoeck, B. Chaudret, *Nat. Mater.* 2005, 4, 750.
[20] J. Park, K. J. An, Y. S. Hwang, J. G. Park, H. J. Noh, J. Y. Kim, J. H..Park, N. M. Hwang, T. Hyeon, *Nat. Mater.* 2004, 3, 891.
[21] S. H. Sun, H. Zeng, D. B. Robinson, S. Raoux, P. M. Rice, S. X. Wang, G. X. Li, J. *Am. Chem. Soc.* 2004, 126, 273.
[22] D. Melville, F. Paul, S. Roath, *Nature* 1975, 255, 706.
[23] M. Uhlen, *Nature* 1989, 340, 733.
[24] C. T. Yavuz, J. T. Mayo, W. W. Yu, A. Prakash, J. C. Falkner, S. Yean, L. L. Cong, H. J. Shipley, A. Kan, M. Tomson, D. Natelson, V. L.Colvin, *Science* 2006, 314, 964.
[25] M. Lewin, N. Carlesso, C. H. Tung, X. W. Tang, D. Cory, D. T.Scadden, R. Weissleder, *Nat. Biotechnol.* 2000, 18, 410.

[26] E. M. Shapiro, S. Skrtic, K. Sharer, J. M. Hill, C. E. Dunbar, A. P., Koretsky, *Proc. Natl. Acad. Sci. USA* 2004, 101, 10901.

[27] J. H. Lee, Y. M. Huh, Y. Jun, J. Seo, J. Jang, H. T. Song, S. Kim, E. J. Cho, H. G. Yoon, J. S. Suh, J. Cheon, *Nat. Med.* 2007, 13, 95.

[28] R. L. Edelstein, C. R. Tamanaha, P. E. Sheehan, M. M. Miller, D. R. Baselt, L. J. Whitman, R. J. Colton, *Biosens. Bioelectron.* 2000, 14, 805.

[29] B. D. Cullity, Introduction to Magnetic Materials, Addison-Wesley, V. Salgueirin˜ o-Maceira, M. A. Correa-Duarte, M. Spasova, L. M. .Liz-Marza´ n, M. Farle, *Adv. Funct. Mater.* 2006, 16, 509.

[30] V. Salgueirin˜ o-Maceira, M. A. Correa-Duarte, M. Farle, A. Lo´ pez-Quintela, K. Sieradzki, R. Diaz, *Chem. Mater.* 2006, 18, 2701.

[31] S. Y. Chou, P. R. Krauss, P. J. Renstrom, *Science* 1996, 272, 85.

[32] B. Heidari, I. Maximov, E. L. Sarwe, L. J. Montelius, *J. Vac. Sci.Technol.* B. 1999, 17, 2961.

[33] R. Beckman, E. Johnston-Halperin, Y. Luo, J. E. Green, J. R. Heath, *Science* 2005, 310, 465.

[34] M. Schena, D. Shalon, R. W. Davis, P. O. Brown, *Science* 1995, 270,467.

[35] S. S. Gambhir, *Nat. Rev. Cancer* 2002, 2, 683.

[36] S. J. Hurst, E. K. Payne, L. D. Qin, C. A. Mirkin, *Angew. Chem. Int. Ed.* 2006, 45, 2672.

[37] D. H. Reich, M. Tanase, A. Hultgren, L. A. Bauer, C. S. Chen, G. J. Meyer, *J. Appl. Phys.* 2003, 93, 7275.

[38] B. Dieny, J. P. Gavigan, *J. Phys. Condens. Matter* 1990, 2, 187.

[39] E. E. Fullerton, D. T. Margulies, M. E. Schabes, M. Carey, B. Gurney, A. Moser, M. Best, G. Zeltzer, K. Rubin, H. Rosen, M. Doerner, *Appl.Phys. Lett.* 2000, 77, 3806.

[40] B. N. Engel, J. Akerman, B. Butcher, R. W. Dave, M. DeHerrera, M. Durlam, G. Grynkewich, J. Janesky, S. V. Pietambaram, N. D. Rizzo, J. M. Slaughter, K. Smith, J. J. Sun, S. Tehrani, *IEEE Trans. Magn.* 2005, 41, 132.

[41] S. S. P. Parkin, N. More, K. P. Roche, *Phys. Rev. Lett.* 1990, 64, 2304.

[42] M. M. Miller, A. A. Lazarides, *J. Opt.* A 2006, 8, S239.

[43] G. X. Li, V. Joshi, R. L. White, S. X. Wang, J. T. Kemp, C. Webb, R.W. Davis, S. H. Sun, *J. Appl. Phys.* 2003, 93, 7557.

[44] H. A. Ferreira, D. L. Graham, P. P. Freitas, J. M. S. Cabral, *J. Appl. Phys.* 2003, 93, 7281.

[45] J. Schotter, P. B. Kamp, A. Becker, A. Puhler, G. Reiss, H. Bruckl, *Biosens. Bioelectron.* 2004, 19, 1149.

[46] a)R. J. Wilson, US Patent 6337215, 2002; b)R. J. Wilson, US Patent 5932097, 1999.

[47] D.M. Pardoll, Cancer vaccines, *Nat. Med.* 4 (5 Suppl) (1998) 525–531.

[48] T. A. Waidmann, J.C. Morris, Development of antibodies and chimeric molecules for cancer immunotherapy, *Adv. Immunol.* 90 (2006) 83–131.

[49] G.P. Dunn, L.J. Old, R.D. Schreiber, *The immunobiology of cancer immunosurveillance and immunoediting, Immunity* 21 (2) (2004) 137–148.

[50] J.N. Biattman, P.D. Greenberg, Cancer immunotherapy: a treatment for the masses, *Science* 305 (5681) (2004) 200–205.

[51] L. Zitvogel, A. Tesniere, G. Kroemer, Cancer despite immunosurveillance: immunoselection and immunosubversion, *Nat. Rev. Immunol.* 6 (10) (2006) 715–727.

[52] L. Gattinoni, D.J. Powell Jr., S.A. Rosenberg, N.P. Restifo, Adoptive immunotherapy for cancer: building on success, *Nat. Rev. Immunol.* 6 (5) (2006) 383–393.

[53] K. Dredge, J.B. Marriott, S.M. Todryk, A.G. Dalgleish, Adjuvants and the promotion of Th1-type cytokines in tumour immunotherapy, *Cancer Immunol. Immunother.* 51 (10) (2002) 521–531.

[54] A. S. Dighe, E. Richards, L.J. Old, R.D. Schreiber, Enhanced in vivo growth and resistance to rejection of tumor cells expressing dominant negative IFN gamma receptors, *Immunity* 1 (6) (1994) 447–456.

[55] E.A. Bach, M. Aguet, R.D. Schreiber, The IFN gamma receptor: a paradigm for cytokine receptor signaling, *Annu. Rev. Immunol.* 15 (1997) 563–591.

[56] D.H. Kaplan, V. Shankaran, A.S. Dighe, E. Stockert, M. Aguet, L.J. Old, R.D. Schreiber, Demonstration of an interferon gamma-dependent tumor surveillance system in immunocompetent mice, *Proc. Natl. Acad. Sci. USA.* 95 (13) (1998) 7556–7561.

[57] S.E. Street, E. Cretney, M.J. Smyth, Perforin and interferon-gamma activities independently control tumor initiation, growth, and metastasis, *Blood* 97 (1) (2001) 192–197.

[58] S.E. Street, J.A. Trapani, D. MacGregor, M.J. Smyth, Suppression of lymphoma and epithelial malignancies effected by interferon gamma, *J. Exp. Med.* 196 (1) (2002) 129–134.

[59] B. Homey, A. Müller, A. Ziotnik, Chemokines: agents for the immunotherapy of cancer? *Nat. Rev. Immunol.* 2 (3) (2002) 175–184.

[60] D.M. Pardoll, Paracrine cytokine adjuvants in cancer immunotherapy, *Annu. Rev. Immunol.* 13 (1995) 399–415.

[61] R. Langer, New methods of drug delivery, *Science* 249 (4976) (1990) 1527–1533.

[62] M.S. Sabel, H. Hill, Y.S. Jong, E. Mathiowitz, R.B. Bankert, N.K. Egilmez, Neoadjuvant therapy with interieukin-12-loaded polylactic acid microspheres reduces local recurrence and distant metastases, *Surgery* 130 (3) (2001) 470–478.

[63] P.T. Goiumbek, R. Azhari, E.M. Jaffee, H.I. Levitsky, A. Lazenby, K. Leong, D.M. Pardoll, Controlled release, biodegradable cytokine depots: a new approach in cancer vaccine design, *Cancer Res.* 53 (24) (1993) 5841–5844.

[64] A.S. Lübbe, C. Bergemann, H. Riess, F. Schriever, P. Reichardt, K. Possinger, M. Matthias, B. Dörken, F. Herrmann, R. Gürtier, P. Hohenberger, N. Haas, R. Sohr, B. Sander, A.J. Lemke, D. Ohlendorf, W. Huhnt, D. Huhn, Clinical experiences with magnetic drug targeting: a phase I study with 4′-epidoxorubicin in 14 patients with advanced solid tumors, *Cancer Res.* 56 (20) (1996) 4686–4693.

[65] P. Tartaj, M.P. Morales, S. Veintemillas-Verdaguer, T. Gonzalez-Carreño, C.J. Serna, in: K.H.J. Buschow (Ed.), *Handbook of Magnetic Materials*, vol. 16, Elsevier, North-Holland, 2006, pp. 403–482.

[66] P. Tartaj, M.P. Morales, S. Veintemillas-Verdaguer, T. Gonzalez-Carreño, C.J. Serna, The preparation of magnetic nanoparticles for applications in biomedicine, *J. Physics D: Appl. Physics* 36 (2003) R182–R197.

[67] X.Q. Xu, H. Shen, J.R. Xu, J. Xu, X.J. Li, X.M. Xiong, Core-shell structure and magnetic properties of magnetite magnetic fluids stabilized with dextran, *Appl. Surface Sci.* 252 (2005) 494–500.

[68] S. Sun, H. Zeng, Size-controlled synthesis of magnetite nanoparticles, *J. Am. Chem. Soc.* 124 (28) (2002) 8204–8205.

[69] S. Sun, H. Zeng, D.B. Robinson, S. Raoux, P.M. Rice, S.X. Wang, G. Li, Monodisperse MFe2O4 (M=Fe, Co, Mn) nanoparticles, *J. Am. Chem. Soc.* 126 (1) (2004) 273–279.

[70] Y.M. Huh, Y.W. Jun, H.T. Song, S. Kim, J. Choi, J.H. Lee, S. Yoon, K.S. Kim, J.S. Shin, J.S. Suh, J. Cheon, In vivo magnetic resonance detection of cancer by using multifunctional magnetic nanocrystais, *J. Am. Chem. Soc.* 127 (35) (2005) 12387–12391.

[71] Y.W. Jun, Y.M. Huh, J.S. Choi, J.H. Lee, H.T. Song, S. Kim, S. Yoon, K.S. Kim, J.S. Shin, J.S. Suh, J. Cheon, Nanoscale size effect of magnetic nanocrystals and their utilization for cancer diagnosis via magnetic resonance imaging, *J. Am. Chem. Soc.* 127 (16) (2005) 5732–5733.

[72] S. Veintemillas-Verdaguer,M.P.Morales, C.J. Serna, Continuous production of γ-Fe2O3 ultrafine powders by laser pyrolysis, *Mater. Lett.* 35 (1998) 227–231.

[73] S. Veintenillas-Verdaguer, M.P. Moraies, C.J. Serna, Effect of the oxidation conditions on the maghemites produced by laser pyrolysis, *Appl. Organometall. Chem.* 15 (2001) 365–372.

[74] G.A. van Ewijk, G.J. Vroege, A.P. Philipse, Convenient preparation methods for magnetic colloids, *J. Magn. Magn. Matt.* 201 (1999) 31–33.

[75] S. Lesieur, C. Grabielle-Madelmont, C. Menager, V. Cabuil, D. Dadhi, P. Pierrot, K.J. Edwards, Evidence of surfactant-induced formation of transient pores in lipid bilayers by using magnetic-fluid-loaded liposomes, *J. Am. Chem. Soc.* 125 (18) (2003) 5266–5267.

[76] M.C. Bautista, O. Bomati-Miguel, M.P. Morales, C.J. Serna, S. Veintemillas-Verdaguer, Surface characterisation of dextran coated iron oxide nanoparticles with applications in nuclear magnetic resonance imaging, *J. Magn. Magn. Mater.* 293 (2005) 20–27.

[77] K. O'Grady, A. Bradbury, Particle size analysis in ferrofluids, *J. Magn. Magn. Mater.* 39 (1983) 91–94.

[78] A.G. Roca, J.F. Marco, M.P. Morales, C.J. Serna, Effect of nature and particle size on properties of uniform magnetite and maghemite nanoparticles, *J. Phys. Chem.* C. 111 (2007) 18577–18584.

[79] P. Persson, N. Niisson, S. Sjöberg, Structure and bonding of orthophosphate ions at the iron oxide-aqueous interface, *J. Colloid Interf. Sci.* 177 (1996) 263–275.

[80] R. Langer, New methods of drug delivery, Science 249 (4976) (1990) 1527–1533.

[81] R.J. Hunter, *Introduction to Modern Colloid Science*, Oxford University Press, New York, 1993, p. 229.

[82] L. Garcell, M.P. Morales, M. Andrés-Vergés, P. Tartaj, C.J. Serna, Interfacial and rheological characteristics of maghemite aqueous suspensions, *J. Colloids Interf. Sci.* 205 (2) (1998) 470–475.

[83] K. Schroder, P.J. Hertzog, T. Ravasi, D.A. Hume, Interferon-gamma: an overview of signals, mechanisms and functions, *J. Leukoc. Biol.* 75 (2) (2004) 163–189.

[84] M.R. Alderson, R.J. Armitage, T.W. Tough, L. Strockbine,W.C. Fanslow, M.K. Spriggs, CD40 expression by human monocytes: regulation by cytokines and activation of monocytes by the ligand for CD40, *J. Exp. Med.* 178 (2) (1993) 669–674.

[85] Islam, T.; Harisinghani, M. G. Overview of nanoparticle use in cancer imaging. *Cancer Biomark.* 2009, *5*, 61–67.

[86] Jarrett, B. R.; Frendo, M.; Vogan, J.; Louie, A. Y. Sizecontrolled, synthesis of dextran sulfate coated iron oxide nanoparticles for magnetic resonance imaging. *Nanotechnology* 2007, *18*, 35603.
[87] Cho, S. J.; Jarrett, B. R.; Louie, A. Y.; Kauzlarich, S. M. Gold-coated iron nanoparticles: A novel magnetic resonance agent for T-1 and T-2 weighted imaging. *Nanotechnology* 2006, *17*, 640–644.
[88] Park, J. Y.; Daksha, P.; Lee, G. H.; Woo, S.; Chang, Y. Highly water-dispersible PEG surface modified ultra-small superparamagnetic iron oxide nanoparticles useful for target-specific biomedical applications. *Nanotechnology* 2008, *19*, 365603.
[89] Chan, D. C. F.; Kirpotin, D.; Bunn, P. A. Synthesis and evaluation of colloidal magnetic iron-oxides for the site-specific radiofrequency induced hyperthermia of cancer. *J. Magn. Magn. Mater.* 1993, *122*, 374–378.
[90] Park, J. H.; Im, K. Y.; Lee, S. H.; Kim, D. H.; Lee, D. Y.; Lee, Y. K.; Kim, K. M.; Kim, K. N. Preparation and characterization of magnetic chitosan particles for hyperthermia application. *J. Magn. Magn. Mater.* 2005, *293*, 328–333.
[91] Kim, D. H.; Lee, S. H.; Im, K. H.; Kim, K. N.; Kim, K. M.; Shim, I. B.; Lee, M. H.; Lee, Y. K. Surface-modified magnetite nanoparticles for hyperthermia: Preparation, characterization, and cytotoxicity studies. *Curr. Appl. Phys.* 2006, *6S1*, e242–e246.
[92] Matsunaga, T.; Sato, R.; Kamiya, S.; Tanaka, T.; Takeyama, H. Chemiluminescence enzyme immunoassay using proteinAbacterial magnetite complex. *J. Magn. Magn. Mater.* 1999, *194*, 126–131.
[93] Lubbe, A. S.; Bergemann, C.; Huhnt, W.; Fricke, T.; Riess, H.; Brock, J. W.; Huhn, D. Preclinical experiences with magnetic drug targeting: Tolerance and efficacy. *Cancer Res.* 1996, 56, 4694–4701.
[94] van der Zee, J. Heating the patient: A promising approach? *Ann. Oncol.* 2002, *13*, 1173–1184.
[95] Engin, K. Biological rationale and clinical experience with hyperthermia. *Control. Clin. Trials* 1996, *17*, 316–342.
[96] Hildebrandt, B.; Wust, P.; Ahlers, O.; Dieing, A.; Sreenivasa, G.; Kerner, T.; Felix, R.; Riess, H. The cellular and molecular basis of hyperthermia. *Crit. Rev. Oncol. Hematol.* 2002, *43*, 33–56.
[97] Chastellain, M.; Petri, A.; Hofmann, H. Particle size investigations of a multistep synthesis of PVA coated superparamagnetic nanoparticles. *J. Colloid. Interf. Sci.* 2004, *278*, 353–360.
[98] Cannas, C.; Gatteschi, D.; Musinu, A.; Piccaluga, G.; Sangregorio, C. Structural and magnetic properties of Fe2O3 nanoparticles dispersed over a silica matrix. *J. Phys. Chem. B* 1998, *102*, 7721–7726.
[99] Kanatani, I.; Ikai, T.; Okazaki, A.; Jo, J. I.; Yamamoto, M.; Imamura, M.; Kanematsu, A.; Yamamoto, S.; Ito, N.; Ogawa, O. et al. Efficient gene transfer by pullulan–spermine occurs through both clathrin- and raft/caveolae-dependent mechanisms. *J. Control. Release* 2006, *116*, 75–82.
[100] Hasegawa, U.; Nomura, S. M.; Kaul, S. C.; Hirano, T.; Akiyoshi, K. Nanogel-quantum dot hybrid nanoparticles for live cell imaging. *Biochem. Biophys. Res. Commun.* 2005, *331*, 917–921.
[101] Gupta, A. K.; Gupta, M. Cytotoxicity suppression and cellular uptake enhancement of surface modified magnetic nanoparticles. *Biomaterials* 2005, *26*, 1565–1573.

[102] Park, K. H.; Song, H. C.; Na, K.; Bom, H. S.; Lee, K. H.; Kime, S.; Kang, D.; Lee, D. H. Ionic strength-sensitive pullulan acetate nanoparticles (PAN) for intratumoral administration of radioisotope: Ionic strength-dependent aggregation behavior and 99mTechnetium retention property. *Colloids Surf. B: Biointerfaces* 2007, *59*, 16–23.

[103] Liu, X. Q.; Kaminski, M. D.; Guan, Y. P.; Chen, H. T.; Liu, H. Z.; Rosengart, A. J. Preparation and characterization of hydrophobic superparamagnetic magnetite gel. *J. Magn. Magn. Mater.* 2006, *306*, 248–253.

[104] Lin, C. L.; Lee, C. F.; Chiu, W. Y. Preparation and properties of poly(acrylic acid) oligomer stabilized superparamagnetic ferrofluid. *J. Colloid. Interf. Sci.* 2005, *291*, 411–420.

[105] Kamruzzaman Selim, K. M; Ha, Y. S; Kim, S. J.; Chang, Y.; Kim, T. J.; Lee, G. H.; Kang, I. K. Surface modification of magnetite nanoparticles using lactobionic acid and their interaction with hepatocytes. *Biomaterials* 2007, *28*, 710–716.

[106] Yoo, M. K.; Kim, I. Y.; Kim, E. M.; Jeong, H. J.; Lee, C. M.; Jeong, Y. Y.; Akaike, T.; Cho, C. S. Superparamagnetic iron oxide nanoparticles coated with galactose-carrying polymer for hepatocyte targeting. *J. Biomed. Biotechnol.* 2007, 94740.

[107] Li, G. Y.; Huang, K. L.; Jiang, Y. R.; Ding, P.; Yang, D. L. Preparation and characterization of carboxyl functionalization of chitosan derivative magnetic nanoparticles. *Biochem. Eng. J.* 2008, *40*, 408–414.

[108] Ma, Z. Y.; Guan, Y. P.; Liu, H. Z. Synthesis and characterization of micron-sized monodisperse superparamagnetic polymer particles with amino groups. *J. Polym. Sci. Polym. Chem.* 2005, *43*, 3433–3439.

[109] Mosmann, T. Rapid colorimetric assay for cellular growth and survival: Application to proliferation and cytotoxic assay. *J. Immunol. Methods* 1993, *95*, 55–63.

[110] Bahadur D, Giri J, Bibhuti B N, Sriharsha T, Pradhan P, Prasad N K, Barick K C and Ambashta R D. *Pramana - J. Phys.* 2005, 65 (4): 663 – 679.

[111] Perez-Castillejos R, Plaza J. A, Esteve J, Losantos P, Acero M. C, Cane C, Serra-Mestres F. *Sensors and Actuators* 2000, 84: 176 – 180.

[112] Piso M.I. Journal of Magnetism and Magnetic Materials 1999, 201 (1): 380 - 384.

[113] Scherer C, and Figueiredo Neto A. M. *Brazilian Journal of Physics*. 2005, 35 (3A): 718 – 727.

[114] Gupta A. K, Gupta M. *Biomaterials*. 2005, 26: 3995 – 4021.

[115] Salata OV. *Journal of Nanobiotechnology*. 2004 [On-Line], 2 (3): 1 – 6. [Cited 21-June-2008] This article is available from: http://www.jnanobiotechnology.com/content/2/1/3.

[116] Jordan A, Scholz R, Maier-Hau K, Johannsen M, Wust P, Nadobny J, Schirra H, Schmidt H, Deger S, Loening S, Lanksch W, Felix Roland. *Journal of Magnetism and Magnetic Materials*. 2001, 225: 118 – 126.

[117] Jordan A, Scholz R, Wust P, Fähling H, Felix Roland. *Journal of Magnetism and Magnetic Materials*. 1999, 201: 413 – 419.

[118] R. Massart. *IEEE Trans. Magn.* 1981. 17: 1247 - 1248.

[119] Tourinho F. Franck R. and Massart R. *J. Mater. Sci.* 1990, 25 (7): 3249 - 3254.

[120] Kholmetskiia A.L, Vorobyovab S.A, T, Lesnikovichb A.I, Mushinskiib V.V, Sobalb N.S. *Materials Letters*. (2005), 59: 1993 – 1996.

[121] Răcuciu M, Creangâ D. E, Câlugâru Gh. *Jounral of Optoelectronics and Advanced Materials*. 2005, 7(6) 2859 - 2864.

[122] Hong R.Y, Zhang S.Z, Han Y.P, Li H.Z, Ding J, Zheng Y. *Powder Technology*.2006, 170: 1 –11.

[123] L.A. García-Cerda, O.S. Rodríguez-Fernández, R. Betancourt-Galindo, R.Saldívar-Guerrero. *Superficies y Vacío*.2003 16(1): 28-31.

[124] Martínez-Mera I, Espinosa-Pesqueira M.E, Pérez-Hernández R, Arenas-Alatorre J.*Materials Letters*. 2007, 61: 4447 – 4451.

[125] Wang X, Zhang C, Wang X, Gu H. *AppliedSurface Science*. 2007 253: 7516 –7521.

[126] Bragg W.L. *Nature*. 1915, 95: 561.

[127] Ayala-Valenzuela O, Matutes-Aquino J, Betancourt-Galindo R, Garcia-Cerda L.A, Rodriíguez F. O, Fannin P.C, Giannitsis A.T. *Journal of Magnetism and Magnetic Materials*. 2005, 294: e37–e41.

[128] Kim Y. Il, Kim D, Lee C.S. *Physica B*. 2003, 337: 42 – 51.

[129] Vaidyanathan G., Sendhilnathan S, Arulmurugan R. *Journal of Magnetism and Magnetic Materials*. 2007 313: 293 – 299.

[130] Betancourt-Galindo R, Rodríguez - Fernández O. S. *Superficies y Vacío* 17 (1): 38 - 41.

[131] Pankhurst Q A, Connolly J, Jones S K, Dobson J. *J. Phys. D: Appl. Phys*. 2003, 36: R167 – R181.

[132] Peia W, Kumadaa H, Natusmea T, Saitoa H, Ishio S. *Journal of Magnetism and Magnetic Materials*. 2007, 310: 2375–2377.

[133] López-López M.T, Durán J.D.G, Delgado A.V, González-Caballero F. *Journal of Colloid and Interface Science*. 2005, 291: 144 – 151.

[134] Apetroaie N, Roca A, Creanga D. E. *Journal of Optoelectronics and Advanced Materials*. 2005, 7(6): 2865 – 2868.

[135] Lacava L. M, B. Lacava M, Azevedo R. B, Lacava Z. G. M, Buske N, Tronconi A. L, Morais P. C. *Journal of Magnetism and Magnetic Materials*. 2001, 225: 79 - 83.

[136] Butter K, Philipse A.P, Vroege G.J. *Journal of Magnetism and Magnetic Materials*. 2002, 252:101 – 103.

[137] Ling Z; Rong He, Hong-Chen G. *Applied Surface Science*. 2006, 253(5): 2611-2617.

[138] Ahn Y, Choi E.J, Kim E.H, *Rev. Adv. Mater*. Sci. 2003, 5: 477.

[139] Farmer V.C, "The Infrared Spectra Of Minerals" Ed. By V.C. Farmer Mineralogical Society, London, 1974 p.18.

[140] Farmer V. C. *Phil. Trans. R. Soc. Lond. A*. 1982; 305: 609-619.

Chapter 6

NANOPARTICLES BASED ON DENDRIMERS

INTRODUCTION

Dendrimers are synthetic tree-shaped macromolecules formed from monomers using a stepwise fabrication process. It is the size of a typical protein, but it does not unfold or come apart easily due to the nature of its stronger chemical bonds. It can be easily attached to a variety of molecules to facilitate the transport and release of medication, contrast agents or DNA for gene therapy. Fullerenes have been investigated for use as a drug delivery platform as well. Fullerenes are a class of carbon molecules that form closed, convex cage molecules containing hexagonal and pentagonal faces. However, there have been problems with this technology as lipophilic uncoated fullerenes were found to selectively translocate into the brains of large-mouth bass and cause oxidative damage [1].

6.1. TARGETING CANCER CELLS WITH DNA-ASSEMBLED DENDRIMERS

Cancer represents an enormous biomedical challenge for drug delivery. Physiological barriers including the vascular endothelial pores, heterogeneous blood supply to the tumor, relatively long transport distances in the interstitium from the vessels to the tumor cells and cellular heterogeneities in different cancers and within single tumors. Once the drug reaches the cell, additional barriers exist such as multi-drug resistance that contributes to the poor delivery of therapeutic agents to cancers. In addition, immunological barriers come into play when the clearance of antibody drug carriers or diagnostic agents by blood monocytes or macrophages of the reticuloendothelial system (RES) makes the efficient targeting of tumor sites problematic. Given this unique biology of cancer, drug delivery requires the development of a multifunctional delivery system which can target subtle molecular alterations that distinguish a transformed cell from the many types of healthy cells found in the body. As a result, there is a growing interest in molecular engineered drug carrier systems that avoid immune recognition and achieve specific targeting ability to the tumor sites [2-10].

Several approaches to cancer drug delivery have been proposed. One is to take advantage of unique structural changes associated with tumor vasculature to allow intra-tumor accumulation of particulate carrier systems. The “leakiness” of the tumor tissue enhances permeability of the polymeric drug to the tumor area through blood flow. The enhanced permeability and retention (EPR) effect reportedly allows for large polymer particles up to approximate 100 nm to exit the vasculature and enter the tumor tissue. However, there are problems with this approach because some vessels are porous whereas others in the same tumor are not; even a single vessel can be "leaky" in one region but not in another. The efficiency and kinetics of particle delivery also varies since the permeability of a molecule is independent of the pore cutoff size of the vasculature as long as the diameter of the molecule is much less than the pore diameter [11]. For example, albumin molecules (particles of less than 10 nm) leave the systematic circulation independent of the fenestrate pore size cut-off. Despite the frequent reports of enhanced permeability to macromolecules in tumor vasculature, little fluid translocation takes place due to the absence of a lymphatic drainage system and the high interstitial pressures in tumors. In addition, the relatively large size of the polymer particles and the heterogeneity of the polymer size have limited this type of drug delivery system. In addition, simply increasing the circulation time of therapeutic agents exploiting this type of passive targeting mechanism may not necessarily improve the therapeutic index [12].

Another approach to efficient cancer targeting is to use biomarkers as selective targets on the tumor cell surface, thereby directing particles to designated but accessible targets. Various targeting ligands such as folic acid, oligopeptides, oligosaccharides or antibodies are currently employed in this active cancer targeting approach. Low and associates have demonstrated that molecules can target the high-affinity folate receptor (hFR) when conjugated to folic acid. When compared to antibodies or other ligands, folic acid is presumed to be non-immunogenic due to its small size, has good solubility, and is relatively specific for a variety of tumors. This is because an increased expression of the folate receptor has frequently been observed in most types of human epithelial cancers including ovarian, choriocarcinomas, choroid plexus, ependymomas, lung, kidney, breast, brain and colon [13-15]. The receptor is generally absent in most normal human tissues with the exception of placenta, lung, thyroid, and kidney. Folate conjugates of radiopharmaceutical agents, MRI contrast agents, low molecular weight chemotherapeutic agents, antisense oligonucleotides, proteins and protein toxins, immunotherapeutic agents, liposomes with entrapped drugs and plasmids have all been reported to be delivered to cancer cells that overexpress the folate receptor. It also has been reported that folic acid conjugated G4 PAMAM dendrimer chelates, complexed with gadolinium, can image tumors expressing the folate receptor. Prior work from our laboratory has also demonstrated the ability of dendrimer-folate conjugates to target and internalize into folic acid expressing cells [16-20]. Dendrimers appear to have advantages for molecular targeting due to their size (less than 5nm), uniformity, and multiple surface velocity. However, conjugating several types of different molecules to a dendrimer may cause synthetic problems that decrease yield, and result in molecules with decreased water solubility due to the hydrophobicity of the attached functional groups. These multiple steps of attachments may also lead to the loss of the intended biological activity of the functional units due to the densely packed terminal groups of dendrimers. Moreover, using a single dendrimer as a targeted drug delivery platform requires that a different targeted drug be produced for each varied tumor type or molecular alterations in each specific cancer. This is synthetically challenging, time-consuming and cost-ineffective. For example, if one needs a therapeutic that targets five different drugs to five different cells it requires at least 25 different syntheses with loss of yield at each step [21-27].

Instead of conjugating all of the molecules to a single dendrimer, synthetic oligonucleotides were used as a tool to self-assemble two dendrimer particles; one bearing folate with another carrying an imaging agent (fluorescein in this case). The dendrimer-based nanoparticles cluster assembled by DNA was found out to specifically target KB cells over expressing folate receptors, and to be internalized into the cells via folate-mediated endocytosis in vitro. Clearly, the ability to design and produce an array of dendrimers in a consistent and reproducible manner is a key to the development of these versatile nanoclusters as a novel platform for a targeted delivery system [28]. In the future, this bi-functional dendrimer cluster system can be readily expanded or adapted to exploit various cancer-specific biomarkers. An array of single-functional dendrimers, such as targets, drugs, and contrast agents, and the ability to link them together efficiently would enable a clinic to readily offer customized nanomedicine based on this dendrimer cluster system (Figure 1).

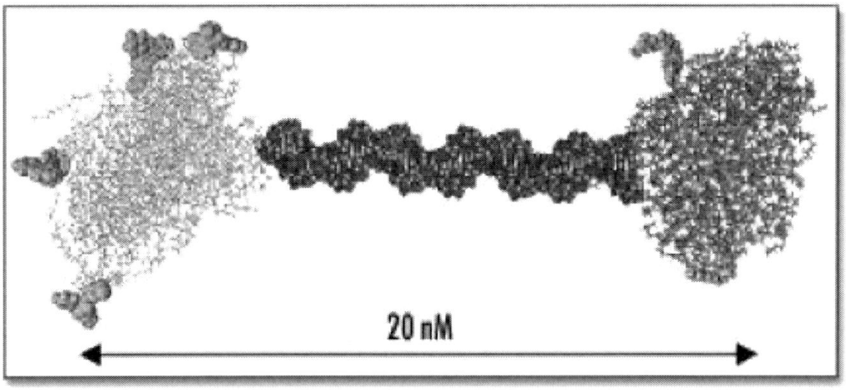

Figure 1. A dendrimer nanocluster with folate attached on one end (left) and a fluorescent dye on the other (right) is held together by complementary strands of DNA.

Polyamidoamine (PAMAM) dendrimers are cationic polymers that have been extensively evaluated as candidates for gene delivery.

6.2. Synthesis and Characterization of Dendriworms

While dendrimers have been used for delivery of genes and antisense oligonucleotides, delivery of siRNA with dendrimers has been challenging and poor efficiencies have been reported. PAMAM dendrimers have been used for siRNA delivery, but most efforts using lower generation dendrimers for siRNA delivery have been unsuccessful and it has been shown that lower generation dendrimers are poor at forming uniform, stable particles with siRNAs. In a recent report, polylysine dendrimers were also found to be inefficient for the delivery of siRNAs. Since dendrimers buffer the endosomes *via* their tertiary and secondary amines, polyvalent conjugation of lower generation dendrimers onto an elongated magnetic nanoparticle host could generate a construct that would induce a high proton sponge effect and enable efficient endosomal escape of siRNAs. The use of lower generation dendrimers may also diminish the likelihood of immune system activation and inflammatory response associated with the use of higher generation dendrimers. The synthesis of dendrimer-conjugated magnetic nanoworms (dendriworms) is shown in Figure 2A. First, generation 4, cystamine core PAMAM dendrimers were reduced in 0.5 M dithiothreitol (DTT), and excess DTT was removed using a G25 chromatography column [29-35].

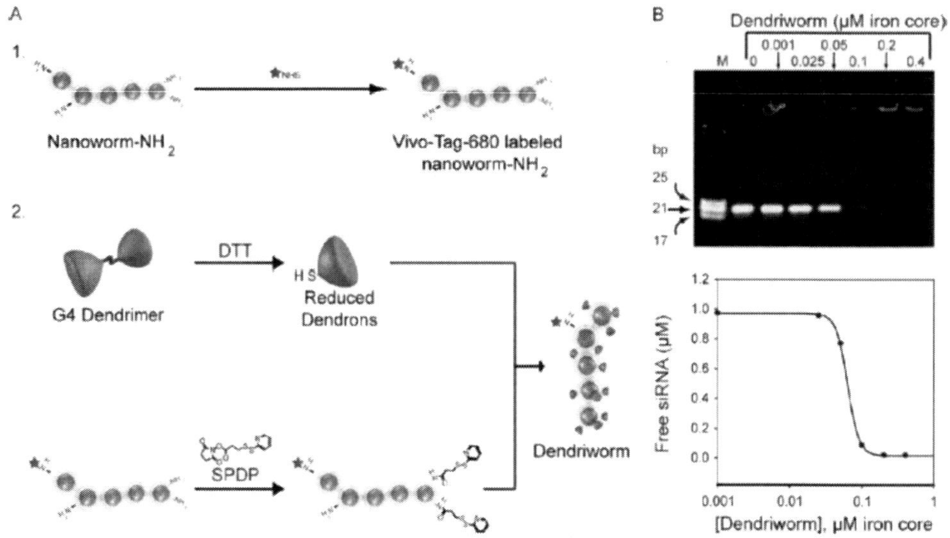

Figure 2. Synthesis and characterization of dendriworms: (A) synthesis scheme (B) siRNA binding characteristics.

Next, amine-modified, cross-linked iron oxide nanoworms, labeled with near-infrared fluorochromes (VivoTag 680, Visen Medical, Inc.), were prepared. Finally, reduced dendrons and nanoworms were conjugated using the heterobifuctional linker *N*-succinimidyl 3-(2-pyridyldithio)-propionate, which links amine groups on the nanoparticle with the central thiols on reduced dendrons (Figure 2A) [36].

The resulting dendriworms were separated from free dendrons by filtering several times using either 100,000 molecular weight cutoff filter (Amicon Ultra, Millipore Corp.), magnetic separation column (MACS, Miltenyi Biotech), or G50 chromatography columns. Dendriworms were stored at 4 °C protected from light and were stable for at least 6 months. The net charge on dendriworms was found to be between 16-24 mV (zetapotential measurement in 1- PBS) and the average hydrodynamic diameter was found to be between 80 and 110 nm (dynamic light scattering measurements in 1- PBS, Zetasizer Nano, Malvern, Inc.). In the transmission electron micrograph, the particles were found to retain a worm-shape, where multiple 5-8 nm iron oxide cores were lined in a series. The number of dendrons per dendriworm was found to be 45-50 using fluorescamine and ninhydrin tests (assuming there are approximately 7 iron cores per dendriworm) [37-40].

Further, using UV_vis absorbance measurements, the number of VivoTags on each dendriworm was found to be between 1 and 5. A key feature of the dendriworms is that dendrons are conjugated via a reducible linker (SPDP) to the nanoparticle surface. This may allow removal of dendrons from the nanoparticle surface in the reducing intracellular environment, resulting in improved siRNA delivery and diffusion inside cells. Variations of magnetic-core particles were also generated using other polymers [polyethyleneimine, myristic acid] and generations 3-6 of PAMAM dendrimers. However, these particles were found to be ineffective for gene silencing either due to a lack of colloidal stability or due to poor siRNA delivery efficiency and were therefore not pursued further. The binding affinity of siRNA and dendriworms was tested by incubating varying amounts of dendriworms (10-400 nM iron core) with 1 -M of siRNA for 10 minutes at room temperature and measuring the amount of siRNA left unbound [41]. The mixture was run on a 15% TBE gel (BioRad Laboratories) at 80 mV for 1 h and siRNAs in the gel were stained with SYBR Gold (Invitrogen Inc., Carlsbad, CA) Figure 1B. Because dendriworms are large and positively charged, they do not electrophorese into the gel toward the cathode and are washed from the top of the gel during the 1 hour run. Furthermore, dendriworms bound to siRNAs remain positively charged (zeta-potential measurements), causing siRNA binding to dendriworms to retard siRNA movement into the gel. At 50 nM dendriworm (iron core concn.), some loss in siRNA band (21 bp) intensity was observed and at 100 nM dendriworm (iron core concn.), nearly complete siRNA retardation was seen. Thus, roughly 9_10 siRNA molecules are bound to each iron core in dendriworm. This implies roughly 50 siRNA molecules loaded on each dendriworm (there are roughly 7 iron cores per dendriworm). To maintain consistency, the amounts of dendriworm and nanoworm are expressed in iron core concentration in the rest of the manuscript [42].

The stability of the siRNA-dendriworm nanoparticle was assessed in fetal bovine serum (FBS) and artificial cerebrospinal fluid (ACSF) by gel electrophoresis and dynamic light scattering. To visualize the residual bound siRNA, 1% SDS was used to release the siRNA prior to electrophoresis. In ACSF, the nanoparticle and the bound siRNA remained stable for up to 6 hours, however, the siRNA showed lower stability in serum [43].

6.3. Dendriworms Enable Efficient Release of Cargo into the Cell Cytosol

Cationic polymers adhere to the negatively charged moieties on the cell membrane and may be taken up *via* fluid phase endocytosis or *via* receptor mediated endocytosis. As a result, cargo that is only active in the cell cytoplasm and is delivered using cationic polymers or cationic nanoparticles needs to escape the endosomes. To test the hypothesis that conjugating multiple cationic dendrimers on a nanoworm scaffold would enhance the proton sponge effect, leading to better cytosolic release of cargos, the efficiency of dendriworm endosomal escape is compared to free dendrimers.

To assess endosomal escape, the excess of a membrane impermeable dye (Calcein at 0.25 mM) along with the siRNA delivery agent (dendrimer, unmodified nanoworms and dendriworms) coupled with siRNAs and counted the fraction of cells with endosomal escape as indicated by diffuse cytoplasmic distribution of the dye[44]. The results of these endosomal escape experiments are shown in Figure 3.

In the presence of Calcein alone, punctate cytoplasmic distribution of dye was consistent with the endosomal uptake. Some degree of endosomal release as judged by diffuse cytoplasmic distribution was seen with aminated nanoworms. Significantly more endosomal escape was seen with both dendrimers and dendriworms, although the extent of endosomal escape was much greater with dendriworms.

Endosomal escape is also dependent on the uptake of the dendrimers or dendriworms into the endosomes which depends on primary amines on the surface of dendrimers that are positively charged in serum (near pH 7). Thus, the addition of more negatively charged siRNAs that neutralize the positively charged primary amines may reduce the number of dendrimers or dendriworms that are able to load into the endosomes, affecting the number of buffering (tertiary) amines per endosome available for inducing endosomal escape.

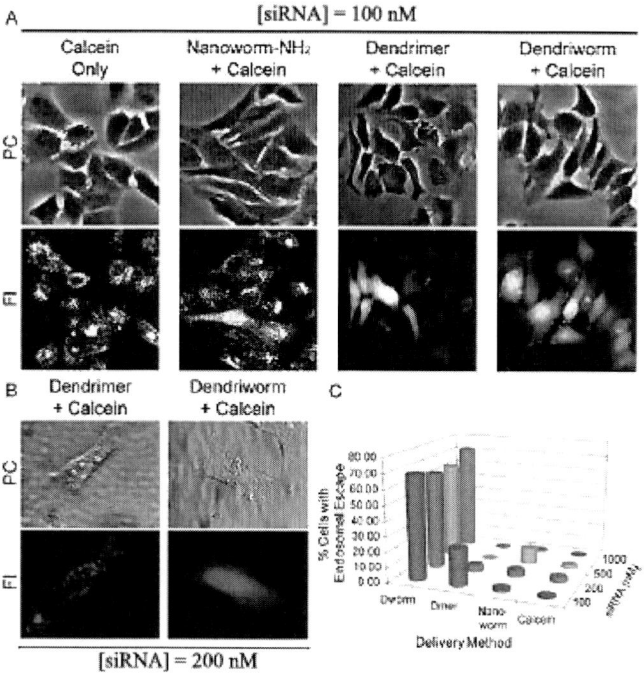

Figure 3. Dendriworms enable efficient escape of endosomal cargos.

Indeed, when higher concentrations of siRNA were mixed with dendrimers or dendriworms, dendrimers failed to induce significant endosomal escape while dendriworms remained effective even when incubated with 1 -M siRNA (Figure 3B and C). Thus conjugation of several dendrons (reduced half of a dendrimer) onto aminated nanoworms allows for more effective endosomal escape [45-50].

The next topic was the cytotoxicity of dendriworms at concentrations used in studies. Two types of toxicity assays relying on either mitochondrial reductase enzyme activity (MTT assay, based on reduction of (3- (4, 5-dimethylthiazol-2-yl)-2,5-diphenyltetrazolium bromide) or on intracellular esterase activity (Calcein AM assay) were used. No significant *in vitro* toxicity was found across various dendriworm and siRNA concentrations used [51].

6.4. Dendriworms Deliver siRNA into Cells and Induce Gene Silencing

Dendriworms are able to deliver cargo into cells efficiently at concentrations that have negligible toxic effects on cells *in vitro*. Next, dendriworms could be used to deliver siRNAs into human primary glioblastoma cells (GBM-6) 47, 48 that express mutant EGFRvIII, and induce gene silencing. EGFRvIII gene silencing (knockdown) in glioblastoma cells was assessed by measuring EGFR protein expression levels and the amount of mRNA transcripts was compared with levels of a housekeeping gene (GAPDH) (Figure 4). Both protein and mRNA levels were suppressed significantly and were comparable to the commercial transfection agent Lipofectamine 2000 when 100 nM siRNA was delivered. Greater relative (dendriworm *vs* Lipofectamine 2000) gene silencing was observed at the protein level than at the mRNA level. One potential cause for this could be the dependence of the mRNA read-out method (qRT-PCR) on several rounds of amplification, which may attenuate the differences if the initial number of mRNAs is different but high in the two samples. Gene silencing was found to be strongly dependent on the siRNA and dendriworm dose (Figure 4) [52-60].

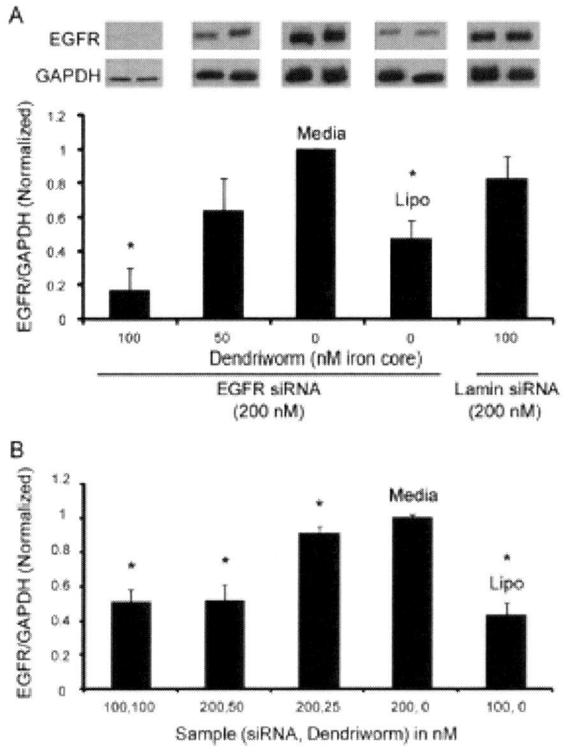

Figure 4. Delivery of EGFR siRNA into glioblastoma cells with dendriworms.

The knockdown was also sequence dependent as lamin siRNA did not lower the expression levels of EGFR (Figure 4). Further, EGFR siRNA delivered using aminated nanoworms or dendrimers did not produce significant EGFR knockdown. In addition to GMB-6 cells, dendriworms also produce significant gene knockdown in HeLa cells although, consistent with the literature on the variability of polymeric formulation dependent siRNA delivery, the extent of knockdown in HeLa cells is lower. Taken together these results imply that dendriworms are a highly effective siRNA delivery vehicle for GBM-6 glioblastoma cells and represent a flexible method for mutilation due to noncovalent siRNA attachment for optimization of siRNA delivery to other cell types [61-70].

6.5. Systemic Delivery of Dendriworms

The utility of dendriworm-based siRNA delivery was studied in mice. When dendriworms (20 _g of iron in 100 _L of 1_PBS) were injected in the bloodstream *via* tail vein injections, preferential accumulation in the lungs, as expected from the cationic nature of the particles, was seen. Swiss Webster mice were injected in the tail vein with fluorescently labeled dendriworms and the amount of fluorescence remaining in the blood at various time points was measured. This was done by drawing blood at different time points, removing cells and debris *via* centrifugation at 800 rpm for 5 min, and measuring the fluorescence from the supernatant using an Odyssey infrared imaging system (Li- Cor Biosciences, Inc.). In this experiment, dendriworms cleared the blood within in 5 minutes [71].

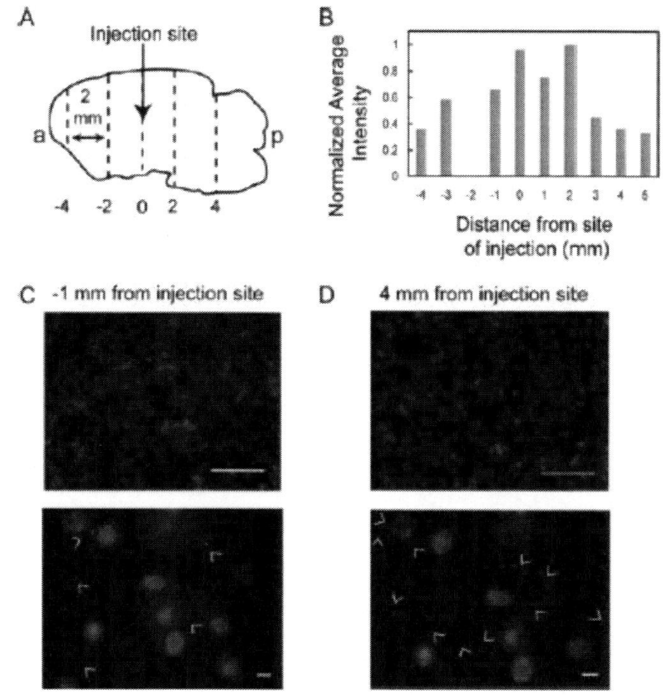

Figure 5. Biodistribution of nanoparticles in the brain.

To further probe where the particles were accumulating, 100 μL of 1μ PBS containing 20 μg of dendriworms or 20 μg of fluorescently labeled CLIO-NH_2 or 1μ PBS alone was injected in the tail vein of mice and the mice were sacrificed after approximately 10 minutes of injections. After an additional 10 minutes, the organs were removed, washed with saline, and imaged with the Odyssey infrared imaging system (Li-Cor Biosciences, Inc.).

Since dendriworms were labeled with infrared emitting dye VivoTag680, fluorescence could be used to infer the degree of uptake of particles in various organs. Images were color coded to reflect the intensity of uptake. After observing significant accumulation in the lungs (also confirmed using histology), in order to test this, one pair of Swiss Webster mice was injected with PBS and the other pair of mice was injected with dendriworms, in the left heart ventricle. The results show that dendriworms still accumulate in the lungs and the reticuloendothelial filtration organs [72-80].

6.6. AN AROMATIC MACROMOLECULE AS A NANO DRUG CARRIER: POLYETHER DENDRIMER/POLY (2-HYDROXYETHYL METHACRYLATE)

Highly branched, functionalized polymers have the potential to act as efficient drug carrier systems. The aromatic polyether dendrimers are spherical, highly ordered, multibranched, nanometer-sized macromolecules having positively charged ether groups on the surface under physiological conditions.

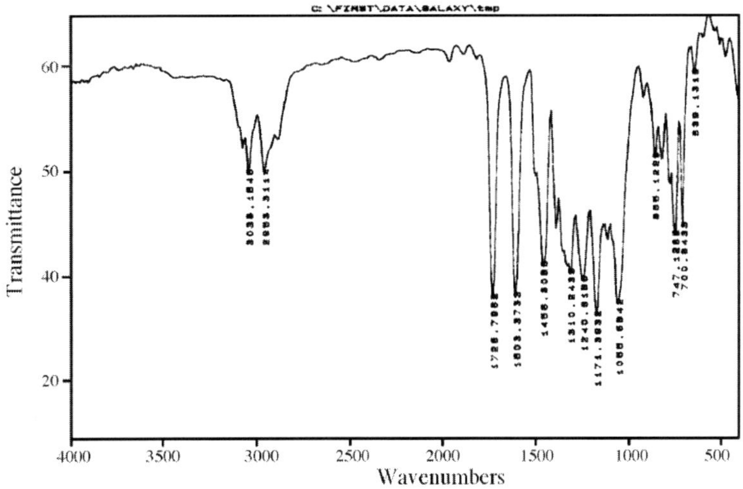

Figure 6. FT-IR spectrum of [G2]-$COOCH_3$ dendrimer.

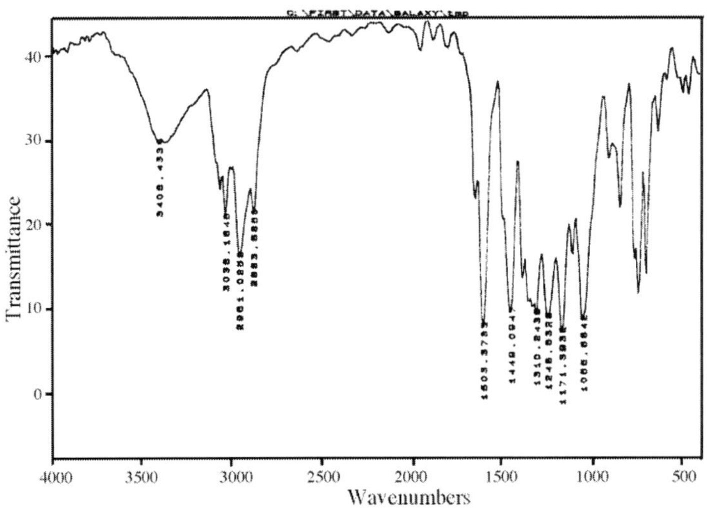

Figure 7. FT-IR spectrum of [G2]-CH_2OH dendrimer.

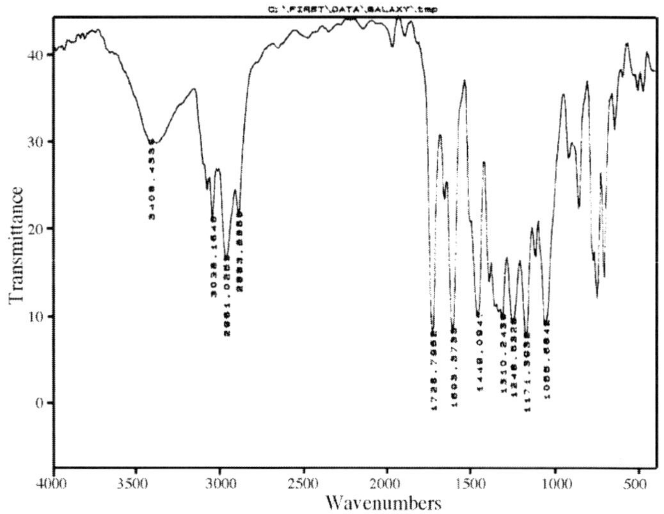

Figure 8. FT-IR spectrum of [G2]-CH2OH dendrimer/poly (2-hydroxy ethyl methacrylate/5-ASA nanocomposites

In this study, we have synthesized a kind of dendrimer/poly (2-hydroxy ethyl methacrylate) nanocomposite/5-aminosalicylic acid (5-ASA) for oral drug delivery. The aromatic polyether dendrimer (generation 2, hyperbranched polyether with -CH_2OH functionality, 3,5-Dihydroxybenzoic acid core) was prepared from generation 2, hyperbranched polyether dendrimer with -$COOCH_3$ form in excellent yield. FT-IR, 1H-NMR and DSC studies suggest that the monomer predominantly forms a complex with polyether dendrimers because of the interaction between the –CH_2OH and groups of dendrimer and the –OH group of 2-hydroxyethyl methacrylate monomer [81-90]. Scanning electron micrography images were obtained from a diluted solution of the nanocomposite particles. The white particles are 5-ASA nanoparticles. The SEM image shows the presence of 5-aminosalicylic acid (5-ASA) particles in the polyfunctional dendrimeric matrix, which were distributed throughout the nanocomposite, which is also confirmed from 1H-NMR studies.

The ability of the dendrimer to form a complex with 2-hydroxyethyl methacrylate monomer depends on the core-surface groups of the dendrimer, electrostatic interactions between the dendrimer and the monomer and the ability of the drug to form a conjugate with the dendrimer through chemical bonding [91-95].

Therefore, it is possible to manipulate the incorporation process for a given monomer by the appropriate selection of the dendrimer and the surface functionality. One might expect that the 2-hydroxy ethyl methacrylate with the carboxylic group may form a complex with surface OH groups of the polyether dendrimer (Figure 9).

Nano and micro polymer-bonded drugs (50 mg) were poured into 3 mL of aqueous buffer solution (SGF: pH 1 or SIF: pH 7.4). The mixture was introduced into a cellophane membrane dialysis bag. The bag was closed and transferred to a flask containing 20 mL of the same solution maintained at 37 °C. The external solution was continuously stirred, and 3 mL samples were removed at selected intervals [96-100]. The volume removed was replaced with SGF or SIF. Triplicate samples were used. The sample of hydrolyzate was analyzed and the quantity of 5-ASA was determined using a standard calibration curve obtained under the same conditions (Figure 10).

Figure 9. SEM of [G2]-CH_2OH dendrimer/2-hydroxyethyl methacrylate/5-ASA nanocomposites.

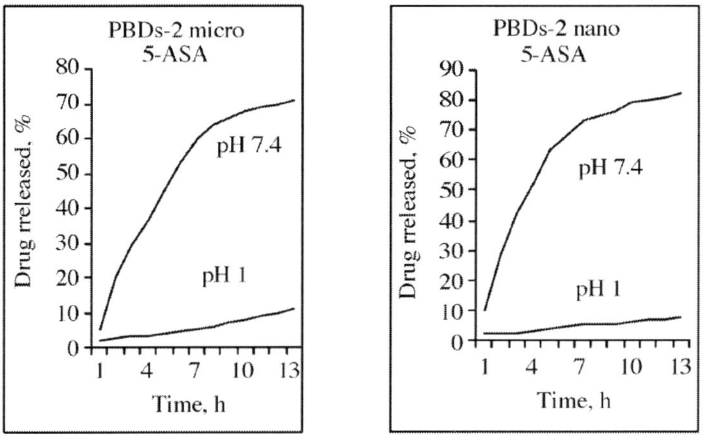

Figure 10. The release of drug from micro and macro as polymeric carriers.

References

[1] Choi YS, Thomas T, Kotlayr A, Islam MT, Baker JR. Synthesis and Functional Evaluation of DNA-assembled Polyamidoamine Dendrimer Clusters for Cancer Cell-Specific Targeting. *Chemistry and Biology* 2005; 12:35-43.

[2] Sampathkumar S-GY, J. Kevin. Targeting Cancer Cells with Dendrimers. *Chemistry and Biology* 2005; 12:5-13.

[3] Jain RK. Delivery of Novel Therapeutic Agents in Tumors - Physiological Barriers and Strategies. *Journal of the National Cancer Institute* 1989; 81:570-6.

[4] Jain RK. Delivery of molecular and cellular medicine to solid tumors. *Advanced Drug Delivery Reviews* 2001; 46:149-68.

[5] Tsuruo T, Naito M, Tomida A, Fujita N, Mashima T, Sakamoto H, Haga N. Molecular targeting therapy of cancer: drug resistance, apoptosis and survival signal. *Cancer* Science 2003; 94:15-21.
[6] Moghimi SM, Hunter AC, Murray JC. Long-circulating and target-specific nanoparticles: Theory to practice. *Pharmacological Reviews* 2001; 53:283-318.
[7] Heuser LS, Miller FN. Differential macromolecular leakage from the vasculature of tumors. *Cancer* 1986; 57:461-4.
[8] Liu DX, Mori A, Huang L. Role of Liposome Size and Res Blockade in Controlling Biodistribution and Tumor Uptake of Gm1-Containing Liposomes. *Biochimica Et Biophysica Acta* 1992; 1104:95-101.
[9] Uchiyama K, Nagayasu A, Yamagiwa Y, Nishida T, Harashima H, Kiwada H. Effects of the Size and Fluidity of Liposomes on Their Accumulation in Tumors - a Presumption of Their Interaction with Tumors. *International Journal of Pharmaceutics* 1995; 121:195-203.
[10] Matsumura Y, Maeda H. A new concept for macromolecular therapeutics in cancer chemotherapy: mechanism of tumoritropic accumulation of proteins and the antitumor agent smancs. *Cancer Research* 1986; 46:6387-92.
[11] Tabata T, Murakami Y, Ikada Y. Tumor accumulation of poly (vinyl alcohol) of different sizes after intravenous injection. *Journal of Controlled Release* 1998; 50:123-33.
[12] Jain RK. Barriers to Drug-Delivery in Solid Tumors. *Scientific American* 1994; 271:58-65.
[13] Hobbs SK, Monsky WL, Yuan F, Roberts WG, Griffith L, Torchilin VP, Jain RK. Regulation of transport pathways in tumor vessels: Role of tumor type and microenvironment. *Proceedings of the National Academy of Sciences of the United States of America* 1998; 95:4607-12.
[14] Goren D, Horowitz AT, Tzemach D, Tarshish M, Zalipsky S, Gabizon A. Nuclear delivery of doxorubicin via folate-targeted liposomes with bypass of multidrug-resistance efflux pump. *Clinical Cancer Research* 2000; 6:1949-57.
[15] Zalipsky S, Mullah N, Harding JA, Gittelman J, Guo L, DeFrees SA. Poly (ethylene glycol)- grafted liposomes with oligopeptide or oligosaccharide ligands appended to the termini of the polymer chains. *Bioconjugate Chemistry* 1997; 8:111-18.
[16] Spragg DD, Alford DR, Greferath R, Larsen CE, Lee KD, Gurtner GC, Cybulsky MI, Tosi PF, Nicolau C, Gimbrone MA. Immunotargeting of liposomes to activated vascular endothelial cells: A strategy for site-selective delivery in the cardiovascular system. *Proceedings of the National Academy of Sciences of the United States of America* 1997; 94:8795-800.
[17] Drummond DC, Meyer O, Hong KL, Kirpotin DB, Papahadjopoulos D. Optimizing liposomes for delivery of chemotherapeutic agents to solid tumors. *Pharmacological Reviews* 1999; 51:691-743.
[18] Lu YJ, Sega E, Leamon CP, Low PS. Folate receptor-targeted immunotherapy of cancer: mechanism and therapeutic potential. *Advanced Drug Delivery Reviews* 2004; 56:1161-76.
[19] Leamon CP, Low PS. Folate-mediated targeting: from diagnostics to drug and gene delivery. *Drug Discovery Today* 2001; 6:44-51.

[20] Leamon CP, Low PS. Delivery of Macromolecules into Living Cells - a Method That Exploits Folate Receptor Endocytosis. *Proceedings of the National Academy of Sciences of the United States of America 1991;* 88:5572-6.

[21] Leamon CP, Low PS. Selective Targeting of Malignant-Cells with Cytotoxin-Folate Conjugates. *Journal of Drug Targeting* 1994; 2:101-12.

[22] Ross JF, Chaudhuri PK, Ratnam M. Differential Regulation of Folate Receptor Isoforms in Normal and Malignant-Tissues in-Vivo and in Established Cell-Lines - Physiological and Clinical Implications. *Cancer* 1994; 73:2432-43.

[23] Weitman SD, Weinberg AG, Coney LR, Zurawski VR, Jennings DS, Kamen BA. Cellular- Localization of the Folate Receptor - Potential Role in Drug Toxicity and Folate Homeostasis. *Cancer Research* 1992; 52:6708-11.

[24] Weitman SD, Lark RH, Coney LR, Fort DW, Frasca V, Zurawski VR, Kamen BA. Distribution of the Folate Receptor Gp38 in Normal and Malignant-Cell Lines and Tissues. *Cancer Research* 1992; 52:3396-01.

[25] Campbell IG, Jones TA, Foulkes WD, Trowsdale J. Folate-Binding Protein Is a Marker for Ovarian-Cancer. *Cancer Research* 1991; 51:5329-38.

[26] Ke CY, Mathias CJ, Green MA. Folate-receptor-targeted radionuclide imaging agents. *Advanced Drug Delivery Reviews* 2004; 56:1143-60.

[27] Konda SD, Aref M, Brechbiel M, Wiener EC. Development of a tumor-targeting MR contrast agent using the high-affinity folate receptor - Work in progress. *Investigative Radiology* 2000; 35:50-7.

[28] Ladino CA, Chari RVJ, Bourret LA, Kedersha NL, Goldmacher VS. Folate-maytansinoids: Target-selective drugs of low molecular weight. *International Journal of Cancer* 1997; 73:859-64.

[29] Li S, Deshmukh HM, Huang L. Folate-mediated targeting of antisense oligodeoxynucleotides to ovarian cancer cells. *Pharmaceutical Research* 1998; 15:1540-5.

[30] Ward CM, Acheson N, Seymour LW. Folic acid targeting of protein conjugates into ascites tumour cells from ovarian cancer patients. *Journal of Drug Targeting* 2000; 8:119-23.

[31] Kranz DR, Manning TC, Rund LA, Cho BK, Gruber MM, Roy EJ. Targeting tumor cells with bispecific antibodies and T cells. *Journal of Controlled Release* 1998; 53: 77-84.

[32] Lee RJ, Low PS. Folate-Mediated Tumor-Cell Targeting of Liposome-Entrapped Doxorubicin in-Vitro. *Biochimica Et Biophysica Acta-Biomembranes* 1995; 1233: 134-44.

[33] Douglas JT, Rogers BE, Rosenfeld ME, Michael SI, Feng MZ, Curiel DT. Targeted gene delivery by tropism-modified adenoviral vectors. *Nature Biotechnology* 1996; 14:1574-8.

[34] Quintana A, Raczka E, Piehler L, Lee I, Myc A, Majoros I, Patri AK, Thomas T, Mule J, Baker JR. Design and function of a dendrimer-based therapeutic nanodevice targeted to tumor cells through the folate receptor. *Pharmaceutical Research* 2002; 19:1310-1316.

[35] Tomalia DA. Starburst(R) dendrimers - Nanoscopic supermolecules according dendritic rules and principles. *Macromolecular Symposia* 1996; 101:243-55.

[36] Cloninger MJ. Biological applications of dendrimers. *Current Opinion in Chemical Biology* 2002; 6:742-8.
[37] KukowskaLatallo JF, Bielinska AU, Johnson J, Spindler R, Tomalia DA, Baker JR. Efficient transfer of genetic material into mammalian cells using Starburst polyamidoamine dendrimers. *Proceedings of the National Academy of Sciences of the United States of America 1996*; 93:4897-902.
[38] Choi YS, Mecke A, Orr BG, Holl MMB, Baker JR. DNA-directed synthesis of generation 7 and 5 PAMAM dendrimer nanoclusters. *Nano Letters* 2004; 4:391-7.
[39] Fire, A.; Xu, S.; Montgomery, M. K.; Kostas, S. A.; Driver, S. E.; Mello, C. C. Potent and Specific Genetic Interference by Double-Stranded RNA in Caenorhabditis Elegans. *Nature* 1998, *391*, 806–11.
[40] Devi, G. R. siRNA-Based Approaches in Cancer Therapy. *Cancer Gene Ther.* 2006, *13*, 819–29.
[41] Takeshita, F.; Ochiya, T. Therapeutic Potential of RNA Interference against Cancer. *Cancer Sci.* 2006, *97*, 689–96.
[42] Elbashir, S. M.; Harborth, J.; Lendeckel, W.; Yalcin, A.; Weber, K.; Tuschl, T. Duplexes of 21-Nucleotide RNAs Mediate RNA Interference in Cultured Mammalian Cells. *Nature* 2001, *411*, 494–8.
[43] Dorsett, Y.; Tuschl, T. Sirnas: Applications in Functional Genomics and Potential as Therapeutics. *Nat. Rev. Drug Discovery* 2004, *3*, 318–29.
[44] Stevenson, M. Therapeutic Potential of RNA Interference. *New. Engl. J. Med.* 2004, *351*, 1772–7.
[45] Dykxhoorn, D. M.; Palliser, D.; Lieberman, J. The Silent Treatment: siRNAs as Small Molecule Drugs. *Gene Ther.* 2006, *13*, 541–52.
[46] Uprichard, S. L. The Therapeutic Potential of RNA Interference. *FEBS Lett.* 2005, *579*, 5996–6007.
[47] Novina, C. D.; Murray, M. F.; Dykxhoorn, D. M.; Beresford, P. J.; Riess, J.; Lee, S. K.; Collman, R. G.; Lieberman, J.; Shankar, P.; Sharp, P. A. siRNA-Directed Inhibition of HIV-1 Infection. *Nat. Med.* 2002, *8*, 681–6.
[48] Soutschek, J.; Akinc, A.; Bramlage, B.; Charisse, K.; Constien, R.; Donoghue, M.; Elbashir, S.; Geick, A.; Hadwiger, P.; Harborth, J.; et al. Therapeutic Silencing of an Endogenous Gene by Systemic Administration of Modified siRNAs. *Nature* 2004, *432*, 173–8.
[49] Zimmermann, T. S.; Lee, A. C. H.; Akinc, A.; Bramlage, B.; Bumcrot, D.; Fedoruk, M. N.; Harborth, J.; Heyes, J. A.; Jeffs, L. B.; John, M.; et al. RNAi-Mediated Gene Silencing in Non-Human Primates. *Nature* 2006, *441*, 111–14.
[50] Song, E.; Zhu, P.; Lee, S. K.; Chowdhury, D.; Kussman, S.; Dykxhoorn, D. M.; Feng, Y.; Palliser, D.; Weiner, D. B.; Shankar, P.; et al. Antibody Mediated *in Vivo* Delivery of Small Interfering RNAs *via* Cell-Surface Receptors. *Nat. Biotechnol.* 2005, *23*, 709–17.
[51] Akinc, A.; Zumbuehl, A.; Goldberg, M.; Leshchiner, E. S.; Busini, V.; Hossain, N.; Bacallado, S. A.; Nguyen, D. N.; Fuller, J.; Alvarez, R.; et al. A Combinatorial Library of Lipid-
[52] Like Materials for Delivery of RNAi Therapeutics. *Nat. Biotechnol.* 2008, *26*, 561–9.
[53] Li, S. D.; Chono, S.; Huang, L. Efficient Oncogene Silencing and Metastasis Inhibition *via* Systemic Delivery of siRNA. *Mol. Ther.* 2008, *16*, 942–6.

[54] Landen, C. N.; Merritt, W. M.; Mangala, L. S.; Sanguino, A. M.; Bucana, C.; Lu, C.; Lin, Y. G.; Han, L. Y.; Kamat, A. A.; Schmandt, R.; et al. Intraperitoneal Delivery of Liposomal siRNA for Therapy of Advanced Ovarian Cancer. *Cancer Biol. Ther.* 2006, *5*, 1708–13.

[55] Yezhelyev, M. V.; Qi, L.; O'Regan, R. M.; Nie, S.; Gao, X. Proton-Sponge-Coated Quantum Dots for siRNA Delivery and Intracellular Imaging. *J. Am. Chem. Soc.* 2008, *130*, 9006–12.

[56] Medarova, Z.; Pham, W.; Farrar, C.; Petkova, V.; Moore, A. *In Vivo* Imaging of siRNA Delivery and Silencing in Tumors. *Nat. Med.* 2007, *13*, 372–7.

[57] Oishi, M.; Nakaogami, J.; Ishii, T.; Nagasaki, Y. Smart Pegylated Gold Nanoparticles for the Cytoplasmic Delivery of siRNA to Induce Enhanced Gene Silencing. *Chem. Lett.* 2006, *35*, 1046–7.

[58] Rosi, N. L.; Giljohann, D. A.; Thaxton, C. S.; Lytton-Jean, A. K.; Han, M. S.; Mirkin, C. A. Oligonucleotide-Modified Gold Nanoparticles for Intracellular Gene Regulation. *Science* 2006, *312*, 1027–30.

[59] Liu, Z.; Winters, M.; Holodniy, M.; Dai, H. siRNA Delivery into Human T Cells and Primary Cells with Carbon-Nanotube Transporters. *Angew. Chem., Int. Ed.* 2007, *46*, 2023–7.

[60] KukowskaLatallo, J. F.; Bielinska, A. U.; Johnson, J.; Spindler, R.; Tomalia, D. A.; Baker, J. R. Efficient Transfer of Genetic Material into Mammalian Cells Using Starburst Polyamidoamine Dendrimers. *Proc. Natl. Acad. Sci. USA*. 1996, *93*, 4897–4902.

[61] Tomalia, D. A.; Baker, H.; Dewald, J.; Hall, M.; Kallos, G.; Martin, S.; Roeck, J.; Ryder, J.; Smith, P. A New Class of PolymersOStarburst-Dendritic Macromolecules. *Polym. J.* 1985, *17*, 117–32.

[62] Tomalia, D. A.; Frechet, J. M. J. Discovery of Dendrimers and Dendritic Polymers: A Brief Historical Perspective. *J. Polym. Sci., Part A* 2002, *40*, 2719–28.

[63] Shi, X. Y.; Thomas, T. P.; Myc, L. A.; Kotlyar, A.; Baker, J. R. Synthesis, Characterization, and Intracellular Uptake of Carboxyl-Terminated Poly(Amidoamine) Dendrimer- Stabilized Iron Oxide Nanoparticles. *Phys. Chem. Chem. Phys.* 2007, *9*, 5712–20.

[64] Luo, D.; Haverstick, K.; Belcheva, N.; Han, E.; Saltzman, W. M. Poly(ethylene glycol)-Conjugated PAMAM Dendrimer for Biocompatible, High-Efficiency DNA Delivery. *Macromolecules* 2002, *35*, 3456–62.

[65] Kono, K.; Akiyama, H.; Takahashi, T.; Takagishi, T.; Harada, A. Transfection Activity of Polyamidoamine Dendrimers Having Hydrophobic Amino Acid Residues in the Periphery. *Bioconjug. Chem.* 2005, *16*, 208–14.

[66] Myc, A.; Patri, A. K.; Baker, J. R. Dendrimer-Based BH3 Conjugate That Targets Human Carcinoma Cells. *Biomacromolecules* 2007, *8*, 2986–9.

[67] Shukla, R.; Thomas, T. P.; Peters, J.; Kotlyar, A.; Myc, A.; Baker, J. R. Tumor Angiogenic Vasculature Targeting with PAMAM Dendrimer-RGD Conjugates. *Chem Commun.* 2005, 5739–41.

[68] Bi, X.; Shi, X.; Majoros, I. J.; Shukla, R.; Baker, J. R. Multifunctional Poly(amidoamine) Dendrimer-Taxol Conjugates: Synthesis, Characterization and Stability. *J. Comput. Theor. Nanosci.* 2007, *4*, 1179–87.

[69] Lee, C. C.; Gillies, E. R.; Fox, M. E.; Guillaudeu, S. J.; Frechet, J. M. J.; Dy, E. E.; Szoka, F. C. A Single Dose of Doxorubicin- Functionalized Bow-Tie Dendrimer Cures Mice Bearing C-26 Colon Carcinomas. *Proc. Natl. Acad. Sci. USA*. 2006, *103*, 16649–54.

[70] Najlah, M.; Freeman, S.; Attwood, D.; D'Emanuele, A. Synthesis and Assessment of First-Generation Polyamidoamine Dendrimer Prodrugs to Enhance the Cellular Permeability of P-Gp Substrates. *Bioconjug. Chem.* 2007, *18*, 937–46.

[71] Choi, Y.; Baker, J. R. Targeting Cancer Cells with DNA Assembled DendrimersOA Mix and Match Strategy for Cancer. *Cell Cycle* 2005, *4*, 669–71.

[72] Landmark, K. J.; DiMaggio, S.; Ward, J.; Kelly, C.; Vogt, S.; Hong, S.; Kotlyar, A.; Myc, A.; Thomas, T. P.; Penner-Hahn, J. E.; *et al.* Synthesis, Characterization, and *in Vitro* Testing of Superparamagnetic Iron Oxide Nanoparticles Targeted Using Folic Acid-Conjugated Dendrimers. *ACS Nano* 2008, *2*, 773–83.

[73] Kobayashi, H.; Jo, S. K.; Kawamoto, S.; Yasuda, H.; Hu, X.; Knopp, M. V.; Brechbiel, M. W.; Choyke, P. L.; Star, R. A. Polyamine Dendrimer-Based MRI Contrast Agents for Functional Kidney Imaging to Diagnose Acute Renal Failure. *J. Magn. Reson. Imag.* 2004, *20*, 512–8.

[74] Lee, C. C.; MacKay, J. A.; Frechet, J. M.; Szoka, F. C. Designing Dendrimers for Biological Applications. *Nat. Biotechnol.* 2005, *23*, 1517–26.

[75] Dufes, C.; Uchegbu, I. F.; Schatzlein, A. G. Dendrimers in Gene Delivery. *Adv. Drug Delivery Rev.* 2005, *57*, 2177–202.

[76] Kang, H.; DeLong, R.; Fisher, M. H.; Juliano, R. L. Tat- Conjugated PAMAM Dendrimers as Delivery Agents for Antisense and siRNA Oligonucleotides. *Pharm. Res.* 2005, *22*, 2099–106.

[77] Juliano, R. L. Intracellular Delivery of Oligonucleotide Conjugates and Dendrimer Complexes. *Ann. N.Y. Acad. Sci.* 2006, *1082*, 18–26.

[78] Inoue, Y.; Kurihara, R.; Tsuchida, A.; Hasegawa, M.; Nagashima, T.; Mori, T.; Niidome, T.; Katayama, Y.; Okitsu, O. Efficient Delivery of siRNA Using Dendritic Poly(L-lysine) for Loss-of-Function Analysis. *J. Controlled Release* 2008, *126*, 59–66.

[79] Zhou, J.; Wu, J.; Hafdi, N.; Behr, J. P.; Erbacher, P.; Peng, L. PAMAM Dendrimers for Efficient siRNA Delivery and Potent Gene Silencing. *Chem Commun.* 2006, 2362–4.

[80] Shen, X. C.; Zhou, J.; Liu, X.; Wu, J.; Qu, F.; Zhang, Z. L.; Pang, D. W.; Quelever, G.; Zhang, C. C.; Peng, L. Importance of Size-to-Charge Ratio in Construction of Stable and Uniform Nanoscale RNA/Dendrimer Complexes. *Org. Biomol. Chem.* 2007, *5*, 3674–81.

[81] Duncan, R.; Izzo, L. Dendrimer Biocompatibility and Toxicity. *Adv. Drug Delivery Rev.* 2005, *57*, 2215–37.

[82] Park, J. H.; von Maltzahn, G.; Zhang, L. L.; Schwartz, M. P.; Ruoslahti, E.; Bhatia, S. N.; Sailor, M. J. Magnetic Iron Oxide Nanoworms for Tumor Targeting and Imaging. *Adv. Mater.* 2008, *20*, 1630–5.

[83] Cho, Y. W.; Kim, J. D.; Park, K. Polycation Gene Delivery Systems: Escape from Endosomes to Cytosol. *J. Pharm. Pharmacol.* 2003, *55*, 721–34.

[84] Hu, Y.; Litwin, T.; Nagaraja, A. R.; Kwong, B.; Katz, J.; Watson, N.; Irvine, D. J. Cytosolic Delivery of Membrane-Impermeable Molecules in Dendritic Cells Using pH-Responsive Core-Shell Nanoparticles. *Nano Lett.* 2007, *7*, 3056–64.

[85] Jones, R. A.; Cheung, C. Y.; Black, F. E.; Zia, J. K.; Stayton, P. S.; Hoffman, A. S.; Wilson, M. R. Poly(2-alkylacrylic acid) Polymers Deliver Molecules to the Cytosol by pH-Sensitive Disruption of Endosomal Vesicles. *Biochem. J.* 2003, *372*, 65–75.

[86] Giannini, C.; Sarkaria, J. N.; Saito, A.; Uhm, J. H.; Galanis, E.; Carlson, B. L.; Schroeder, M. A.; James, C. D. Patient Tumor EGFR and PDGFRA Gene Amplifications Retained in an Invasive Intracranial Xenograft Model of Glioblastoma Multiforme. *Neuro. Oncol.* 2005, *7*, 164–76.

[87] Sarkaria, J. N.; Carlson, B. L.; Schroeder, M. A.; Grogan, P.; Brown, P. D.; Giannini, C.; Ballman, K. V.; Kitange, G. J.; Guha, A.; Pandita, A.; et al. Use of an Orthotopic Xenograft Model for Assessing the Effect of Epidermal Growth Factor Receptor Amplification on Glioblastoma Radiation Response. *Clin. Cancer Res.* 2006, *12*, 2264–71.

[88] Derfus, A. M.; Chen, A. A.; Min, D. H.; Ruoslahti, E.; Bhatia, S. N. Targeted Quantum Dot Conjugates for siRNA Delivery. *Bioconjug. Chem.* 2007, *18*, 1391–6. 50. von Gersdorff, K.; Sanders, N. N.; Vandenbroucke, R.; De Smedt, S. C.; Wagner, E.; Ogris, M. The Internalization

[89] Route Resulting in Successful Gene Expression Depends on Both Cell Line and Polyethylenimine Polyplex Type. *Mol. Ther.* 2006, *14*, 745–53.

[90] Schaffert, D.; Wagner, E. Gene Therapy Progress and Prospects: Synthetic Polymer-Based Systems. *Gene Ther.* 2008, *15*, 1131_38.

[91] Lai, S. K.; Hida, K.; Chen, C.; Hanes, J. Characterization of the Intracellular Dynamics of a Non-degradative Pathway Accessed by Polymer Nanoparticles. *J. Controlled Release* 2008, *125*, 107–11.

[92] Uyechi, L. S.; Gagne, L.; Thurston, G.; Szoka, F. C., Jr. Mechanism of Lipoplex Gene Delivery in Mouse Lung: Binding and Internalization of Fluorescent Lipid and DNA Components. *Gene Ther.* 2001, *8*, 828–36.

[93] Zhu, H.; Acquaviva, J.; Ramachandran, P.; Boskovitz, A.; Woolfenden, S.; Pfannl, R.; Bronson, R. T.; Chen, J. W.; Weissleder, R.; Housman, D. E.; *et al.* Oncogenic EGFR Signaling Cooperates with Loss of Tumor Suppressor Gene Functions in Gliomagenesis. *Proc. Natl. Acad. Sci. USA.* 2009, *106*, 2712_6.

[94] Nicholas, M. K.; Lukas, R. V.; Jafri, N. F.; Faoro, L.; Salgia, R. Epidermal Growth Factor ReceptorOMediated Signal Transduction in the Development and Therapy of Gliomas. *Clin. Cancer Res.* 2006, *12*, 7261–70.

[95] Belda-Iniesta, C.; de Castro Carpeno, J.; Sereno, M.; Gonzalez-Baron, M.; Perona, R. Epidermal Growth Factor Receptor and Glioblastoma Multiforme: Molecular Basis for a New Approach. *Clin. Transl. Oncol.* 2008, *10*, 73–7.

[96] Friedman, H. S.; Bigner, D. D. Glioblastoma Multiforme and the Epidermal Growth Factor Receptor. *N Engl J Med.* 2005, *353*, 1997–9.

[97] Halatsch, M. E.; Schmidt, U.; Behnke-Mursch, J.; Unterberg, A.; Wirtz, C. R. Epidermal Growth Factor Receptor Inhibition for the Treatment of Glioblastoma Multiforme and Other Malignant Brain Tumours. *Cancer Treat Rev.* 2006, *32*, 74–89.

[98] Mellinghoff, I. K.; Wang, M. Y.; Vivanco, I.; Haas-Kogan, D. A.; Zhu, S.; Dia, E. Q.; Lu, K. V.; Yoshimoto, K.; Huang, J. H.; Chute, D. J.; *et al.* Molecular Determinants of the Response of Glioblastomas to EGFR Kinase Inhibitors. *New Engl. J. Med.* 2005, *353*, 2012–24.

[99] Jain, K. K. Nanobiotechnology-Based Drug Delivery to the Central Nervous System. *Neurodegener. Dis.* 2007, *4*, 287–91.

[100] Josephson, L.; Tung, C. H.; Moore, A.; Weissleder, R. High- Efficiency Intracellular Magnetic Labeling with Novel Superparamagnetic-Tat Peptide Conjugates. *Bioconjug. Chem.* 1999, *10*, 186–91.

[101] Palmacci, S.; Josephson, L. *Synthesis of Polysaccharide Covered Superparamagnetic Oxide Colloids.* U.S. Patent 5,262,176, 1993.

[102] Mohammad Reza Saboktakin, Roya M.Tabatabaie, Abel Maharramov, and Mohammad Ali Ramazanov," An Aromatic Macromolecule as a Nano Drug Carrier: Polyether Dendrimer/Poly(2-hydroxyethyl methacrylate)", *Journal of Polymers & Polymer Composites, Vol. 18, No. 8, 2010.*

Chapter 7

CONDUCTIVE NANOCOMPOSITES FOR COATING

7.1. INTRODUCTION

The ability of polymers to become electrically conductive was shown when a scientific breakthrough demonstrated that to become electrically conductive, a polymer must imitate a metal, which means that electrons in polymers must be free to move and not be bound to atoms. Most polymeric materials are poor conductors of electricity because a large number of free electrons are not available to participate in the conduction process. In principle, an oxidation or reduction reaction is often accompanied by the addition or removal of electrons. Therefore, a polymer might become electrically conductive by removing electrons, a process described as doping. Doping means an increase in carriers in conjugated polymers where the process involves electron acceptors and electron donors, respectively. This unexpected discovery not only challenged the traditional concept that organic polymers are only insulators, but also established a new class of conducting polymers called 'synthetic metals'. Since the discovery of electrical conductivity in ionic polymers, various ionically conducting polymers or polymer electrolytes have been prepared for a wide range of applications. The most interesting property of conducting polymers is their high (almost metallic) conductivity, which can be altered by simple oxidation or reduction and also by bringing the material into contact with different compounds [1-6].

The first and most widely used conducting polymeric systems were composites in which an insulating polymer matrix was filled with particulate or fibrous conductive filler, such as a carbon or metal, to impart high conductivity. Applications for such composites are widespread; these composites are used for interconnections, printed circuit boards, encapsulations, die attaches, heat sinks, conducting adhesives, electro-magnetic interference (EMI) shielding, electrostatic discharge (ESD) and aerospace engineering. Unfortunately, these conductive fillers will impart heavy weight, poor surface finish, poor mechanical properties and easy oxidative degradation to the end product [7].

Polyaniline (PANI) is a promising conducting polymer due to its easy synthesis, environmental stability and high electrical conductivity. The preparation of PANI composites with various materials has received great attention because of their unique properties and applications in various electrical devices. However, the main problem associated with the effective utilization of all intrinsically conducting polymers (ICPs), including PANI, is inherent in their lower level of conductivity compared to metal and their infusibility and poor solubility in all available solvents.

Polymer composites containing PANI (matrixes) have received considerable attention because of the resultant combination of improved processability and fairly good mechanical properties coupled with good conductivity. Research on creating blends of PANI with conventional polymers concluded that these blends possess relatively high conductivity with good mechanical properties. PANI/polyamide blended films have been successfully prepared, and their excellent electrical and mechanical properties have contributed to the film industry. In addition, the preparation and characterisation of PANI materials with ferromagnetic properties has also been primarily performed. Although the first goal of advanced studies was to fabricate composites with good room temperature conductivity, only low values of conductivity of 10^{-6}–10^{-3} S/cm were obtained [8-15].

Because PANI is one of the most promising conducting polymers with enhanced properties, numerous research papers are available concerning the synthesis and application of PANI composites. Many papers dealing with the preparation of conducting composites of PANI, such as TiO_2/PANI, ZrO_2/PANI, Fe_3O_4/PANI, zeolite/PANI, MoO_3/PANI, MnO_2/PANI, WO_3/PANI and others, have been published. Another area of research is based on $CaCO_3$, for example, the correlation between hardness and yield stress of $CaCO_3$ filled polyethylene.

The evaluation of CaCO3 dispersion in polypropylene composite fibers shows that: 1) Fourier transform infrared (FTIR) spectroscopy may be reliably applied for the evaluation of the relative degree of dispersion of calcite in composites, 2) the impact strength of high density polyethylene (HDPE) is increased upon $CaCO_3$ reinforcement and 3) the presence of the filler influences the crystallization process and leads to an increase of the amount of imperfect crystalline phase. Other results indicated that $CaCO_3$ nanoparticles could induce nucleation but slow the mobility of polymer chains [16-20].

Traditionally, BET surface area measurements have been used to determine the specific surface area of colloidal inorganic oxides and polymer latexes. There have been very few BET studies on conductive polymer-based materials. The surface area and porosity values obtained show that the application of ESP into a PANI matrix enhanced its composite properties because it created a different phase relative to the matrix materials. These are important characteristics that affect the quality and utility of the PANI-eggshell composite and are important in understanding its structure, formation and conductivity potential, which corresponded well to its applications in various fields [21].

As shown in Figure 1, the relative pressure above 0.8 reflects a dramatic adsorption and desorption behavior, meaning that above this point, the applied pressure is sufficient to allow the surface to be saturated by the adsorbate. Adsorption is usually described through isotherms, that is, the amount of adsorbate on the adsorbent as a function of its pressure at a constant temperature. The determination of a specific surface by means of the BET theory is based upon the phenomenon of the physical adsorption of N2 gases on the external and internal surfaces of a sample and is able to show the porosity and pore surface of the ESP, which is different in its physical properties from the matrix of conductive polymer. The amount of adsorbed gas is dependent on its relative vapor pressure and is proportional to the total external and internal surface of the material.

The porous structure of ESP will be an added advantage for electron transfer when current is applied to the conductive polymer [22].

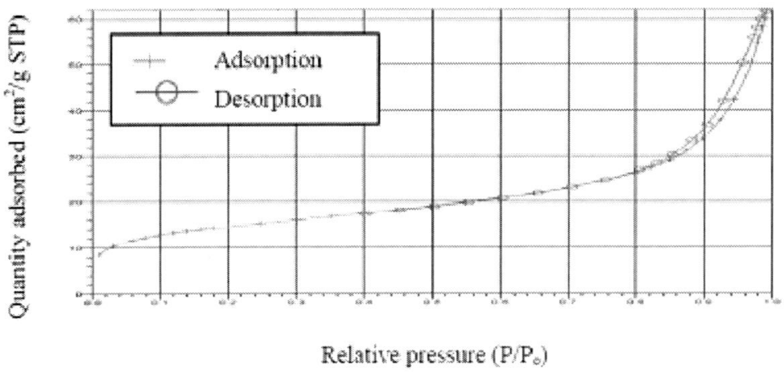

Figure 1. BET analysis shows quantity absorbed versus relative pressure.

Figures 2 and 3 show the pore volume and area of the ESP. The pore volume and pore area decrease with increasing pore width. There are a few differences between BET and Langmuir surface area values, but the main difference is that Langmuir values can only be used for surfaces that are covered by one layer of gas and BET values are calculated using a multilayer model. The Langmuir adsorption isotherm was determined for single layer adsorption and produced a curve that describes the fraction of the surface area of the adsorbent covered with solute as a function of the concentration of the solute in the contacting liquid phase[23].

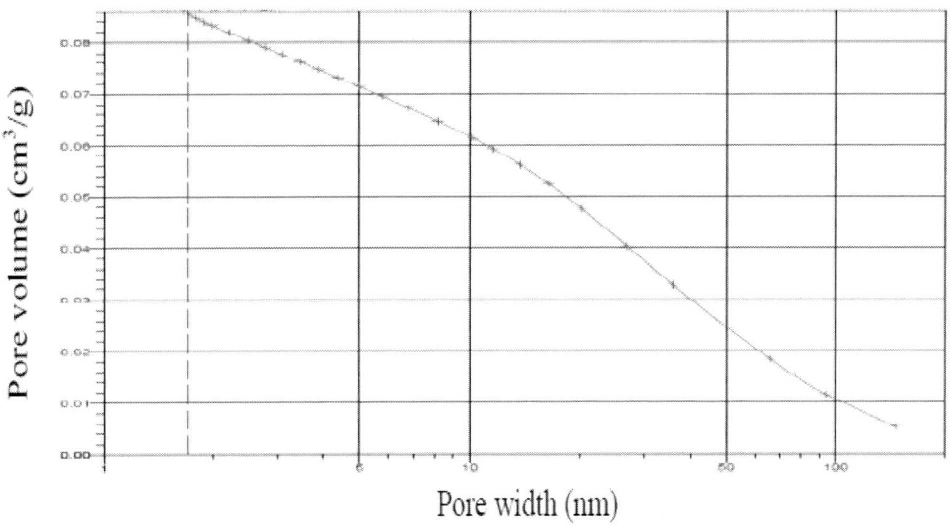

Figure 2. BET analysis shows the pore volume versus pore width.

The main goal of infrared (IR) spectroscopy analysis was to determine the chemical functional groups in the PANI-eggshell conducting polymer sample. Different functional groups absorb characteristic frequencies of IR radiation. From the result shown in Figure 4, high wave numbers were observed that presented very strong spectrum absorption intensity peaks at 3525.01 cm–1 corresponding to the NH group. These peaks are perfectly positioned to correspond to an aromatic amine (NH2). Because the peak intensity was 1658.22 cm–1, this molecule must be a primary aromatic NH2. Aromatic C=O bonds and C=C bonds also fall into this interval. The peak at 2910.16 cm–1 is within the range of C-H bonds and -CH2 asymmetric bonds [24].

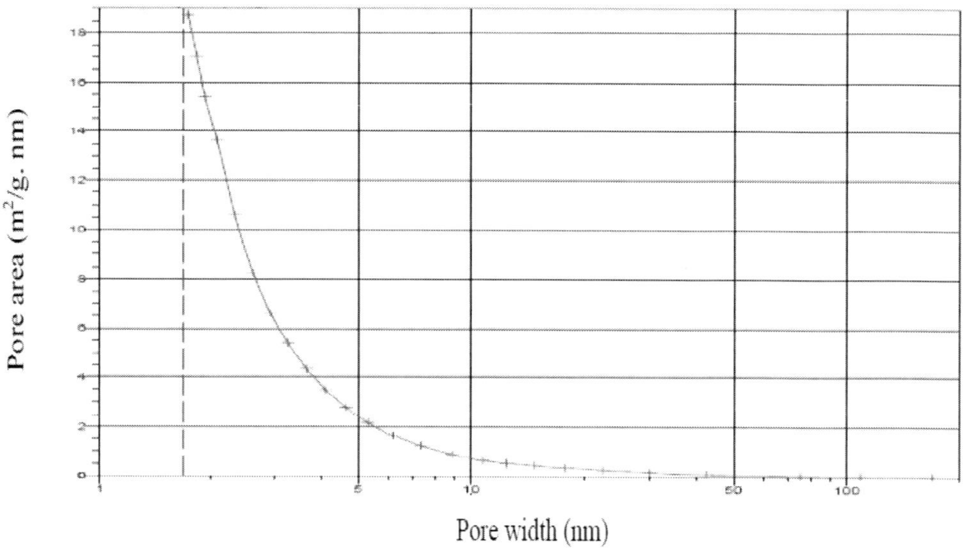

Figure 3. BET analysis shows pore area versus pore width.

Small peaks located in the region between 600 and 700 cm–1 were also observed. The bands at 618.30 cm–1 can be assigned as a bend of a benzene ring. These bands confirm that the source of unsaturation is a benzene ring. Thus, it is a mono-substituted benzene ring. Because we know that we have a primary aromatic NH2 and a mono-substituted benzene ring, the molecule must be aniline [18]. Since the aniline is not an individual substance in this composite, S-O bends and inorganic sulphates were found between 610 and 680 cm^{-1}.

In addition, a peak at 2875.38 cm–1, which falls in between 2835 and 2878 cm–1, shows that the functional group of this conductive polymer structure contains a CH3 symmetric and CH2-O symmetric C-H strength. Moreover, additional peaks at 1558.30 and 1359.93 cm^{-1} indicate an N-H in-plane bend and the presence of N-O strength, respectively. From the results obtained, all expected functional groups were revealed, further confirming that the structure of the PANI-eggshell composite is exhibited in the composite. Functional groups play an important role in improving the thermal or conductivity properties of conductive composites.10 In the PANI-ESP composites, no chemical change was found within the structure. The only interaction occurs between the filler and the reinforcement of CB with the matrix [25-28].

Figure 5(a) shows the surface morphology of the control sample without ESP at 100 X magnification. In this morphology, a chain configuration is indicated as the PANI chain is embedded within its matrix. An expanded coil of the PANI structure was also observed.19 A major part of the PANI is compatible with 1-methyl-2-pyrrolidinone, which makes it soluble in the matrix, and hence, a flat surface was observed [29].

Figure 5(b) shows the surface morphology with a filler of 0.25 g of ESP at 100 X magnification. The distribution of the ESP particles in the polymer matrix is noticeably uniform. There are no large agglomerations of particles within the matrix. The chain of PANI is dissolved in the matrix and is compatible with ESP [5]. The average ESP particle size distribution obtained from the sieve instrument was 63 μm. The ESP particles are distributed far from each other because only a small amount of reinforcement was used [30].

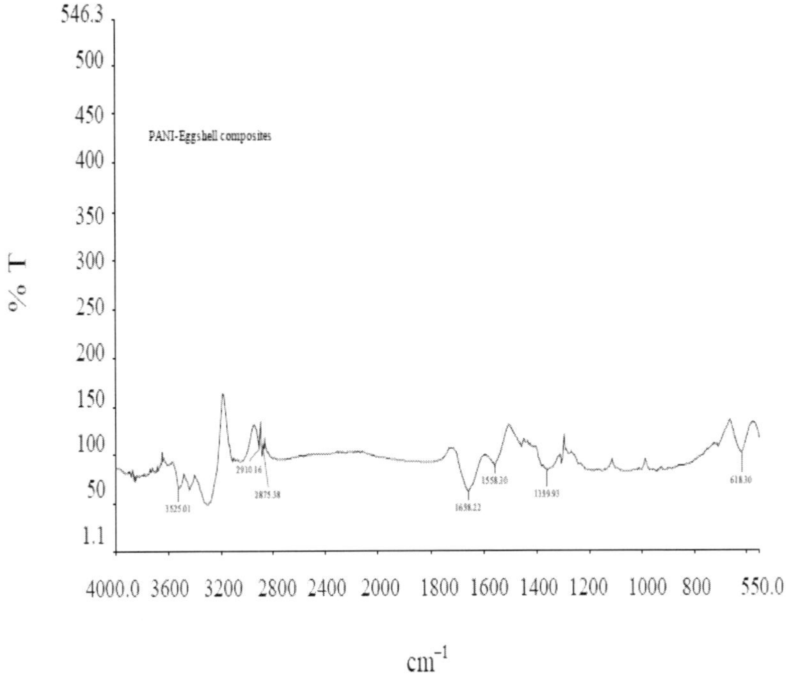

Figure 4. FTIR spectrum of conductive polymer base on PANI-ESP composites.

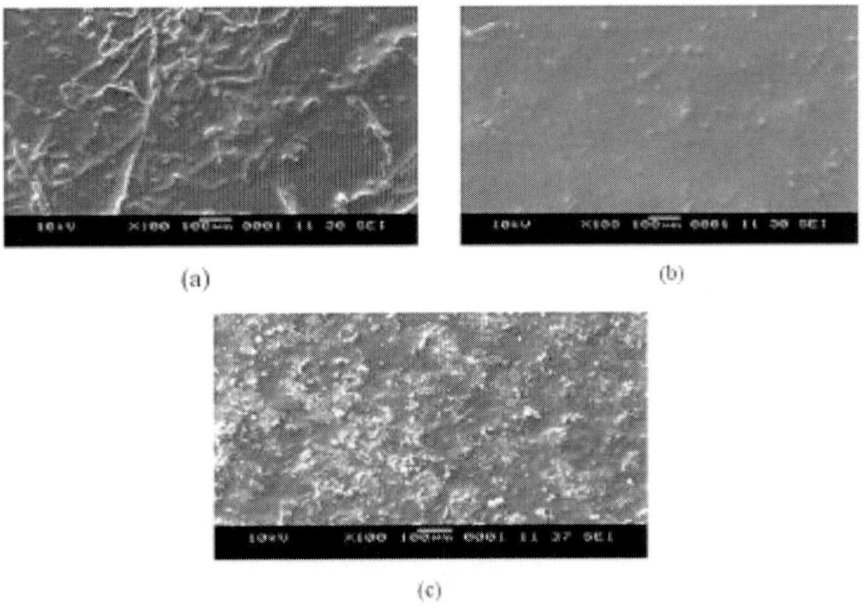

Figure 5. SEM micrographs of (a) without ESP; b) 0.25g of ESP; c) 0.75g of ESP.

Figure 5(c) shows the surface morphology with a filler of 0.75 g of ESP at 100 X magnification. The distribution of the ESP particles in the polymer matrix is noticeably uniform, but note that there are agglomerations of ESP on the surface of the sample. Agglomeration with PANI makes the connection of the conductive path more complete and improves conductivity. A rougher surface was observed as filler loading was increased; the formation of microfiller was found due to difficulties of achieving a homogeneous dispersion of filler at high filler loading [31-38].

XRD was used to determine the index of the diffraction pattern of the PANI-eggshell composites by assigning the correct Miller indices (hkl) to each peak (reflection) in the diffraction pattern, as shown in Figure 6. An XRD pattern is properly indexed when all of the peaks in the diffraction pattern are labeled and no peaks expected for the particular structure are missing [22]. All major peaks were accounted and assigned the correct Bravais lattice to calculate the lattice parameters for the PANI-eggshell composite. In cubic systems, the first XRD peak in the XRD pattern is due to diffraction from planes with the lowest Miller indices [i.e., simple cubic, (100), h2 + k2 + l2 = 1; body centered cubic, (110), h2 + k2 + l2 = 2; and face centered cubic (FCC), (111), h2 + k2 + l2 = 3]. Because h, k and l are always integers, we can obtain h2 + k2 + l2 values by dividing the 2sinθ values for the different XRD peaks with the minimum peak in the pattern (i.e., the 2sinθ value from the first XRD peak) and multiplying that ratio by the proper integer (either 1, 2 or 3). This calculation should yield a list of integers that represent the various values. The correct Bravais lattice can then be identified by recognizing the sequence of allowed reflections for cubic lattices (i.e., the sequence of allowed peaks written in terms of the quadratic form of the Miller indices) [39].

Table 4. Calculation of interplanar spacing from the results interpreted from XRD

Peak	2θ	θ	$\sin^2\theta$	$\sin^2\theta / \sin^2\theta_{min}$	$3 \times \sin^2\theta / \sin^2\theta_{min}$	$h^2 + k^2 + l^2$	*hkl*	*d*
A	21.0	10.50	0.03321	1.0000	3.0000	3	111	7.318
B	23.9	11.95	0.04287	1.2909	3.8726	4	200	7.438
C	28.4	14.20	0.06018	1.8121	5.4363	5	210	7.019
D	30.3	15.15	0.0683	2.0566	6.1698	6	211	7.217
E	33.9	16.95	0.08499	2.5592	7.6775	8	220	7.471
F	37.5	18.75	0.10332	3.1111	9.3333	9	300	7.187
G	41.8	20.90	0.12726	3.8320	11.4959	11	311	7.159
H	43.0	21.50	0.13432	4.0446	12.1337	12	222	7.278

Notes: λ = 1.54 Å
n = 1
h = 1/x, k = 1/y, l = 1/z (from the Miller index)
Bravais lattice is FCC
Average lattice parameter = 7.26071 Å

Figure 6 shows the X-ray diffraction pattern of amorphous materials in partially crystalline mixtures of conductive polymer. From the curve obtained and the calculation of interplanar spacing, Bravais lattice was found to be followed by a FCC pattern as shown in Tables 3 and 4. Therefore, the conductive polymer based on PANI was concluded to possess a FCC structure. The powders exhibited peaks at 2 angles, around 200 and 450, which indicates that the conductive polymer contains some crystalline domains [10]. The morphological structures of PANI derivatives are similar, closed-packed structures, which results in similar conductivity values [40].

The polymer backbone in the PANI-ESP conducting polymers consists of a π-conjugated chain. The overlap of π-electrons is called conjugation because it leads to a sequence of alternating double and single bonds, resulting in unpaired electrons that are delocalized along the polymeric chain. PANI exhibits conductivity due to this conductive theory.

The conductive polymer was created for use as a sensor in several applications, such as in the detection of NH_3 gas, coffee aromas and ethanol. From the results shown in Figures 7(a), (b) and (c), the sensors were able to detect all three gases because there was either a significant increase or decrease in conductivity after application of those gases. This enabled the detection of these gases due to the differences in conductivity that were obtained. When the sensor was applied to ethanol, the highest resistivity change was obtained at 0.75 g of ESP, whereas the highest resistivity difference was obtained at 0.5 g ESP when coffee aroma was applied [41].

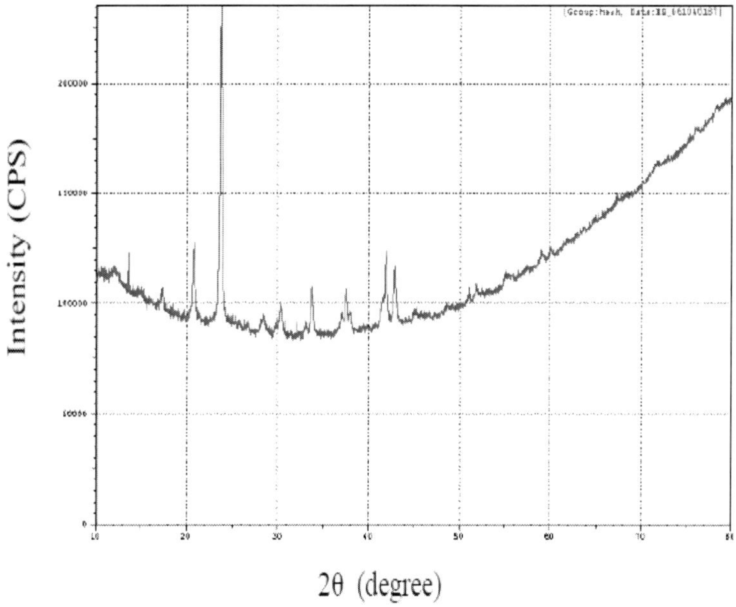

Figure 6. Graph shows the XRD diffractogram on PANI-eggshell powded conductive polymer.

Table 3. Sequential of Bravais lattice for cubic system

Primitive	1,2,3,4,5,6,8,9,10,11,12,13,14,16…
Body-centered	2,4,6,8,10,12,14,16…
Face-centered	3,4,8,11,12,16,19,20,24,27,32…
Diamond cubic	3,8,11,16,19,24,27,32…

This finding shows that the resistivity increased when these two gases were applied. However, when NH_3 gas was applied to the sensor, the resistivity value decreased significantly to about 25 M ohm meaning that the conductivity increased significantly. Therefore, for comparison, the sensor is best used as an NH_3 gas detector because the conductivity was increased, which was the aim of this experiment. Resistivity, which is simply the inverse of the conductivity, was measured because the multimeter only shows resistivity values when ethanol becomes ethylene (C_2H_4) gas, whereas ammonium hydroxide will become NH_3 when heated as it gives off a water molecule. When C_2H_4 gas, NH_3 gas and caffeine ($C_8H_{10}N_4O_2$) were brought into contact with the sensor, an interaction occurred between the structure of NH_3, C_2H_4 and caffeine that distorted the chemical structure of the PANI-ESP. Therefore, the conductive behavior changes due to changes of the structure of the PANI-ESP. Due to structural distortion, the highest conductivity was shown upon application of NH_3 where it has the lowest resistivity [Figure 7(c)].

The conductivity of the conductive polymer was calculated according to the following formula using ASTM D257:

$$\text{Conductivity of a material,} \qquad \sigma = \frac{L}{R \times A} \tag{3}$$

In which the conductivity is calculated from the resistance, *R,* between two sections separated by distance, *L,* having an average area, *A.* Table 5 shows the conductivity value calculated from this formula.

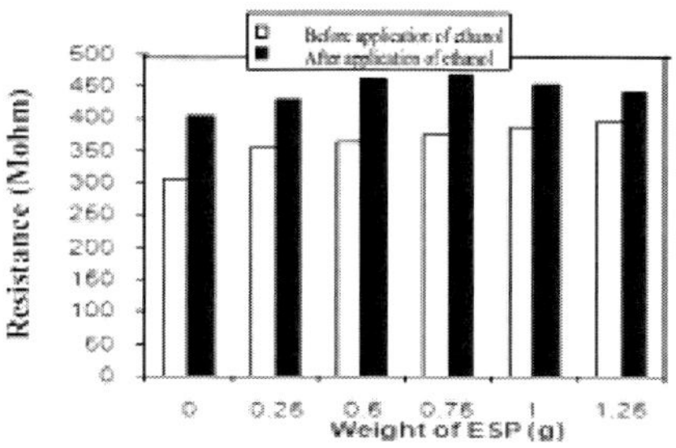

(a) Ethanol

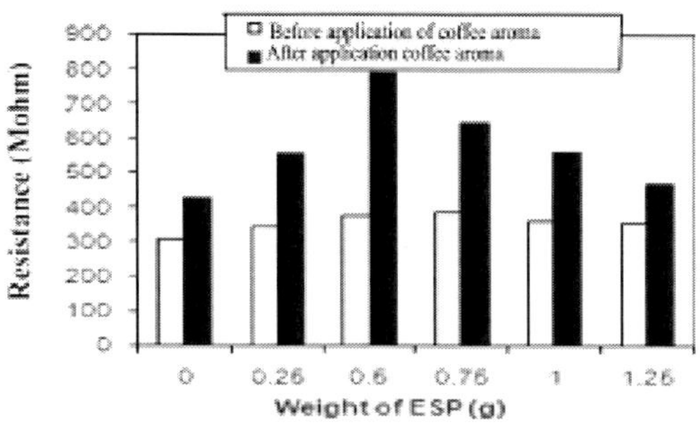

(b) Aroma coffee

Figure 7. (Continued)

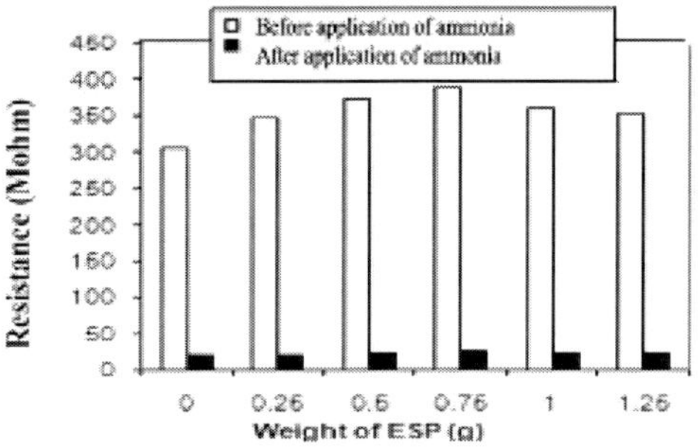

(c) Ammonia gas

Figure 7. Application of conductive test based on (a) ethanol (b) coffee aroma (c) ammonia.

7.2. Catalytic Properties of Conductive Polymer Supported Catalysts

Polymers represent a class of materials that play a role of growing importance in catalysis, due to the possibility of controlling in a simple way their morphology and their physico-chemical properties. Until now, most of the studies have been devoted to the use of two types of polymers for the preparation of metallic catalyst, namely (i) synthetic cross-linked macromolecular materials, such as ion exchange resins, and (ii) hydrosoluble polymers. Resins have been used as support for catalytically active polydipersed noble metal phases since 1969 [42].

Table 5. Conductivity calculated from the experimental

Eggshell (g)	Conductivity, M (ohm^{-1} cm^{-1})					
	Ammonia		Coffee aroma		Ethanol	
	Before	After	Before	After	Before	After
0	7.44E-05	0.00113	7.44E-05	5.32E-05	7.44E-05	5.63E-05
0.25	6.59E-05	0.00106	6.59E-05	4.08E-05	6.42E-05	5.28E-05
0.5	6.12E-05	0.00093	6.12E-05	2.88E-05	6.25E-05	4.92E-05
0.75	5.89E-05	0.00089	5.89E-05	3.54E-05	6.09E-05	4.81E-05
1	6.32E-05	0.00097	6.32E-05	4.07E-05	6.38E-05	5.03E-05
1.25	6.46E-05	0.00103	6.46E-05	4.85E-05	6.63E-05	5.17E-05

These bifunctional catalysts, comprising acid as well as hydrogenation active centers, are generally prepared by ionic-exchange between the resin and the metal precursor followed by a chemical reduction [43]. They are employed, for example, in the industrial synthesis of methyl-isobutyl ketone [44]. Hydrosoluble polymers are used to prepare suspension of metal nanoparticles in order to prevent coagulation and precipitation of the metal nanoparticles. These catalysts, named metal colloids, are generally prepared by reduction of metal salts in the presence of protective polymers such as poly(vinylalcohol) and poly(N-vinyl-2-pyrrolidone) [45]. The noble-metal colloids are active and selective in the hydrogenation of olefins and dienes.

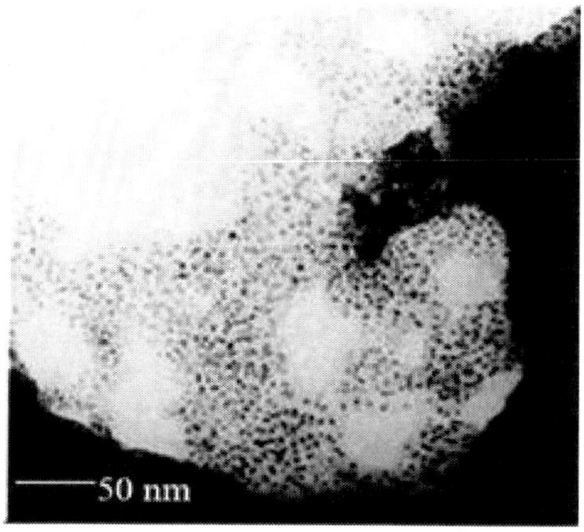

Figure 8. TEM image of pd/Cu-PPy.

Chloride probably acts as counterions to the N_+ in the polymer. The activity and selectivity of a catalyst for nitrite and nitrate was determined and compared with those of classical Pd or Pd-Cu catalysts supported on alumina. These novel catalysts are more active than the classical one for nitrite reduction with a low selectivity (<1%) towards ammonia production, which is an undesired product. With nitrate reduction, the activity of the polymer-supported catalyst is equal to that of the classical one ($PdCu/Al_2O_3$), but no intermediate nitrite is observed and the selectivity towards ammonia is lower. In conclusion, the results show that electroactive polymers can be used as support for noble metals and that the resulting materials have promising catalytic properties for reactions carried out at ambient temperature, such as nitrate and nitrite reduction in water[46].

7.3. Development and Modeling of Thermally Conductive Polymer/Carbon Composites

Thermally conductive polymer composites can replace metals in many applications. This technology is a substantial improvement since polymers are commonly used due to their thermal isolating properties. The advantages of thermally conductive polymers over metals are reduced density; increased corrosion, oxidation, and chemical resistance; increased processibility; and properties are adjustable to fit the application. However, polymers have many disadvantages; for example, creep, thermal instability, and a limited number of processing techniques. The main application for thermally conductive polymers is heat sinks. Other possible benefits are faster injection molding cycle times and improved thermal stability [47].

Carbon black, synthetic graphite, and carbon fiber were added to nylon 6,6 or polycarbonate matrices. For this work, the transverse and longitudinal thermal conductivities were measured. In addition, optical microscopy and image analysis were also performed to characterize the structure of the composites. From these studies, it was found that all three fillers positively affect the transverse thermal conductivity. Synergisms in transverse thermal conductivity were found between synthetic graphite and carbon fiber, between synthetic graphite and carbon black in both polymer matrixes and between carbon black and carbon fiber in polycarbonate only. The results were used to improve the predictions of a transverse thermal conductivity model. The improved model was based on the Nielsen Model. This model includes the aspect ratio, thermal conductivity, packing of the filler(s), and the thermal conductivity of the polymer. The model was revised by updating a parameter, which was originally determined for each filler. Now the value is the same for each filler used. The updates to the model significantly improved the predictive nature of the model for the systems studied [48-50].

7.4. Electrically Conductive Composites of Poly(Vinyl Alcohol)-G-Poly(Acrylic Acid) Hydrogels Impregnated with Polyaniline (PANI)

A thermodynamic system capable of transforming chemical energy directly into mechanical work is known as a chemo-mechanical system. Many artificial chemo-mechanical systems made of environmentally sensitive polymer gels are being developed for sensor/actuator systems such as a controllable membrane separation system and an electrically regulated drug delivery system [51]. A gel containing ionic groups can be actuated isothermally by an electric field. When the gel is negatively charged, it swells near the anode and contracts near the cathode, the contraction rate being proportional to the external electric current [52]. The absolute change in volume is not in-significant. The dimensional changes of some percentages are quite usual. A gel bar made of polyelectrolyte materials can bend backwards and forward by the application of an electric field. Water and ions migrate towards the electrode bearing a change opposite in sign to the net change in the gel, and this coupling of electro-osmosis and electrophoresis is thought to be responsible for the observed chemo-mechanical behavior [53]. Electrically conducting polymers (ECPs) and in particular polyaniline, polypyrrole, polythiophene and their derivatives have been the subject of intensive research for their unique electrical, electrochemical and/or optical properties [54-56] and uncountable applications [57-60]. ECPs are also named ‘conjugated polymers’ because they are macromolecules containing a spatially extended π bonding system, which is the reason for their intrinsic semiconducting nature. Polyaniline is the one of the most promising conducting polymers due to a good combination of properties, stability, price and ease of synthesis by different routes [61]. It exists in a variety of reversible protonated and oxidized forms, differing in their electrical, electrochemical and/or optical properties like many ECPs [62-70]. PANI is difficult to process because it is soluble only in a limited number of organic solvents. Electrically conducting polymers can be prepared by the polymerization of aniline in acidic medium, resulting in the formation of polyaniline (PANI) [71]. The influence of polymerization parameters on the molecular weight of polyaniline has been studied [72] and a new aggregation mechanism for shape and aggregation control of PANI nanoparticles was suggested [73]. PANI is known as being stable, promising, and of low cost so different synthetic routes have been developed like the one phase emulsion polymerization process [74], Triton Y-100 [75], potentiometric method [76], polymerization by different oxidizing agents like cupric sulphate [77], iron (III) and chloride [78]. PANI nanowires have also been potentiostatically deposited on a stainless steel electrode [79]. The polymerization process for PANI and poly (2-methoxy-aniline) (POMA) nanotubes formation was also investigated [94]. Electrically conducting hydrogel composites made of PANI nanoparticles and poly (Hvinyl- O-/m amino acetophenone of different compositions were synthesized by oxidative polymerization [80-85]. PANI powder with crystalline semi-conducting ZnS composite by mechanical mixing was also found to increase conductivity. Similarly, PANI/Co_3O_4 was synthesized to overcome the solubility problem of PANI. Lee et al. reported direct patterning of conducting water-soluble polymers. Other aspects of preparation methods and properties of polyaniline blends and composites with organic polymer have also been reviewed. Thus, being motivated by the significant and diversified applications of PANI and related conducting polymers, the selection of PVA and PAA as the hydrogel components of the polymer matrix lie in their unlimited applications in medicine, biology and technology.

7.5. Microwave Absorbing Foams Loaded Silica Nanoparticles As Electromagnetic Shields

The electromagnetic radiation interference is one of the by-products of the rapid proliferation of electronic devices [86]. The nano-structured materials have attraction for microwave radiation absorbing and shielding materials in the GHz frequency range due to their unique chemical and physical properties [87]. The volume to weight ratio of shielding material is very important in microwave absorbing materials for lightweight and strong absorption properties. The use of plastic materials for the housing of computer and electronic devices has grown very rapidly due to their advantages over metals, such as light weight, design flexibility, low cost and easy mass production[87-96]. The plastic casing of electronic equipment does not provide protection from external fields. Composites with discontinous conducting fillers, such as metal particles are extensively employed in electromagnetic interference (EMI) shielding [97]. The conducting polymers now offer great technological application potential in several areas [98] such as static films for transparent packaging of electronic components, electromagnetic shielding, rechargeable batteries, light – emitting diodes, nonlinear optical devices, sensors for medicine and pharmaceutics apparatus, membranes for separation of gas mixture, protection against corrosion, conducting paints and glues and others [99-106]. The most important application of these polymers is microwave adsorbing materials [107]. These polymers are generally prepared by adding fillers in a polymeric matrix. One very common way, among the several methods for preparing conducting polymer blends or composites, is by mechanical mixing of the components [108, 109]. Great interest has been focused on PANI and polyurethane resin (PU) matrices, within the field of conducting polymers, due to the important characteristics that it present. Its conductive form has excellent chemical stability combined with a relatively easy, inexpensive and high yield. These blends may combine the desired properties of two components, the electrical conductivity of PANI/ PU with the physical and mechanical properties of the polymeric matrix [110, 111]. The microwave absorption properties of doped PANI / PU are combined with SiO_2 [112]. Frequency conductivity and dielectric relaxation measurements have proven to be especially valuable in giving additional information on the conduction mechanism so that conductivity measurement alone is not proposed [113]. A new potential use SiO_2 has been proposed. The high dielectric constant SiO_2 has been used for the synthesis of PANI / PU / SiO_2 nanocomposites [114]. The morphology and conducting properties of doped PANI / PU / SiO_2 nanocomposites, synthesized by the chemical oxidation process. Scanning electron microscopy, conductivity and X-Ray measurements provide support to improve the system understanding [115, 116]. SEM of SiO_2 nanoparticles loaded PANI – camphor sulfonic acid (CSA) foams synthesized by chemical oxidation is shown in Figure 9.

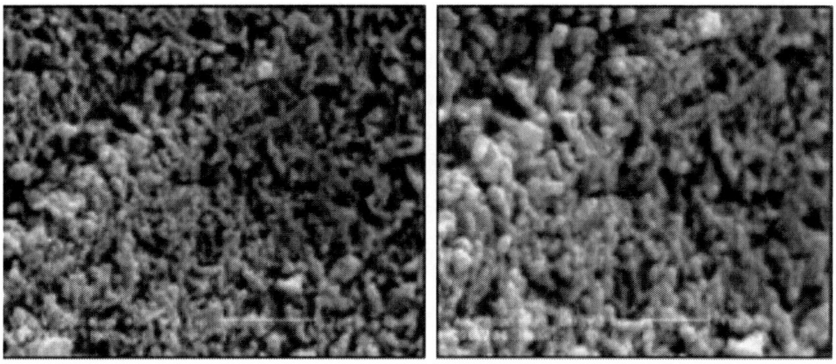

Figure 9. SEM of doped polyaniline with SiO_2 nanoparticles foams.

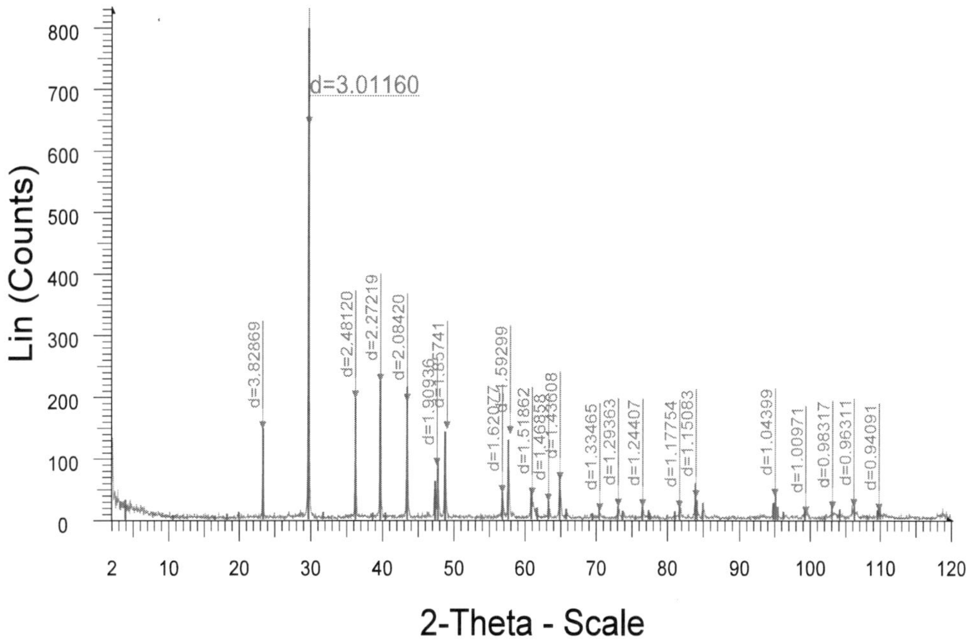

Figure 10. XRD pattern for conductive polyaniline foam with SiO_2 (15%) nanoparticles.

The SiO_2 nanoparticles loaded PANI - CSA foams are very sensitive to the temperature. The scanning electron micrography images were obtained from a diluted solution of the nanocomposite particles. The white spots are SiO_2 nanoparticles. These nanoparticles show a large variation in their dimensions.

Since the particles of SiO_2 are spherical in shape, the observed porosity in these composites is less than the other PANI composites. The crystallinity of the formed composites was followed with X-Ray diffraction (XRD) as a function of weight percent inorganic component. Studies on XRD patterns of PAni are scarce in the literature. Figure 10 shows the XRD pattern for SiO_2 nanoparticles loaded PANI - CSA foams (15%).

The diffraction pattern of SiO_2 nanoparticles loaded PAni - CSA foams show a peak at about $2\Theta = 29.89°$. In order to verify the influence of the doping process in PANI – CSA crystallinity, x-ray measurements were made. A semicrystalline structure of PANI - CSA is observed and does not change with the processing temperature suggesting that the morphology of the main polymeric chain is unaffected. However, concentration of CSA infer on crystallinity behavior as shown on Figure 10. An amorphous structure is obtained for PAni – CSA doped by reactive processing due to the crosslinking reaction in agreement with the torque and conductivity results.

Figure 11a shows the FT- IR spectrum of SiO_2 nanoparticles loaded PANI - CSA foam, where the % of transmittance is plotted as a function of wave number (cm^{-1}). The characteristic FT-IR peak at 1579 and 1423 cm^{-1} are due to the presence of quinoid and benzenoid rings, respectively and are clear indication of these two states in the polymer chain. The peaks at 1135 , 1249 , 1579cm^{-1} are due to the C-N bond stretching vibration, N=O, respectively.

Figure 11b shows the FT-IR spectrum of doped PANI with CSA in the presence of SiO_2 particles exhibit new absorption peaks distinctly at 1579,1423,1249,1135 and 773 cm^{-1} which are assignable to the presence of various metal oxides in the composite. The FT-IR spectra of other nanocomposites (PANI – CSA with 15% SiO_2) do not show much variation in the characteristic peaks.

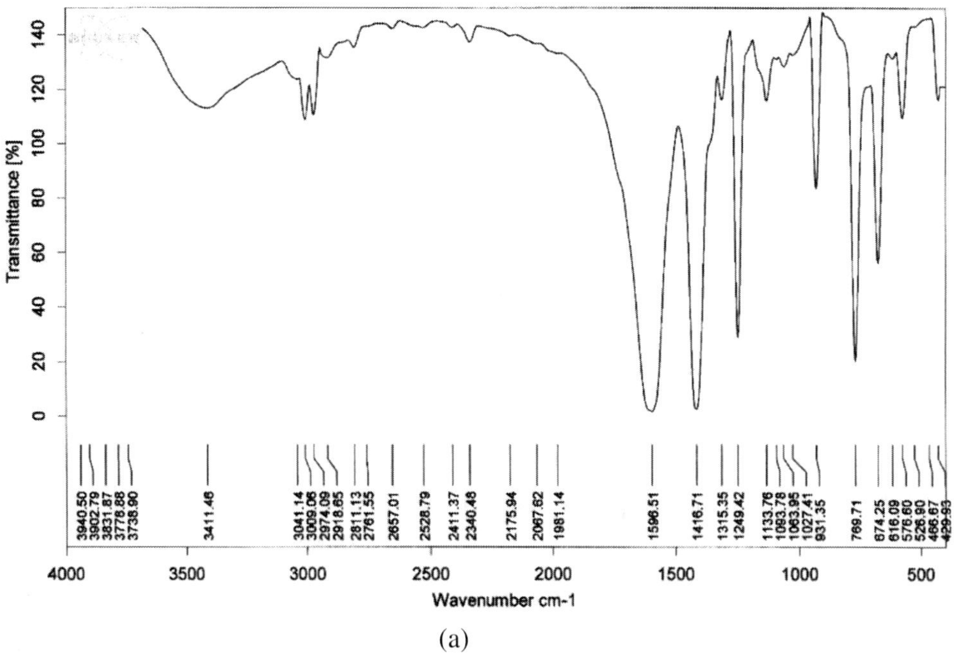

(a)

Figure 11. (Continued)

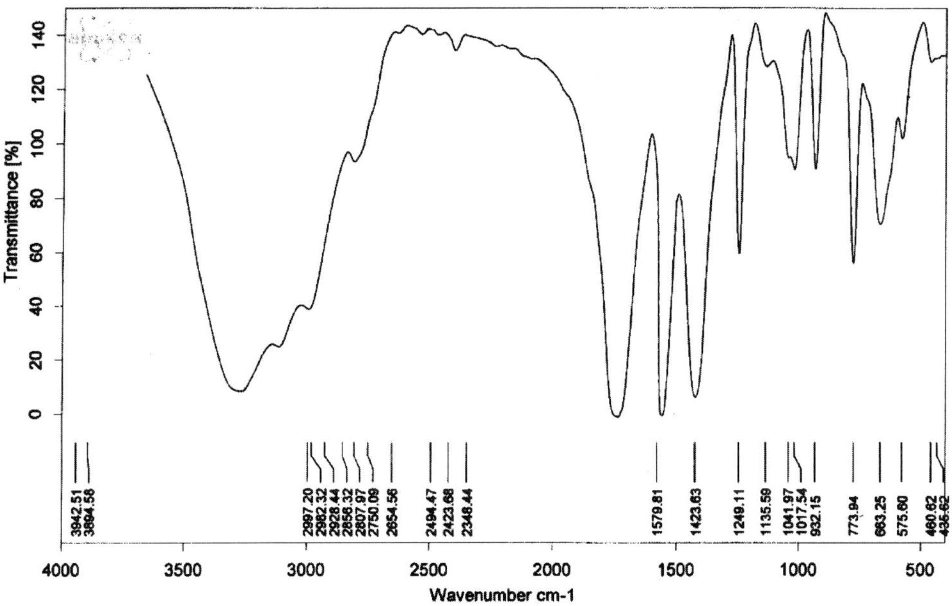

(b)

Figure 11. FT- IR spectrum of a) doped polyaniline, b) doped polyaniline in the presence of SiO2 particles.

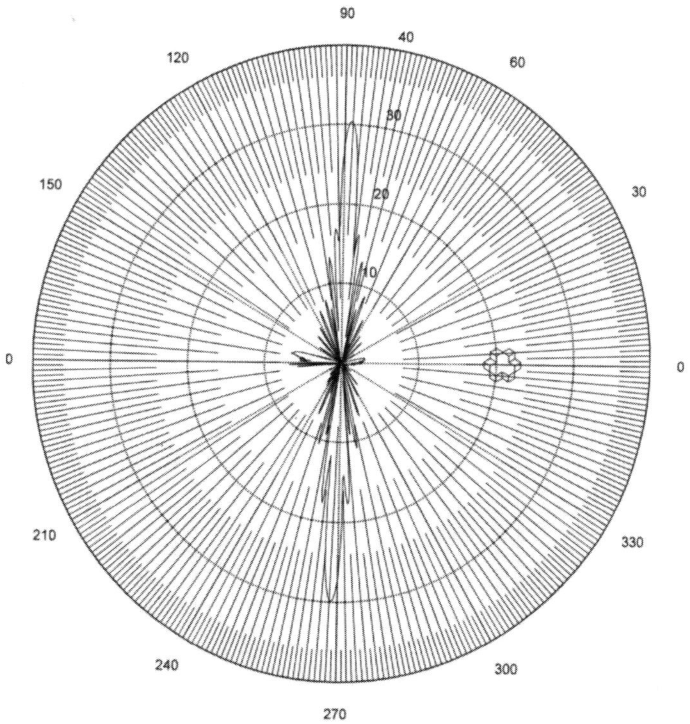

Figure 12. The RCS diagram of nanocomposite film on 200×200 mm Aluminium plate at 8-12GHz band.

The effect of SiO_2 particles content on the electrical conductivity of SiO_2 nanoparticles loaded PAni - CSA foam is very important. The frequency behavior (8-12 GHz) of all nanocomposites looks like a straight line typical of hopping conduction.

The absolute conductivity for individual samples increases as a function of frequency except for the nanocomposite with 15 % SiO_2 For the SiO_2 nanoparticles loaded PANI - CSA foam, the conductivity values change the order of magnitude.(10^{-5} S/cm). As the mole % of SiO_2 in PANI – CSA increases the absolute conductivity shifts to lower scales except for the 30% SiO_2 nanoparticles loaded PANI - CSA foam.

This shift is due to the electrical charges being displaced inside the nanopolymer and their lower concentration. The decrease in conductivity, by increase in the mole % of SiO_2 may be due to the particle blockage conduction path by SiO_2 (nano size particle) embedded in the PANI – CSA /PU foam.

The increase in mole% of SiO_2 leads to an increasing inter chain distance, which makes hopping between chains more difficult, resulting in the reduction of conductivity. At higher frequencies the conductivities of all the samples (with a different mole % of SiO_2) merge to show similar frequency dependence, indicating the formation of excess charge carriers.

The variation of the microwave absorption of SiO_2 nanocomposite in polyurethane resin matrix is over the frequency range of 8-12 GHz. It is interesting to note the absorption of nanocomposite over X band. Figure 12 shows the RCS diagram of nanocomposite film on 200×200 mm Aluminium plate at 8-12GHz band. Conducting coating based on PU/ PAni – CSA / SiO_2 blends were examined as microwave absorbing materials. Measurements of absorbed energy in the frequency range of 8 to 12 GHz were performed using the wave-guide method. The influence of the CSA concentration is clearly larger than the doping process.

REFERENCES

[1] Callister, W. D. (2005). *Fundamentals of materials science and engineering,* 2nd ed. USA: John Wiley & Sons Inc.

[2] Riande, E. & Diaz-Calleja, R. (2004). *Electrical properties of polymers.* New York: Marcel Dekker, Inc.

[3] Askeland, D. R. & Phule, P. P. (2006). *The science and engineering of materials*, 5th ed. USA: Cengage Learning.

[4] Min, G. (1999). Conducting polymers and their applications in the film industry – Polyaniline/polyamide blended films. *Synth. Met.*, 102(1–3), 1163–1166.

[5] Bhadra, S., Khastgir, D., Singha, N. & Lee, J. H. (2009). Progress in preparation, processing and applications of polyaniline. *Prog. Polym. Sci.*, 34(8), 783–810.

[6] Laska, J. & Pron, A. (1997). Conducting blends of polyaniline with conventional polymers. *Synth. Met.*, 84(1), 117–118.

[7] Jacobo, S. E., Aphesteguy, J. C., Anton, R. L., Schegoleva, N. N. & Kurtyandskaya, G. V. (2007). Influence of the preparation procedure on the properties of polyaniline based magnetic composite. *Eur. Polym. J.*, 43(4), 1333–1346.

[8] Jang, J. (2006). *Conducting polymer nanomaterials and their applications.* Seoul: Spinger-Verlag Berlin Heidelberg.

[9] Gurunathan, K., Amalnerker, D. P. & Trivedi, D. C. (2003). Synthesis and characterization of conducting polymer composite (PAn/TiO2) for cathode material in rechargeable battery. *Mater. Lett.*, 57(9), 1642–1648.

[10] Gok, A., Omastora, M. & Prokes, J. (2007). Synthesis and characterization of red mud/polyaniline composites: Electrical properties and thermal stability. *Eur. Polym. J.*, 43(6), 2471–2480.

[11] Suwanprateeb, J. (2000). Calcium carbonate filled polyethylene: Correlation of hardness and yield stress. *Composites Part A*, 31(4), 353–359.

[12] Durcova, O., Michlik, P. & Knotek, L. (1989). Evaluation of the calcium carbonate dispersion in polypropylene composite fibres. *Polym. Test.*, 8(4), 269–275.

[13] Tanniru, M. & Misra, R. D. K. (2005). On enhanced impact strength of calcium carbonate reinfored high-density polyethylene composites. *Mater. Sci. Eng. A,* 405(1–2), 178–193.

[14] Nowaczyk, G., Glowin Kowski, S. & Jurga, S. (2004). Rheological and NMR studies of polyethylene/ calcium carbonate composites. *Solid State Nucl. Magn. Reson.*, 25(1), 194–199.

[15] Wang, X., Tong, W., Li, W., Huang, H., Yang, J. & Li, G. (2006). Preparation and Properties of nanocomposite of poly (phenylene sulfide) / calcium carbonate. *Polym. Bull.*, 57(6), 953–962.

[16] Maeda, S., & Armes, S. P. (1995). Surface area measurements on conducting polymer-inorganic oxide nanocomposites. *Synth. Met.*, 73(2), 151–155.

[17] Fagerlund, G. (2006). Determination of specific surface by the BET method. *Mater. Struct.,* 6(3), 239–245.

[18] Smith, B. C. (1999). *Infrared spectral interpretation: A systematic approach*, 1st ed. USA: CRC Press LLC.

[19] Laska, J. (2004). Conformations of polyaniline in polymer blends. *J. Mol. Struct.*, 701(1–3), 13–18.

[20] Kulshreshtha, A. K. (2002). *Handbook of polymer blends and composites*, 1st ed., vol. 2. UK: Rapra Technology Limited.

[21] Supri, A. G. (2009). LDPE-isophthalic acid-modified eggshell powder composites, (LDPE/ESP). *J. Phy. Sci.*, 20(1), 87–98.

[22] Smith, W. F. & Iltashemi, J. (2006). *Foundation of materials science and engineering*, 4th ed. USA: McGraw Hill.

[23] Gercek, B., Yaruz, M., Yilmaz, H., Sari, B. & Unal, H. I. (2007). Comparison of electrorheological properties of some polyaniline derivatives. *Colloids Surf, A,* 299, 124–132.

[24] W.O. Haag, D.D. Whitehurst, German Patents 1800371 (1969).

[25] Biffis, B. Corain, Z. Cvengrošová, M. Hronec, K. Jeřábek, M. Králik, *Appl. Catal. A,* 124.

[26] (1995) 355; D. Gašparovičová, M. Králik, M. Hronec, Collect. *Czech. Chem. Commun.* 64 (1999) 502.

[27] Mitschker, R. Wagner, P.M. Lange, in: M. Guisnet et al. (eds.), *Heterogeneous Catalysis and Fine Chemicals*, Elsevier, Amsterdam, 1988, p. 61.

[28] H. Hirai, J. *Macromol. Sci.-Chem.* A13 (1979) 633; H. Hirai, N. Yakura, Y. Seta, S.

[29] Hodoshima, *Reactive and Functionnal polymers* 37 (1998) 121.

[30] F. Epron, F. Gauthard, C. Pineda, J. Barbier, *J. Catal.,* 198 (2001) 309.

[31] 29. F. Epron, F. Gauthard, J. Barbier, *J. Catal.,* 206 (2002) 363.
[32] Ashwell G J 1992 *Molecular electronics* (New York: John Wiley & Sons Inc.).
[33] Azaroff L V and Duerger M J 1958. *The powder method in Xray crystallography* (New York: McGraw-Hill).
[34] Benerjee P and Mandal B M 1995 *Synth. Metals* 47 257.
[35] Carmono F, Barreau F, Delhaes P and Canet R 1980 *J. Phys. Lett.* 41 L531
[36] Choi C H, Tracz A, Jeszka J K, Boiteux G, Seytre G, Ulanski J and Kryszewski M 1990 *Synth. Metals* 37 181.
[37] Foster R 1969 *Organic charge-transfer complexes* (New York: Academic Press)
[38] Jeszka J K and Tracz A 1992 *Polym. Adv. Technol.* 3 139.
[39] Jeszka J K, Ulanski J and Kryszewski M 1981 *Nature* 289 390.
[40] Kang T J, Miyaki Y, Han J H, Motobe T, Whang Y E and Miyata S 1994 *Proc. Pacific Polym. Conf.* (New York: Springer- Verlag) 3rd ed., p. 307.
[41] Kryszewski M and Jeszka J K 1998 *Synth. Metals* 94 99.
[42] Manoharan S P, Birlasekarn S and Suryanarayana C V 1986 *Bull. Electrochem.* 2 509.
[43] Margolis J M (ed) 1989 *Conducting polymers and plastics* (New York: Chapman & Hall).
[44] Omartova M, Kosina S, Pionteck J, Janke A and Pavlince J 1996 *Synth. Metals* 81 49.
[45] Singh R and Singh R A 1997 *Mol. Mater.* 8 187.
[46] Singh R A and Srivastava D N 1999 *Mol. Mater.* 11 223.
[47] Singh R A, Singh R and Verma S M 1991 *Bull. Chem. Soc. Jpn* 64 1938.
[48] X. Ren, T. E. Springer, T. A. Zawodzinski, S. Gottesfeld, *J. Electrochem. Soc.*, 174, 466 (2000).
[49] Y. Yin, J. Fang, Y. Cui, K. Tanaka, H. Kita, K. Okamoto, *Polymer*, 44, 4509 (2003).
[50] T. Yamaguchi, F. Miyata, S. Nakao, *Adv. Mater.,* 15, 1198 (2003).
[51] C. Hasiotis, V. Deimede, C. Kontoyannis, *Electrochimica Acta*, 46 (2001) 2401-2406.
[52] F. Lufrano, G. Squadrito, A. Patti, E. Passalacqua, *J.Appl. Poly. Sci.*, 77 (2000) 1250-1256.
[53] J. F. Blanco, Q. T. Nguyen, P. Schaetzel, *J. Appl. Poly. Sci.*, 84 (2002) 2461-2473.
[54] V. E. Gul', *Structure and Properties of Conducting Polymer Composites,* VSP, Utrecht, 1996.
[55] X. S. Yi, G. Wu, D. Ma, J. *Appl. Polym. Sci.,* 1998, *67*, 131.
[56] D. Stauffer, *Phys. Rep.,* 1979, *54*, 1.
[57] R. Tchoudakov, O. Breuer, M. Narkis, A. Siegmann, *Polym. Eng. Sci.,* 1996, *36*, 1336.
[58] H. Stephen, J. Polym. Sci., Part B: *Polym. Phys.,* 1999, *37*, 1899.
[59] G. Yu, M. Q. Zhang, H. M. Zeng, *J. Appl. Polym. Sci.,* 1998, *70*, 559.
[60] J. F. Zhang, Q. Zheng, Y. Q. Yang, X. S.Yi, *J. Appl. Polym. Sci.,* 2002, *83*, 3112.
[61] C. G. Jaime, W. G. William, F. F. Lorraine, *J. Appl. Polym. Sci.,* 2001, *80*, 692.
[62] Kajiwara K., Ross-Murphy S. B.: Synthetic gels on the move. *Nature*, 355, 208–209 (1992).
[63] Kurkuri M. D., Lee J-R., Han J. H., Lee I.: Electroactive behaviour of poly(acrylic acid) grafted poly(vinyl alcohol) samples, their synthesis using a Ce(IV) glucose redox system and their characterization. *Smart Materials and Structure*, 13, 417–423 (2006).
[64] Ma J. T., Liu L. R., Yang X. J., Yao K. D.: Bending behaviour of gelatin/poly(hydroxyethyl methacrylate) IPN hydrogel under electric stimulus. *Journal of Applied Polymer Science*, 56, 73–77 (1995).

[65] Blanchet G. B., Fincher C. R., Lefenfeld M., Roger J. A.: Contact resistance in organic thin film transistors. *Applied Physics Letters*, 84, 296–298 (2004).

[66] Tittu M., Hiekkataipale P., Kainen J. H., Makela T. Ikkala O.: Viscoelastic and electrical transitions in gelation of electrically conducting polyaniline. *Macromolecules,* 35, 5212–5217 (2002).

[67] Akbayir C., Bulut F., Farrell T., Goldschmidt A., Günther R., Kam A. P., Miclea P., Scherf U., Seekamp J., Solovyev V. G., Sotomayor Torres C. M.: Nanostructured conjugated polymeric system for photovoltaic applications. *Reviews on Advanced Materials Science,* 5, 205–210 (2003).

[68] Tiitu M., Talo A., Forsén O., Ikkala O.: Aminic epoxy resin hardeners as reactive solvents for conjugated polymers: Polyaniline base/epoxy composites for anticorrosion coatings. Elsevier Science (2005). *Polymer,* 46, 6855–6861 (2005).

[69] Li M., Guo Y., Wei Y., MacDiarmid A. G., Lelkes P. I.: Electrospinning polyaniline-contained gelatin nanofibers for tissue engineering applications. *Biomaterials,* 27, 2705–2715 (2006).

[70] Kim D-H., Abidian M., Martin D. C.: Conducting polymers grown in hydrogel scaffolds coated on neural prosthetic devices. *Journal Of Biomedical Materials Research, Part A*, 71, 577–585 (2004).

[71] Qiang Y., Antony J., Sharma A., Nutting J., Meyer D.: Iron/iron oxide core shell nano clusters for biomedical application. *Journal of Nanoparticles Research*, 8, 489–496 (2005).

[72] Schmid A. L., Torresi. S. I. C., Bassetto A. N., Carlos I. A.: Structural, morphological and spectroelectrochemical characterization of poly(2-ethyl aniline). *Journal of the Brazil Chemical Society,* 11, 317–323 (2000).

[73] Ansari R., Keivani M. B.: Polyaniline conducting electroactive polymers, thermal and environmental stability studies. *Journal of Chemistry,* 3, 202–217 (2006).

[74] Ryu K. S., Chang S. H., Kang S-G., Oh E. J., Yo C. H..: Physicochemical and electrical characterization of polyaniline induced by crosslinking, stretching and doping. *Bulletin of the Korean Chemical Society*, 20, 333–336 (1999).

[75] Jang S-Y., Seshadri V., Khil M-S., Kumar A., Marquez M., Mather P. T., Sotzing G. A.: Welded electrochromic conductive polymers nanofibres by electrostatic spinning. *Advanced Material*, 17, 2177–2180 (2005).

[76] Deore B. A., Yu I., Anguiar P. M., Siedler C. R., Freund M. S.: Highly crosslinked, self doped polyaniline exhibiting unprecedented hardness. *Chemical Materials*, 17, 3803–3805 (2005).

[77] Huang J.: Synthesis and applications of conducting polymer polyaniline nanofibers. *Pure and Applied Chemistry International*, 78, 15–27 (2006).

[78] Tang B. Z., Geng Y., Sun Q., Zhang X. X., Jing X.: Processible nanomaterials with high conductivity and magnetizability: Preparation and properties of magnemite/ poly aniline nanocomposites film. *Pure Application Chemistry,* 72, 157–162 (2000).

[79] Hussain A. M. P., Kumar A.: Electrochemical synthesis and characterization of chloride doped polyaniline. *Bulletin of Material Science*, 26, 329–334 (2003).

[80] Ibrahim M., Koglin E.: Spectroscopic study of polyaniline emeraldine base: Modelling approach. *Acta Chimica Slovenica*, 52, 159–163 (2005).

[81] Liu G., Freund M. S.: New approach for the controlled crosslinking of polyaniline: Synthesis and characterization. *Macromolecules,* 30, 5660–5665 (1997).

[82] Oh K. W., Kim S. H., Kim E. A.: Improved surface characteristic and the conductivity of polyanilinenylon fabrics by plasma treatment. *Journal of Applied Polymer Science*, 81, 684–694 (2001).

[83] Wang Y., Jing X.: Intrinsically conducting polymers for electromagnetic interference shielding. *Polymer for Advanced Technologies*, 16, 344–351 (2005).

[84] da Silva P. J. E., Temperini M. L. A., Torresi S. I. C.: Characterization of conducting polyaniline blends by resonance. *Journal of Brazil Chemical Society*, 16, 322–327 (2005).

[85] Plesu N., Hiescu S., Ilia G., Popa A., Muntean C.: Rheology of polyaniline aispersims in acrylic resin. *Turkey Journal of Chemistry*, 30, 155–163 (2006).

[86] Draman S. F. S., Daik R., Ahmad M.: Fluorescence characteristics of sulfonated polyaniline solution in DMF when exposed to oxygen gas. *Malaysian Journal of Chemistry*, 8, 37–44 (2006).

[87] Steyskal J., Gilberg, K. G.: Polyaniline, preparation of a conducting polymer (IUPAC Technical Report). *Pure and Applied Chemistry*, 74, 857–867 (2002).

[88] Vilènik M., igon M., Zupan M., Šebenik A.: Influence of polymerization parameters in the molecular weight of polyaniline. *Acta Chimica Slovenica*, 45, 173–183 (1998).

[89] Li D., Kaner R. B.: Shape and aggregation control of nanoparticles: Not shaken not stirred. *Journal of American Chemical Society*, 128, 968–975 (2006).

[90] Pérez R., Pinto N. J.: Electrical characterization of polyaniline synthesized using one phase emulsion polymerization process. in '*Proceeding of the National Conference on Undergraduate Research (NCUR) 2004,* Indianapolis, USA' 15–17 (2004).

[91] Arsalani N., Khavei M., Entezami A. A.: Synthesis and characterization of novel-N-substituted polyaniline by Triton X-100. *Iranian Polymer Journal*, 12, 237–242 (2003).

[92] Gaikwad P. D., Shirale D. J., Gade V. K., Savale P. A., Kharat H. J., Kakde K. P., Hussaini S. S., Dhumane N. R., Shirsat M. D.: Synthesis of H2SO4 doped polyaniline film by potentiometric method. *Bulletin of Material Science*, 29, 169–172 (2006).

[93] Mallick K., Witcomb M. J., Binsmore A., Scurrell M. S.: Polymerization of aniline by cupric sulfate: A facile synthetic route for producing polyaniline. *Journal of Polymer Research,* 13, 397–401 (2006).

[94] Nestorovic G. D., Jeremic K. B., Jovanovic S. M..: Kinetics of aniline polymerization initiated with Iron III chloride. *Journal of Serbian Chemical Society*, 71, 895–904 (2006).

[95] Gupta V., Miura N.: Electrochemically deposited polyanilines network: A high performance electrode materials for redox super capacitors. *Electrochemical and Solid State Letters*, 8, A630–A632 (2005).

[96] Mazur M., Tagowska M., Palys B., Jackowska K.: Template synthesis of poly aniline and poly(2- methoxy aniline) nanotubes: Comparison of the formation mechanism. *Electrochemistry Communications*, 5, 403–407 (2003).

[97] Dispenza C., Lo Presti C., Belfione C., Spadaro G., Piazza S.: Electrically conductive hydrogel composites made of polyaniline nanoparticles and poly(Nvinyl- 2-pyrrolidone). *Polymer,* 47, 961–971 (2006).

[98] Savitha P., Swapna Rao P., Sathyanarayana D. N.: Highly conductive new aniline copolymers: poly(aniline- co-amino acetophenone)s. *Polymer International*, 54, 1243–1250 (2005).

[99] B.D. Ratner; "*Comprehensive Polymer Science – The Synthesis, Characterisation, Reactions & Applications of Polymers*", Aggarwal, S. K., ed., vol 7, Pergamon Press, Oxford; (1989); 201-247.
[100] M.R. Saboktakin, A. Maharramov, M.A. Ramazanov;" Synthesis and characterization of hybride polyaniline / polymethacrylic acid/ Fe_3O_4 nanocomposites"; *Nature and Science*, 5(3) ; (2007); 67-71.
[101] M.R. Saboktakin , A. Maharramov ,M.A. Ramazanov ;" The synthesis and properties of Fe_3O_4 / Sodium acetate / CMS ternary nanocomposites as electrorheological fluid" ; *Journal of American Science*,3(4) ; (2007); 30-34.
[102] M.R. Saboktakin , A. Maharramov ,M.A. Ramazanov;" Synthesis and characterization of aromatic polyether dendrimer / Mesalamine (5-ASA) nanocomposite as drug carrier system"; *Journal of American Science*; 3(4) ;(2007);40-45.
[103] W. Li; S. Tutton; AT. Vu ; et al.; "First-Pass Contrast-Enhanced Magnetic Resonance Angiography in Humans Using Ferumoxytol, a Novel Ultrasmall Superparamagnetic Iron Oxide (USPIO)-Based Blood Pool Agent"; *Journal of Magnetic Resonance Imaging* ; 21 ; (2005) ; 46-52.
[104] D. Simberg; T. Duza; Ji Hu Park; et al.; " Biomimetic amplification of nanoparticle homing to tumors"*; PNAS* ;104(3); (2007);932-936.
[105] Zh. Tang; N. Alvarez; SZe. Yang; "Organic /Inorganic Hybride Material for Coating on Metals"; *Material Research Society, Symp.Proc.* ;734 ;(2003); 57.1-57.8.
[106] C.C. Berry; S.G. Curtis;" Functionalisation of magnetic nanoparticles for applications in biomedicine"; *Journal of Physics D: Applied Phys*ics; 36;(2003);198-206.
[107] R.; David; M.; Anne Wallace; K. Carl; et al.;" A synthetic Macromolecule for sentinel Node Detection"; *Journal of nuclear medicine*; 42(6);(2001);951-958.
[108] N. Hildebrandt; D. Hermsdorf; A. Signorell;et al.;"Superparamagnetic iron oxide nanoparticles functionalized with peptides by electrostatic interactions"; (2007); *ARKIVOC*,79.
[109] M. Andrzej; P. Thommey, J. Baker, et al; "Dendrimer- Based Targeted Delivery of an Apoptotic Sensor in Cancer Cells", *Biomacromolecules*; 8;(2007);13-18.
[110] S.C. Raghavendra; S. Khasim; M.V.N. Revanasiddappa, et al; "Synthesis, characterization and low frequency a.c. conduction of polyaniline/fly ash composites"; *Journal of Bulletin Material Science*;26(7); (2003); 733-739.
[111] H.S. Nalwa; " Conductive polymers: synthesis and electrical properties in handbook of organic conductive molecules and polymers"; V.2, *John Wiley & Sons; New York* (1997).
[112] P. Hourquebie; B. Blondel and S. Dhume;" Microwave an optical properties properties of soluble conducting polymers"; *Journal of Synthetic Metals*, 85; (1997);1437-1438.
[113] R. Faez and M. De Paoli; "Elastic polyaniline with EPDM and Dodecylbenzene sulfonic acid as plasticizers, *Journal of Applied Polymer Science*; 82; (2001); 1768-1775.
[114] W.A. Gazotti; M. De Paoli; "High yield preparation of a soluble polyaniline derivative"; *Journal of Synthetic Metals*; 80; (1996) ; 263-269.
[115] W. Luzny ; E.J. Samuelsen ; D. Djurado ; Y.F. Nicolau ; " Polyaniline protonated with camphor sulfonic acid : modeling of its crystalline structure " ; *Journal of Synthetic Metals* ; 90 ; (1997) ; 19-33.

[116] Y. Cao; P. Smith and A.J. Heeger; "Counter-ion inducted processability of conducting polyaniline"; *Journal of Synthetic Metals*; 55-57;(1993);3514-3519.
[117] R. Faez and M. De Paoli; "A conductive rubber based on EPDM and polyaniline: I. Doping method effect"; *European Polymer Journal*; 37; (2001); 1139-1143.
[118] Y. Cao; P. Smith and A.J. Heeger; "Counter-ion inducted processability of conducting polyaniline and of conducting polyblends of polyaniline in bulk polymers, *Journal of Synthetic Metals*;48 ; (1992) ; 91-97.

Chapter 8

METALLIC NANOCOMPOSITES

8.1. INTRODUCTION

Plasmon resonances of metallic nanoparticles have been of historical interest, first quantified by Mie in the case of spherical particles [1]. They are currently the subject of great interest due to the multitude of applications enabled by locally enhanced electric fields and the nonlinear optical phenomena that they mediate. The optical properties of nanoparticles, summarized by their extinction spectra, depend on size, shape, dielectric medium, and interfacial structure. Beyond the Mie theory, numerical solutions of Maxwell's equations are used to compute the spectra of arbitrary structures. The success of such treatments has been validated in light-scattering measurements on single particles that are simultaneously characterized through microscopy. Time-resolved measurements, although carried out on ensembles of particles, provide the dynamical connection between structure and optical response, and can lead to a more detailed characterization of properties. Ultrafast pump-probe measurements have been used to monitor the damping of surface plasmons in real-time. It is well established that the damping rate of the collective electron motion, Γ, (the sum of dephasing and dissipation) determines the width of the spectral response. Accordingly, the extinction spectra, $\sigma(\omega,t)$, can be approximated as Lorentzian, defined by the two moments: line-center, Ω, and line-width, Γ. It is also well established that the dissipation of plasmons through electron-phonon scattering is completed on the time scale of 10-12 seconds. The optical excitation of nanoparticles with subpicosecond pulses leads to a sudden jump in lattice temperature and drives the particles into shape oscillations [2]. This in turn leads to the modulation of Plasmon resonances on time scales of 1 ps $< t <$ 1 ns, as observed in transient-scattering spectroscopy. The period of modulation and its damping rate can be gainfully employed to extract mechanical and thermal properties of particles, as demonstrated on gold spheres and nanorods, silver spheres, ellipsoids, prisms, and triangular nanoplates. The only assumption necessary for such analyses is for the resonance, Ω, to be modulated by the mechanical vibrations of the particle. The microscopic connection is suggested by noting that because the free plasmon frequency, $\omega p = (ne^2/\varepsilon_0 me)^{1/2}$, is determined by the electron density, then volume oscillations should modulate the resonance frequency, $\delta\omega p/\omega p = -\delta V/2V$. This is clearly an oversimplification of the dynamics; it ignores the dielectric response of the medium and does not leave room for the interpretation of modulations in Γ, which is observed in most reported measurements. Spectrally resolved transient scattering provides two-dimensional (2D) images of $\sigma(\omega, t)$, mapping the dynamical correlation between Ω and Γ. The results indicate that the spectral response contains a component that tracks the particle strain rate. The effect of the dynamical response of the interfacial double-layer is present in all colloidal preparations and estimates the magnitude of the effect to be in accord with the observation. Some of the implications of this novel finding are addressed [3].

A great deal of effort has been devoted to the development of platinum-ruthenium alloy electrocatalysts for the oxidation of methanol in fuel cells. More recent efforts have focused on improving the so-called CO tolerance of these catalysts, *i.e.*, their activity in the oxidation of hydrogen that is obtained by reforming methanol, which inevitably contains a certain concentration of carbon monoxide. Despite considerable advances in the development of these electrocatalysts, their activity and CO tolerance are still unsatisfactory. This issue is one of the two major obstacles for a large-scale application of the polymer electrolyte membrane fuel cell (PEMFC), which seems to be the most promising energy system for electric vehicles and for on-site power generation. The second obstacle is the price of platinum whose loading is too high even in the best available electrocatalysts [4].

A new approach is described for obtaining electrocatalysts that can alleviate both of the above impediments. The new method is based on the spontaneous deposition of Pt on Ru that for Pt 2 and Pd 3 deposition on a Ru(0001) single crystal surface, which involves a reduction of H_2PtCl_6 coupled with the oxide formation on Ru. A selective Pt deposition on Ru ~no deposition on carbon! is attainable without the application of an external potential. The preparation of bimetallic catalysts by surface modification of metal substrates has attracted considerable attention in catalysis and electrocatalysis. Most studies involve alloy formation and modifications of Pt surfaces, which include underpotential deposition (UPD) of metal adlayers, the so-called spontaneous or irreversible adsorption of non-noble metal cations, and spontaneous deposition of noble metals adlayers. The latter type of modification requires a subsequent application of potential and yields small coverages, as recently shown by Wieckowski and co-workers for the deposition of Ru adlayers on Pt (111) and Pt (100) surfaces. The present deposition method can produce monolayer-to-multilayer coverages, of Pt 2 and Pd on Ru single crystals. If extended to Ru nanoparticles, it would open up an important possibility for tailoring Pt-Ru catalysts with an atomic level control [5].

The Pt-Ru bimetallic alloys are the best available electrocatalysts for the oxidations of reformate hydrogen. It is generally believed that those catalysts operate through a bifunctional mechanism, although the electronic effects cannot be ignored. Spontaneous deposition of Pt on Ru nanoparticles can be used to control the Pt cluster size and to tune the electronic and catalytic properties of Pt-Ru catalysts. In addition, this approach facilitates a considerable reduction of Pt loadings by depositing Pt at the surface of Ru nanoparticles rather than having Pt throughout the Pt-Ru nanoparticles.

8.2. Properties and Structure Formation of Cadmium Sulfide Nanocomposites with Polypropylene

The research of photoluminescent polymeric nanocomposites (PNC) is a question of great scientific and practical importance for the understanding of energy transfer mechanism and transport carriers in multiphase polymeric systems. Luminescent screens, transducers, sensors and other facilities with improved physical-chemical characteristics and broadened intervals of phosphorescence in visible region of spectra are created on a PNC basis. Development and preparation of new photoactive nanocomposites is closely connected with the understanding of the interrelation of the "structure-technology properties application" of these materials. The creation of polymeric nanocomposites, composed of two or more phases, challenge the studies of physical and chemical processes of modification structures and properties of new active elements. The polymer matrix provides the high physical-mechanical properties of PNC (flexibility, machinability) and active fillers including the high sensitivity and sufficient photoluminescent properties. It should be noted that by variation of the components' properties, it is possible to change the properties of nanocomposites, to study the interface effects, to research intermolecular transfer processes and migration of electron excitation in polymeric mediums and the influence of interface interactions on photoluminescent properties of filler. The research of polymeric nanocomposites structures with inorganic semi-conductors (sulfides, transition metals oxides) is of interest in terms of the new electrochemical and photochemical characteristics [5].

The structure and properties of these composites associate with each other. The data of structure (size and character of distribution of the particles of dispersed phase in the matrix of polymer) allows forecasting the properties of this system and the investigation of the properties allows the forecasting of the structure of nanocomposites. It is clear that there is a strong change in the IR spectra, especially in the wavelength region 2846 cm^{-1}, 1456-1186 cm^{-1}.

Depending on the radiation dose, the increasing of absorption band strength in wavelength region 2950 cm^{-1} and 2839 cm^{-1} was observed, resulting from the activation of CH valence vibrations in a spectra of polypropylene strength of CH valence and deformation vibrations and vibrations mutual influence of CH and CH_2 groups were observed [6].

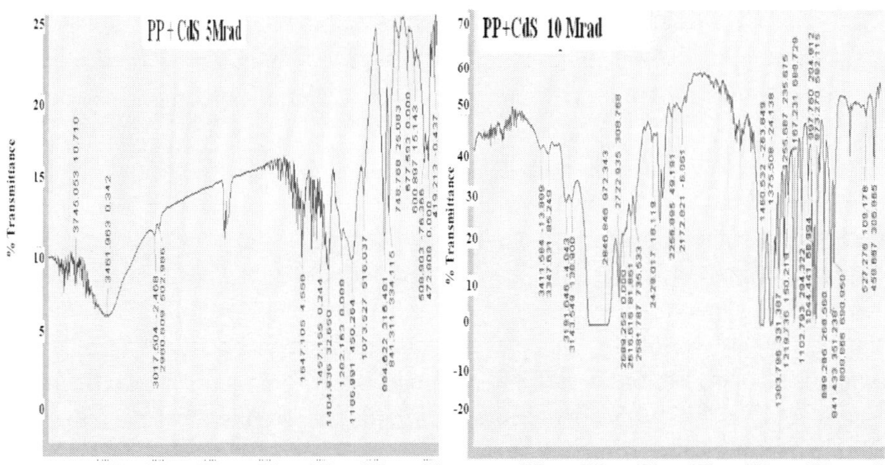

Figure 1. IR spectra of PP and nanocomposite samples, treated by γ-radiation (5+10 Mrad).

These were scanned by the atomic-force microscope (AFM) and the relief of nanocomposites PP + CdS samples, obtained from PP powder, was treated by γ-radiation

(10 Mrad). The AFM investigation of the relief of nanocomposites PP + CdS samples testify to their increasing numbers. These reactions lead to the formation of different oxygen containing and trans-vinyl groups and deep trapping sites (up to 1.2 electron volt), dielectrical cavities and free quasi-stable radicals. The ability of a polymeric matrix to form complexes increases with γ-radiation treatment, i.e., the majority of dispersed component forms around the oxidation centers in polymer. The AFM-scanning of PP + CdS samples shows that the sizes of CdS nanoparticles do not change with a change of radiation dose and the concentration of CdS nanoparticles in the polymeric matrix increases. The further increasing of the radiation dose leads to polymer structure decomposition. The concentration change of CdS in PP with γ-radiation treatment is seemingly correlated with the forming of oxidizing centers in the polymer, which are the nuclease center for CdS.

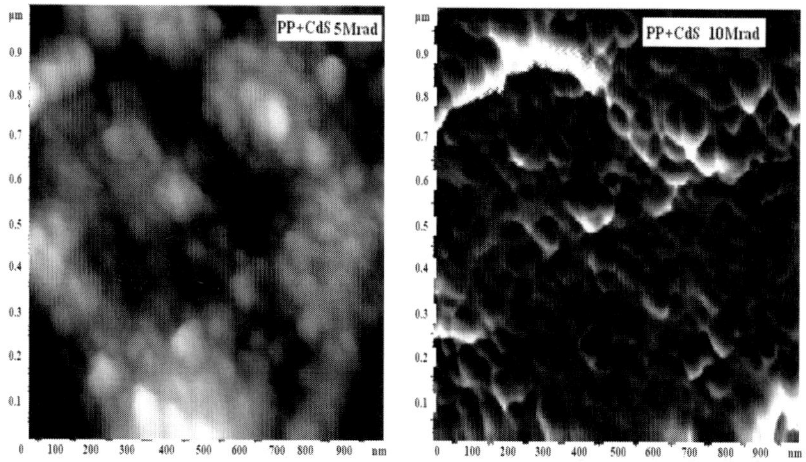

Figure 2. 2D images of PP+CdS nanocomposites samples treated by γ-radiation (5mrad and 10 mrad).

It was found that the CdS nanoparticles size is 25-35 nm and does not depend on the γ-radiation dose. The photoluminiscent spectra of PP + CdS nanocomposites treated and untreated by various doses of γ- radiation are presented in Figure 3. There are two maximum peaks observed at the wavelength interval 534-560 nm and the maximum band width changes by the changing of the γ-radiation dose at the extremum of 10 Mrad. The intensity of photoluminescence of the above mentioned peaks increases at the 10 Mrad γ-radiation dose and then begin to decline. It is most likely connected with the decomposition processes of the nanocomposite in the polymeric matrix. The photoluminiscent spectra and AFM-scanning shows that the dimensions of the nanoparticles do not depend on the dose of γ-radiation treatment, and the concentration of CdS nanoparticles in the polymeric matrix increase with γ- radiation treatment. The further increasing of γ-radiation probably destroys the chemical structure of the polymer [7].

The concentration change of CdS in PP with changing γ- radiation dose is correlated with the experimental results, getting from the photoluminescent spectra and evidently is connected with the forming of oxidizing centers in the polymer, which are the nuclease center for CdS formation [8].

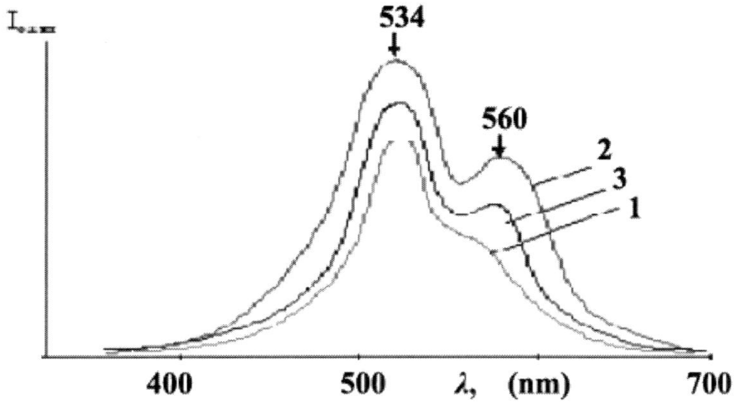

Figure 3. Photoluminescence spectra of PP+CdS and nanocomposite samples treated by various doses of γ-radiation 1) 5 mrad, 2) 10 Mrad, 3) 15 Mrad.

8.3. DERIVATOGRAPHIC ANALYSIS OF PE POLYMER NANOCOMPOUNDS WITH $Co(AlO_2)_2$ FILLER

It is known that the use of different methods of polymer material modification leads to pronounced development of the field of their application. In this case, directional change of structure and polymer properties are carried out either during the synthesis process or by the effect on the finished product with the introduction of different chemical character fragments into the macromolecule. Recently, when creating polymer materials with given properties, special attention was given to modification of their surface as the structure of the surface layer determines their behavior under working conditions in many ways. The modification of electret and electrophysical properties of two – phase systems of polymers with different fillers is of greater importance. In polymers with strong electric fields there have been ageing processes covering deeper layers of polymer. As a result, there have been changes in the structural part of the polymer related to the changes in the crystal and amorphous part of the polymer, as well as with the formation of the interlaminar crystal phase in the polymer matrix etc. The study of polymer composite material thermal properties provides the scope for evaluating the quality of filled material and working out recommendations on their use and production technology [9].

In Figure 4, PE derivatographs are given. As it is seen on the DTA curve at E=0, two endothermal effects at T105°C and T405°C and two exothermal effects at T187 °C and T327°C have been found. The endothermal effect at T105°C is in agreement with polymer matrix crystal phase melting. The exothermal effect within T155°-255 °C with the peak at T187 °C is in agreement with the formation of hydroperoxide groups, i.e., the first stage of oxidation.

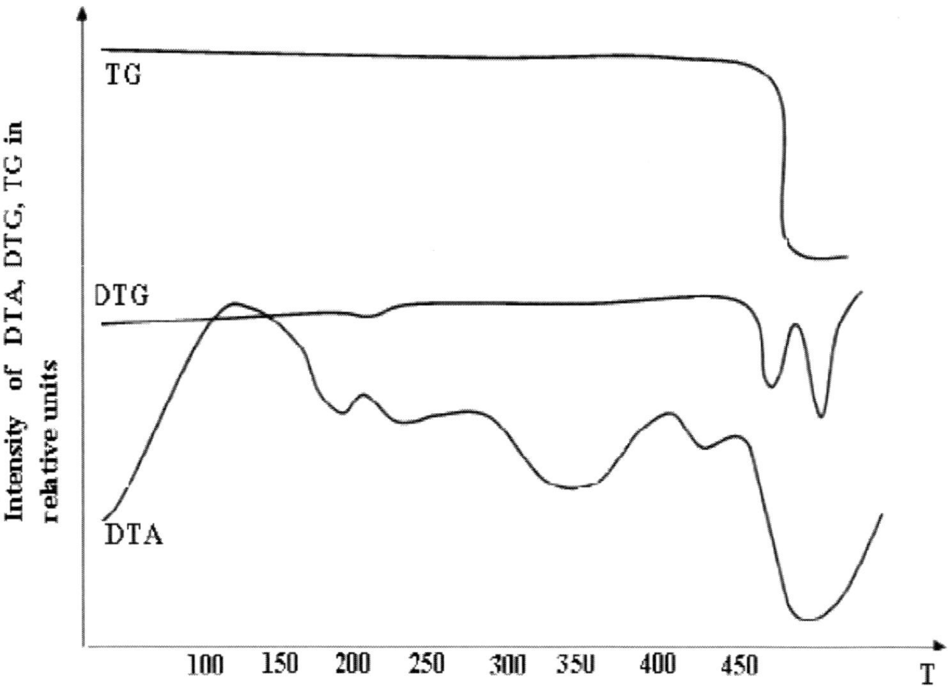

Figure 4. Derivatograph of PE films obtained under water quenching conditions without treatment in constant electric field.

The second blurred exothermal effect at T327 °C is in agreement with the breaking of weak bonds. The vigorous endothermal effect at T405 °C is in agreement with polymer depolymerization [10].

In Table 1, the results of PE thermogravimetric and differential-thermal analyses filled by Co $(AlO_2)_2$ in amounts of 0.5 vol. % and 1 vol. % are shown. On the DTG curve, the process of weak bond break and depolymerization appears as the endothermal effects at T400 °C and T430 °C are accompanied by the loss of mass on the TG curve in the amount of 15.6%. The balance of the PE unburnt mass by TG calculation is 84.4% [11]. On the PE derivatograph (Figure 5) exposed to electrothermopolarization at E=7×106V/m the melting temperature of the polymer matrix crystal phase shifts to low temperatures by 5°C, i.e., is equal to 100°C. The first stage of oxidation occurs at T240°C. The onset of weak bond break on both curves DTA and DTG fits T272°C. The peak on curve DTG is observed at T290 °C but on curve DTA it is observed at T305 °C. The process of weak bond break is followed by mass volatility in amounts of 20% but the balance of unburnt mass is 80%. From the formation of volatile components of TG and DTG curve course calculation it follows that by applying electric power E=7×106V/m, polymer thermodestruction comparing with PE at E=0 is more than 4%.

By the addition of PE+0.5%Co(AlO2)2, (E=0) three endothermal effects on DTA curve have been revealed at T120 °C, T210 °C and T270 °C (Figure 6). The endothermal effect at T120°C is due to the melting of the polymer matrix crystal phase at T210 °C the first stage of polymer oxidation and at 270 °C polymer depolymerization has taken place.

Table 1. TG and GTA analyses of polyethylene films

Composition on PE-based compounds	Differential-thermal analysis - DTA				Differential-thermo gravimeter analysis - DTG, TG		
	Meling temperature °C of crystal phase	I[st] stage of oxidation T°C	Break of wear bonds T°C	Depolymerization T °C	Endothermal effect-DTG, T°C	Amount of volatile component by TG in %	Balance as a remit of depolymerization by heating up to 360-380°C, in %
PE, E=0	105	187	380	405	400,430	15,6	84,4
PE, E=7×0^6 V/m	–	–	240	305	290	20	80
PE+0,5% Co(AlO_2)$_2$, E=0	120	190	240	290	240	17	83
PE+0.5% Co(AlO_2)$_2$, E=7×10^6 V/m	–	170	205	230	230	30	40
PE+1% Co(AlO_2)$_2$, E=0	70	90	188	290	280	23	77
ПЭ+1% Co(AlO_2)$_2$, E=7×10^6 V/m	70	140	195	230	230	33	67

These processes on the DTG curve appear as one endothermal effect at T260 °C accompanied along the TG curve by the volatility of volatile components in amounts of 17%.

As a result of polymer depolymerization, the balance of polymer unburnt part by heating up to T380 °C is 83%. It should be noted that Co(AlO2)2 in filler amounts of 0.5% increases the melting temperature of the polymer matrix crystal phase by 15 °C but the temperature of the 1st oxidation stage by 3 °C compared with the initial PE[12-15].

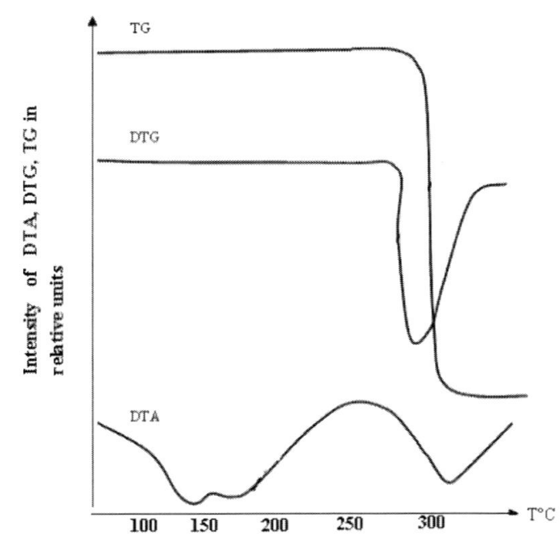

Figure 5. Derivatograph of PE films obtained under water quenching conditions treated in constant electric field.

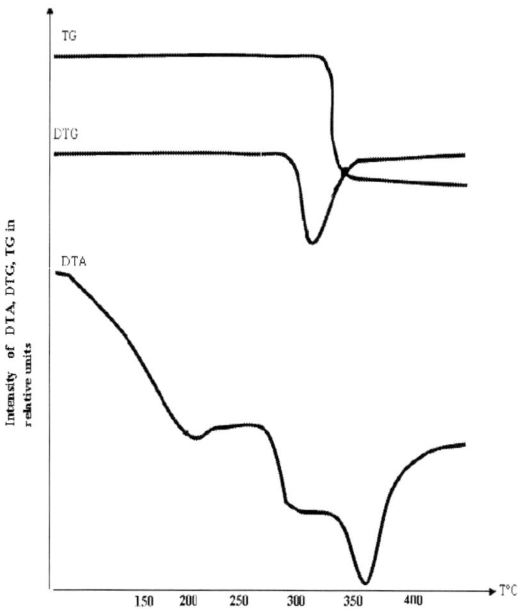

Figure 6. Derivatograph of PE+ 0.5%Co $(ALO_2)_2$ nanocompounds obtained under water quenching conditions without treatment in constant electric field.

The breakage of weak bonds and depolymerization have taken place at a lower temperature than in the initial polymer, i.e., by 97°C and 115°C, respectively. The latter indicates that thermal stability of the polymer decreases. In the derivatograph PE+0.5%Co(ALO2)2 composition nanocompound has been exposed to electro thermopolarization at E=7×106V/m, (Figure 4). On the DTA curve, one exothermal effect at T170 °C has been found appropriate to the 1st stage of nanocompound oxidation. This was followed by an endothermal effect at T230°C appropriate to compound depolymerization. The given processes on DTG curves are accompanied by one endothermal effect at T230 °C due to the loss of volatile components in amounts of 30% calculated according to the TG curve. The balance of nanocompound unburnt mass by heating up to T360°C is 70%.

By adding to the PE+1% Co(ALO2)2 composition when E=0 there has taken place, the polymer amorphysm process (Figure 7) on the DTA curve appears as a decrease of the matrix crystal phase melting temperature up to T50°C and a blurred endothermal effect at T50°C. The first stage of polymer oxidation manifests itself by exothermal effect at T70 °C, breaking of weak bonds takes place at T160°- 188 °C with the peak of endothermal effect at T188 °C but depolymerization is followed by the endothermal effect at T290 °C[16].

The processes on DTG curve like these appear as one endothermal effect at T290°C accompanied by the volatility of volatile components in the amount of 23% calculated according to the TG curve. The solid balance of the depolymerization processes unburnt part by heating up to 360°C is 77%. In the derivatograph of PE+1%$Co(ALO_2)_2$, the composition nanocompound exposed to electrothermopolarization at $E=7\times10^6$ V/m, on DTA curve (Figure 8) endothermal effect appropriate to polymer matrix crystal phase melting (amorphism of compound matrix crystal phase appeared to happen because of a high electric discharge) has not been determined. The first oxidation stage of the nanocompound takes place at 140°C but the break of weak bonds is revealed at 195 °C and the effect of depolymerization is revealed at 230°C [17].

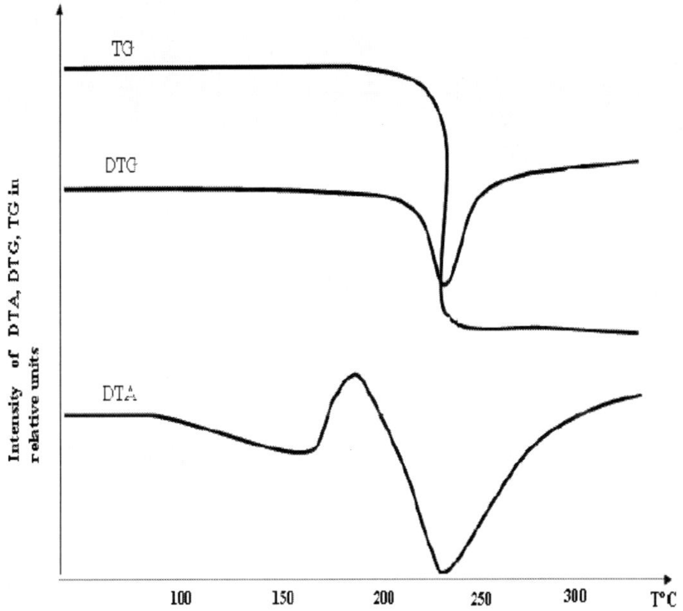

Figure 7. Derivatograph of PE+ 0.5% Co (ALO2)2 nanocompounds obtained under water quenching conditions without treatment in constant electric field E.

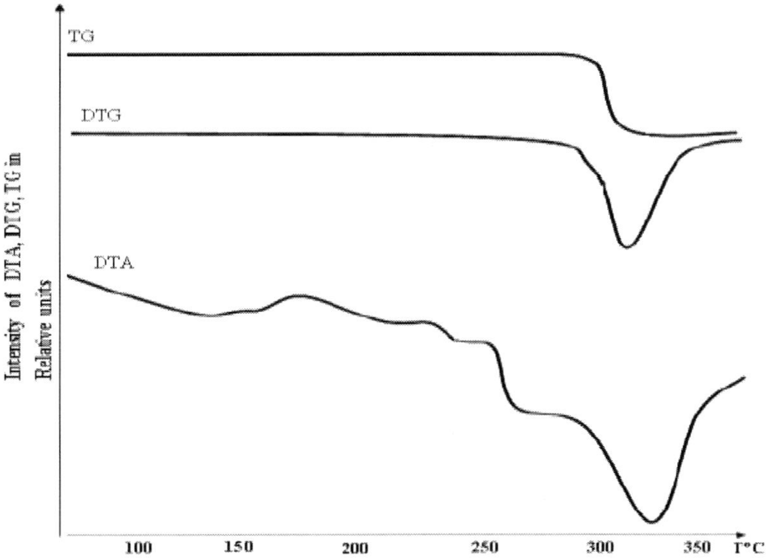

Figure 8. Derivatograph of PE+ 1% Co (ALO2)2 nanocompounds obtained under water quenching conditions without treatment in constant electric field E.

The processes on DTA curves like these manifest themselves by one endothermal effect at 230 °C that by TG curve calculation is in agreement with loss of volatile components in amounts of 33% while the balance of depolymerization by heating up to 360 °C is 67%. From thermal data given in Table 1 and Figure 1-6, it follows that compared with the initial PE Co(ALO2)2 filler addition in amounts of 0.5% in PE leads to an increase of the matrix crystal phase melting temperature by 15 °C. In this case, in all PE+0.5%Co(ALO2)2 and P+1%Co(ALO2)2 compositions under investigation exposed to electrothermopolarization at E=7×106V/m, maximum thermal destruction is produced. This points to the low temperature depolymerization and large amount of volatile component volatility (30– 33%)[18].

8.4. Structure and Magnetic Properties of Nanocomposites on the Basis PE+Fe3O4 и PVDF+ Fe3O4

Over the past decade, the composites on the basis of polymers and ferromagnetic have been widely used in radio engineering, TV, communication engineering, storage devices and computers, because of their relatively high magnetic and exploitation properties. Polymeric composition materials are composed of magnetic particles, other functional fillers and polymeric matrices, in which these particles are situated. The matrix defines the structure of the nanoparticle assembly (size distribution of nanoparticles, distance between them, their shape, ordering etc.), and the structure of the formed nanocomposite defines the properties of the nanocomposite. The polymeric matrix stabilizes the active nanoparticles and the properties of these nanoparticles principally differ from the bulk material. The nanocomposite properties are defined by the chemical nature of the polymeric matrix, the structure of the large interphase part and the interaction between the nanoparticles and the polymeric matrix. They are interesting from the point of view of their structure and properties as well as for their practical applications. In case of low concentrations of nanoparticles in the matrix, nanocomposites are superparamagnetic, due to the directions of weak magnetization axes, which are distributed randomly. A high concentration of nanoparticles in the matrix results in a reaction between them that forms an infinite conductive cluster, and the nanocomposite exhibits ferromagnetic properties. The size of the magnetite nanoparticles has been defined by the magneto-granulemetric method and AFM scanning. The results of the different methods are harmonized well. It is established that the size of the magnetite nanoparticles depends on their concentration in the polymeric matrix and polymer crystallization rate [19]. It is known that magnetization of the medium can be described by the Langeven equation:

$$M(H) = M_s \cdot \varphi_m \int_0^\infty f(m) \cdot L\left(\frac{mH}{kT}\right) \cdot dm \qquad (1)$$

where $L\left(\frac{mH}{kT}\right) = cth\left(\frac{mH}{kT}\right) - \frac{1}{\frac{mH}{kT}}$ - function of Langeven, $f(m)$ - function of distribution of magnetite nanoparticles on magnetic moment, ϕ*m*– concentration of magnetite nanoparticles in the medium. The magnetic moment of magnetite nanoparticles is defined as) (*mfisVMm*·=, where *Ms* – magnetization of satiation of massive magnetite which is equal 491,6 кА/м, *Vi* – volume of magnetite nanoparticle.

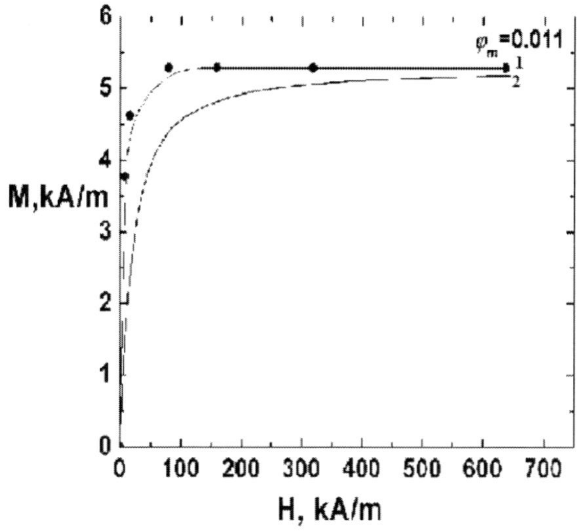

Figure 9. The magnetization of the sample PVDF-Fe_3O_4 (10%).

The dependence of the magnetic diameter of the nanoparticles coagulant from their concentration in the polymeric matrix is shown in Figure 10. The dependence of *dm* from *φm* is linear, with an increasing concentration increasing the magnetic diameter. This means that the nanoparticles coagulate can be observed. However, the coagulation of the particles depends not only on its concentration, but also depends on its supermolecular structure and the degree of crystallization.

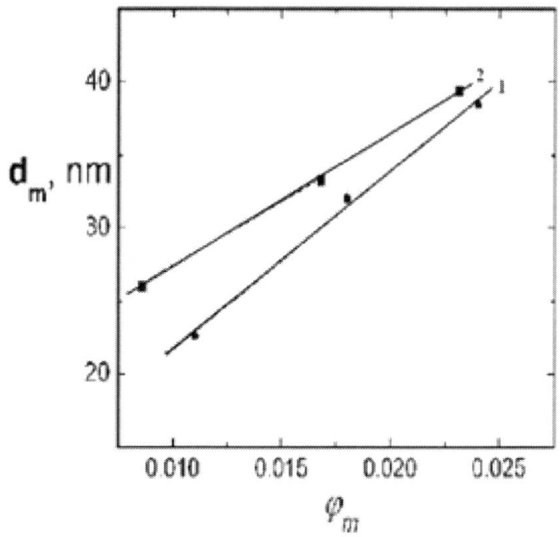

Figure 10. The dependence of magnetic diameter of nanoparticles agglomerations on its concentration in polymeric matrix.

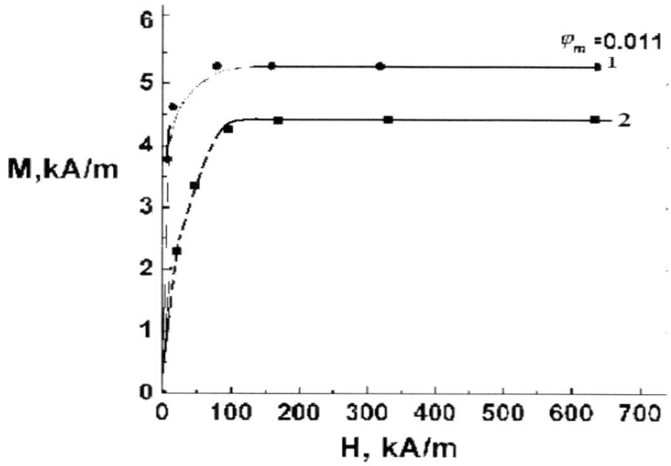

Figure 11. The magnetization of the sample PE+ Fe3O4 in various cooling regimes.

The change of the supermolecular structure of PVDF is larger than the one for PE. The values of magnetic diameters are various for the same concentration of nanoparticles in PVDF and PE (φm =0.017)[20]. The curves of magnetization PMNC on the basis PVDF and PE with part per volume concentration of magnetite nanoparticles φm =0.011. The magnetization of the sample PVDF+Fe3O4 and PE+Fe3O4 has a superparamagnetic character (Figure 11).

Specific magnetization for the nanocomposite on the basis of PVDF is higher than the one for PE basis. It should be noted that PVDF and PE have diamagnetic properties – they are considered diamagnetic. However, the magnetic properties of the C-F and C-H bonds in macromolecules PVDF and PE are different. Magnetic perception depends on the supermolecular structure of the polymer and crystallization degree. The change of specific magnetization depending on the magnetic field and the polymeric matrix type can be linked with difference of diamagnetic anisotropy, supermolecular structure and crystallization degree of polymeric matrix PVDF and PE. It is established experimentally that specific magnetization of the Fe3O4 nanoparticle in the polymer also depends on temperature-duration regime of polymer crystallization.

The value of specific magnetization of PE+Fe3O4 changes depend on the temperature-duration regime of polyethylene crystallization, which means that the specific magnetization of PE+Fe3O4 increases with the decline of the temperature-duration regime of the polyethylene crystallization. The analysis of magnetization curves show the maximum size of nanoparticles coagulants forms in various cooling regimes of the polymer (1. β=30deg/min (BO) 2.β=4 deg/min (MO)) are equal to correspondingly 21 nm, 15,7 nm. They may be the size of magnetite nanoparticles coagulants depending on the rate of polymer cooling. In the case of great cooling rates, it forms small size coagulants [21].

The Fe_3O_4 nanoparticle sizes and the topography of the distribution of nanoparticles in the polymeric matrix for the samples, is obtained in regimes β=4 deg/min and β=30 deg/min(Figure 12). The size of Fe_3O_4 nanoparticles, formed in various cooling regimes of polymer (1. β=30deg/min 2.β=4 deg/min) are equal to correspondingly 21 nm and 15 nm. The analysis of the properties of the nanocomposite surface PE+ Fe_3O_4 and the histogram of the values of elements images is shown below.

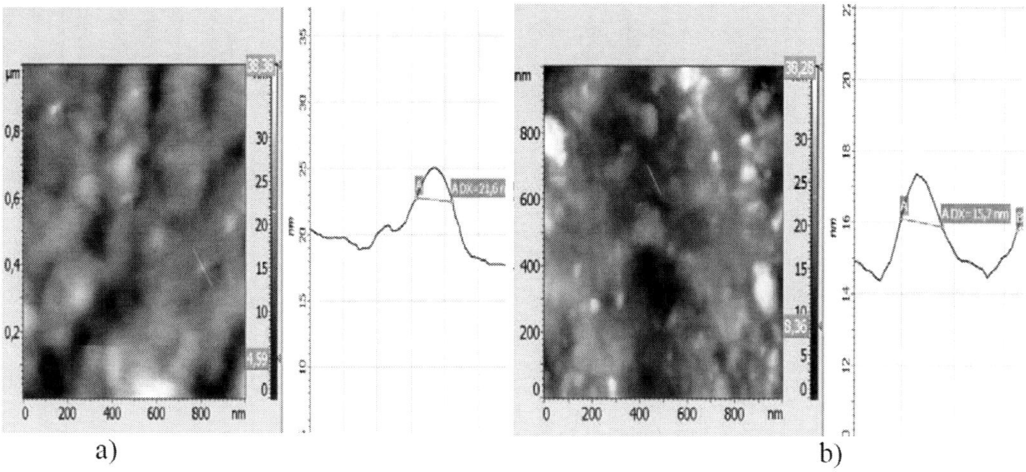

Figure 12. AFM image of nanocomposites surface PVDF+ Fe_3O_4 and size of Fe_3O_4 nanoparticles.

The mean-square surface roughness of the nanocomposite is approximately 15-30 nm as seen in Figure 13. AFM scanning shows that nanoparticles Fe_3O_4 are equally distributed in a polymeric matrix. It should be noted that the formation of nanocomposite properties plays a great role in the interphase interactions between matrix and filler. The microscopic heterogeneity of the polymeric matrix structure is conditioned by the peculiarities of interactions of matrix and filler during the composite crystallization process.

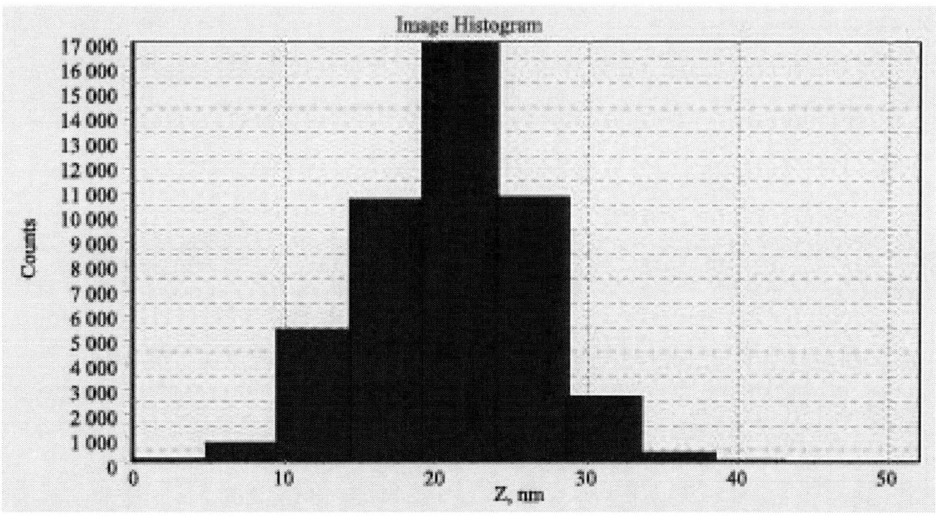

Figure 13. Analysis of properties of nanocomposites surface and histogram of values of elements images.

There is some change of structural micro heterogeneity for crystallized polymers PE and PVDF as a result of the interactions on the border, which differ from micro heterogeneity in the polymeric matrix and the various changes of structural micro heterogeneity for PE and PVDF. The scans of the PE+ Fe_3O_4 nanocomposition obtained by Atomic-Force Microscope and Magnetic Force Microscope is shown below.

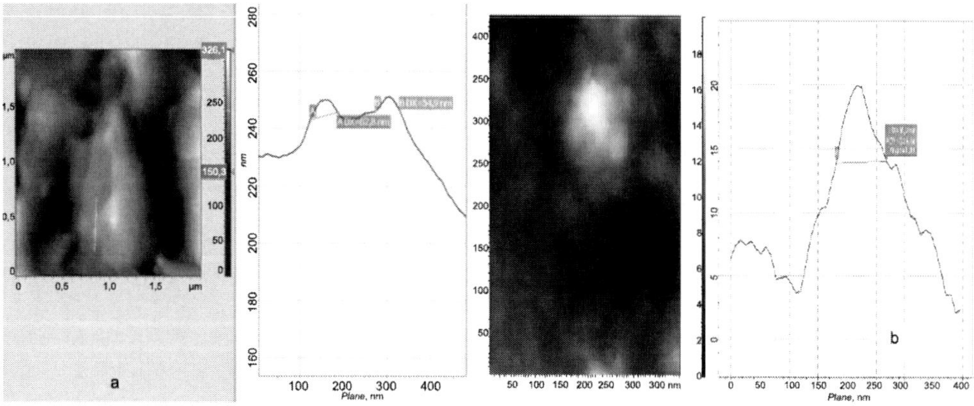

Figure 14. 2D scans and sizes of PE+ Fe_3O_4 nanocomposites.

The magnetic addition is taken here as the volume of the amount γm=0,025 is shown in Figure 14. Investigations of the magnetite conglomerate composed of nanoparticles showed that its geometrical size obtained by AFM is substantially smaller than the geometric size obtained by MFM. The magnitude of the coefficient K by experimental way and is appropriate with experimental results. Thus, it was found that a coagulation of Fe_3O_4 nanoparticles in the polymeric matrix varies depending not only on the volumetric quantity of nanoparticles but also on the surface-molecular structure and crystallization degree of the polymer.

8.5. Magnetoresistance Effect in a PE+Fe_3O_4 Based Polymer Nanocomposite System

A study of nanocomposites based on the ferromagnetic metals and oxides is an area of intense interest. In those materials, a giant magnetic resistance (GMR) is observed. As shown, it was observed that large positive magnetic resistance appeared after the preliminary storage of samples in magnetic field and it reaches saturation at lower values of the field. The role of magnetostriction in the appearance of giant magnetoresistance (GMR) in nickel-polyarylenphtalide- copper system was studied. It has been found that the GMR phenomenon at the boundary of polymer-nickel is not related to the considerable variation of magnetostrictional deformations as the transverse and longitudinal magnetostrictions on the plane of substrate practically is not varying in the interval of magnetic fields causing junction. The perfect nature of GMR in the polymeric nanocomposite was not finally explained yet.

Fe_3O_4 nanoparticles were obtained as follows: salts of $FeSO_4 \times 7H_2O$ and $FeCl_3 \times 3H_2O$ were individually solved in water then solutions were associated in a vessel and stirred by slowly adding of 25% solution of hydroxide ammonium. As a result, the thin layer of dispersed deposit of Fe_3O_4 was produced on the bottom of the vessel. To enhance the sedimentation rate, the vessel it was equipped with had a permanent magnet. The illuminated solution was decanted from the surface of the deposited layer and the product of sedimentation was washed with distilled water. The potassium oleate and aqueous solution of sodium dodesulphate were added to the washed deposit under continuous stirring and heating up to 50^0C. The temperature of the reaction mixture was raised up to 90^0C. The peptizing process was carried out for 30 minutes. As a result of the processes carried out, the Fe_3O_4 nanoparticles were obtained within the solution. Polymer magnetic nanocomposites were obtained as follows: nanoparticles of Fe_3O_4 were added to the solution of polyvinylidenfluoride (PVDF) in dimethyl formamide and the mixture was stirred until the emulsion formation at the temperature of 343 K. After adding the aqueous solution, the Fe_3O_4 containing PVDF was separated. It was then dried in a vacuum box. The samples of nanocomposites were obtained by the thermal pressing method at the melting point of the polymer matrix under 15 MPa pressure for 10 minutes with further cooling at a rate β=30 grad/sec until it reached room temperature. The sizes of magnetic nanoparticles have been studied by AFM and MFM of Integra Prima type. Figure 15 presents the AFM images of surfaces of Fe_3O_4 nanoparticles for various volume contents. AFM studies showed that the sizes of nanoparticles increase with the concentration of Fe_3O_4. The results show that increasing the content of Fe_3O_4 nanoparticles in the polymer matrix, the sizes of particles also increase. It leads to an increase in the probability of the formation of larger particles. In Figure 16 a) and b) the dependencies of the intrinsic magnetization of the Fe_3O_4 nanocomposite on the intensity of the magnetic field are demonstrated. It has been found that intrinsic magnetization increases with concentration of Fe_3O_4 in a polymeric matrix of polyethylene. At the lower values of the magnetic field, the saturation is observed as the ferromagnetic nanoparticles have a single domain structure and the particles are easily oriented under the action of the external permanent magnetic field. It seems to be due to the random distribution of orientations of axes for easily magnetization. At sufficiently small sizes, the nanoparticles have a single domain structure and their magnetic moments are oriented due to external magnetic fields. These particles are stabilized by magnetic anisotropy.

As can be seen from Figure 16 b), the increase in the intrinsic magnetization and saturation are observed with an increase in the intensity of the magnetic field. In the process of demagnetization, i.e., after the magnetizing field has been reduced, the residual magnetization is observed. The repeating of the cycle in the change of the magnetic field direction results in the observation of the hysteresis loop. It is known that the important magnetic characteristics of ferromagnetic materials are coercive force (H_0) and residual intrinsic magnetization (M_H). From Figure 16 it is seen that with an increase in the concentration of the Fe_3O_4 in the polymer matrix, the intrinsic magnetization also increases. An increase of the Fe_3O_4 concentration in the polymer matrix of PE leads to the change in the properties of the nanocomposite, i.e., nanocomposite demonstrates ferromagnetic properties. At sufficiently small sizes, the particles possess a single domain structure.

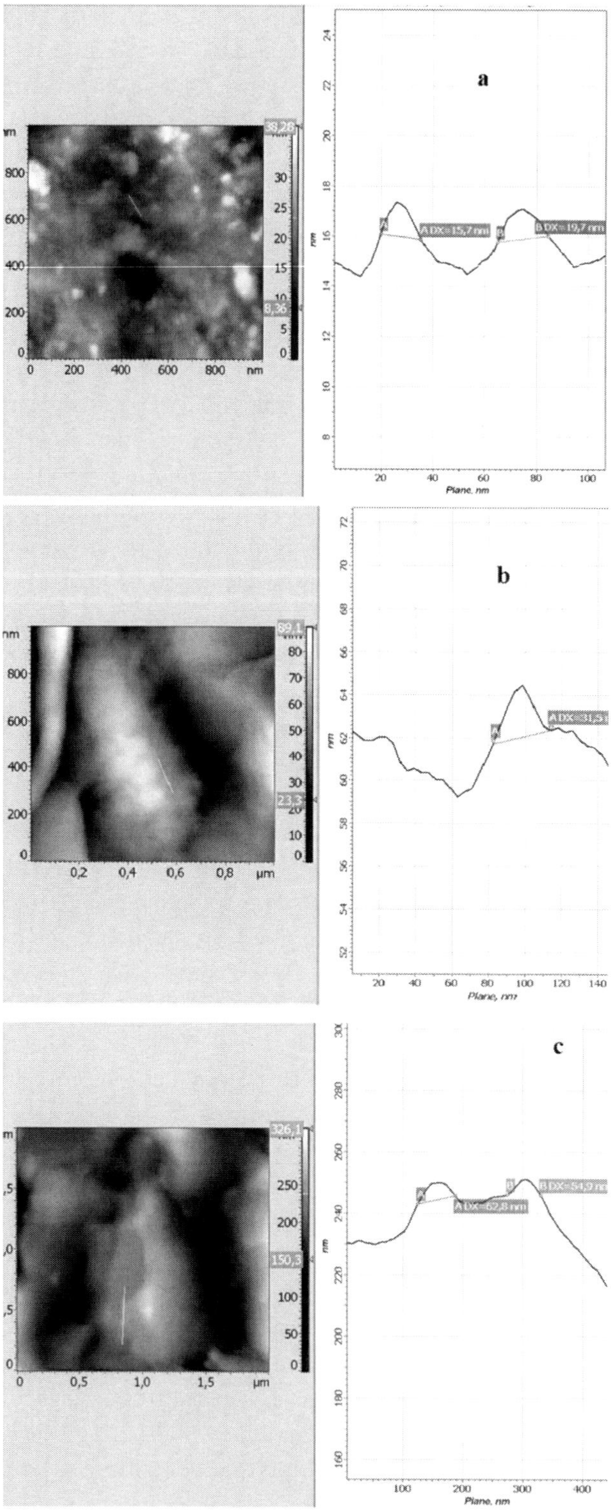

Figure 15. AFM images of surfaces of $PE+Fe_3O_4$ nanocomposites and sizes of Fe_3O_4 nanoparticles $PE+Fe_3O_4$ 5 ml, b) $PE+Fe_3O_4$ 10 ml, c) $PE+Fe_3O_4$.

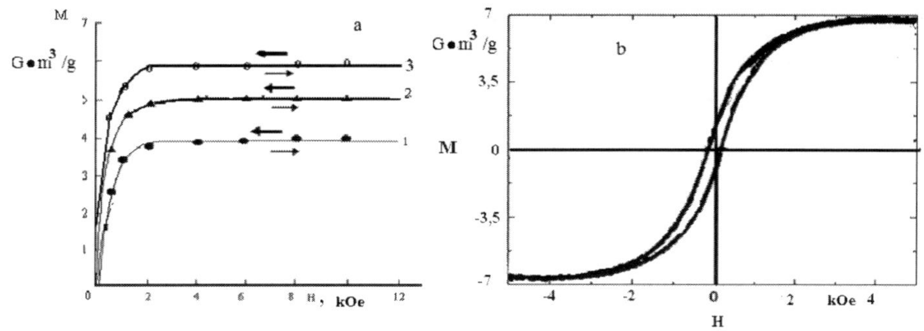

Figure 16. Intrinsic magnetization as a function of intensity of magnetic field for the PE+Fe_3O_4 nanocomposites with different mass contents (MC) :a) 1. PE+5% MC Fe_3O_4, 2.PE+10% MC Fe_3O_4 3. PE+10% MC Fe_3O_4 b) PE+20% MC Fe_3O_4

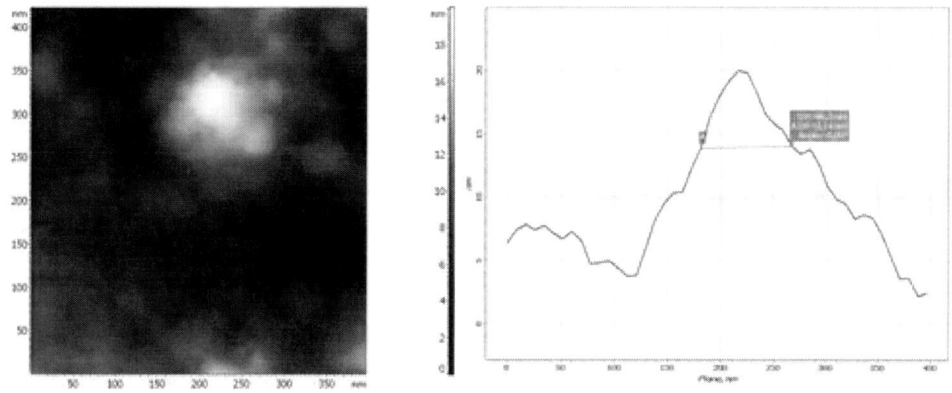

Figure 17. An MFM image of Fe_3O_4 nanocomposite surface and graph between sizes of Fe_3O_4 nanoparticles and their contents.

The direction of their magnetic moments are defined by the "game" between the orientating action of the external magnetic field and the stabilizing behavior of either crystalline or geometric magnetic anisotropy. An increase in the sizes of Fe_3O_4 also leads to changes in the properties of the nanocomposite, i.e., Fe_3O_4 particles enter into the multi-domain structure and the material gets ferromagnetic properties. Figure 17 presents the MFM image of PE+Fe_3O_4 nanocomposite surface and the sizes of Fe_3O_4 for different contents. Both AFM and MFM studies show that the real geometrical sizes are distinct from the magnetic sizes. These experimental results are in accordance with the theoretical calculation. It was experimentally known that the coagulation of Fe_3O_4 nanoparticles in a polymeric matrix depends not only on the concentration of magnetite nanoparticles but also on the supermolecular structure of the polymer.

The periodic change in the resistance of the PE+Fe_3O_4 nanocomposite has been studied in the absence of a magnetic field and when it was present. It has been found that the magnetoresistance effect for nanocomposite decreases with increasing in sizes of Fe_3O_4 nanoparticles. The periodic change in resistance as a function of time is shown in Figure 18. In the absence of a magnetic field (H=0), a decrease in resistance is observed. However, in the presence of a magnetic field (H=1 kOe), magnetic resistance shows an increase with time. The observed variation in magnetic resistance can be explained by taking into account the role of ferromagnetic cluster formations composed of nanoparticles surrounded by the superparamagnetic (SPM) Fe_3O_4 particles [22]. The mutual orientation of magnetic moments of clusters and SPM particles more probably leads to tunneling of charge carriers and a variation in the resistance of the magnetic composite.

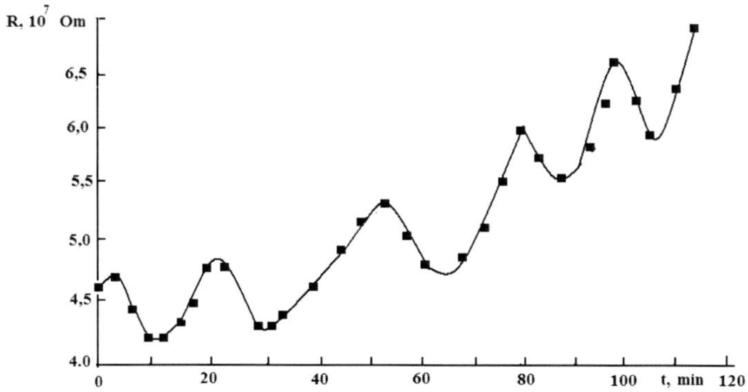

Figure 18. Magnetic resistance of PE+Fe_3O_4 nanocomposite as a function of time at periodic change of magnetic field in the interval 0÷1кОэ.

References

[1] D. Pomoqailo, A. S. Rozenberg, U. E. Uflyand, Nanoscale metal particles in polymers. M.: *Chemistry,* 671 (2000).

[2] M. Magerramov, M. A. Ramazanov, F. V. Gadjieva, *J. of American Science* 3(4), 62 (2007).

[3] M. A. Ramazanov, *J. Applied Physics*, 6, 8 (2007).

[4] Dudziak, A. Jaskowska, R. Borc, E. Spielwla, *Current Topics in Biophysics*, 25(2), 89 (2001).

[5] E. Peeters, M. Christiaans, R. Janssen, H. Schoo,P. Dekkers, E. Meijer, *J.Amer.Chem.Soc.*, 119, 9909 (1997).

[6] Ruhstaller, J. C. Scott, P. J. Brock, U. Scherf, S. A. Carter, *Chem. Phys. Letters*, 317, 238 (2000).

[7] M Magerramov, M. A Ramazanov, F. V. Gadjieva J. Fizika i Ximiya obrabotka materialov, 1, 71 (2008).

[8] A Ishchenko., N. A. Derevyanko, V. A. *Svidro Dyes and Pigments*, 19(3), 169 (1992).

[9] N. Davidenko, A. Ischenko, A. Kadaschuk, N. Kuvshinsky, N. Ostapenko, Yu, Skryshevsky, *Functional Materials,* 5(3), 353 (1998).

[10] M. A. Ramazanov, A. A. Ismailov, O. B. *Tagiyev Proc.SPIE*, 5257, 198 (2003).

[11] M. Magerramov, A. M. Bakirov, A. C. Saidov, A. A Jafarov, *J. Physics of the Solid State,* 30(11), 3511 (1988).

[12] N. A. Eyubova, A. M. Magerramov, V. P. Malin, Y. N. Gazaryan, Collected Volume. *Baku, Elm,* ed. 1, 9-13.

[13] M. A. Bagirov, N. A. Eyubova, V. P. Malin, A. A. Aliyev, A. M. Magerramov, *High-molecular compounds*. M., 29a (5), 917 (1987).

[14] M. A. Ramazanov, A. S. Huseynova, N. A. Eyubova, S. A. Abasov, *J. Optoelectron. Adv. Mater.-Rapid Commun.* 4, (12), 2003 (2010).
[15] M. A. Ramazanov, A. S. Huseynova, S. I. *Mekhtiyeva, Electron treatment of materials,* № 3, 77 (2008).
[16] D. Pomogailo, A. S. Rozenberg, U. E. Uflyand Nanoscale metal particles in polymers. Moscow, "*Chemistry*", 2000, 671p.
[17] Y. I. Petrov Cluster and small particles. Moscow, *Science*, 1986, 124p.
[18] L.V. Nikitin, L. S. Mironova and etc. Magnetic elasticity, structure and magnetic deformation properties of magnetic-elastici. *J.VMS.* v.46(A), №3. c498.
[19] A. M. Magarramov, M.A.Ramazanov, R.Z. Sadikov, R.A. Ali-zade Magnetic polymer nanocomposite on the base of PVDF+Fe3O4 Nanotekhnika, 2005, 4, стр. 111-114.
[20] A. M. Magarramov, M.A.Ramazanov, P.B.Agakishiyeva, F.V.Hajiyeva Influence microstructure to magnetic properties of polymer nanocomposite on the base of Fe3O4. *Plastic Mass* 2008, 10, стр.14-16.
[21] D. Stadnik, G. V. Kirik Polymer composite and nanocomposite in magnetic field. *Sumi,* 2005, 239c.
[22] R. A. Ali-zade Magnetite nanoparticles with an antiferromagnetic surface layer. Inorganic *Materials*, 42(11), 1215 (2006).

Chapter 9

BIOLOGICAL AND BIOMEDICAL NANOPARTICLES

9.1. BIOFUNCTIONALIZED TARGETED NANOPARTICLES FOR THERAPEUTIC APPLICATIONS

Advances in nanotechnology have impacted the field of therapeutics delivery significantly. This is evidenced by the increase in the number of nanoparticle- based therapeutic products in development over the last two decades. A 2006 global survey conducted by the European Science and Technology Observatory (ESTO) revealed that more than 150 companies are developing nanoscale therapeutics, and 24 nanoparticle therapeutics are currently in clinical use [1]. These drugs are being developed to treat a wide range of diseases, such as fungal or bacterial infections, HIV infections, diabetes and cancers. There are several advantages of using nanoparticles for therapeutics delivery. The use of materials on the nanoscale level provides unprecedented freedom to modify some of the most fundamental properties of therapeutic carriers, such as solubility, diffusivity, biodistribution, release characteristics and immunogenicity. Precise nanoparticle engineering has yielded longer circulation half-lives, superior bioavailability and lower toxicity. For example, the liposomal encapsulation of doxorubicin significantly reduces its most serious and dose-limiting side effect, cardiac toxicity [2]. One strategy to further improve the therapeutic index of nanoparticle therapeutics is to functionalize nanoparticles with targeting ligands. The addition of targeting ligands allows the delivery of drug-encapsulated nanoparticles to uniquely identified sites while having minimal undesired effects elsewhere. Since biologically targeted nanoparticles have the potential to be the optimal drug delivery vehicles, there has been tremendous amount of interest in developing novel targeted nanoparticles for therapeutic applications. This chapter reviews recent advances in the development of biofunctionalized targeted nanoparticles. The authors discuss the existing nanoparticle platforms for therapeutic applications, targeting ligands that can be used to functionalize the nanoparticles, and the various targeted nanoparticles in development. Over the last several decades, numerous nanoparticle platforms have been studied for their use in therapeutic applications. These nanoparticle platforms include liposomes, polymer therapeutic conjugates, polymeric micelles, dendrimers, nanoshells and nucleic acid-based nanoparticles. The two dominant classes of nanoparticles, liposomes and polymer-drug conjugates, account for more than 80% of the available nanoparticle therapeutics in clinical use. Liposomes have been used widely as pharmaceutical carriers in the past decade, with 11 formulations approved for clinical use and many more in clinical development [3]. Some of the commonly used therapeutics include: liposomal amphotericin, liposomal doxorubicin and liposomal daunorubicin. Liposomes are spherical vesicles that contain a bilayered membrane structure composed of natural or synthetic amphiphilic lipid molecules. Their biocompatible and biodegradable composition as well as their unique ability to encapsulate both hydrophilic and hydrophobic therapeutic agents, make liposomes excellent therapeutic carriers. Liposomes can also be coated with biocompatible and antibiofouling polymers, such as polyethylene glycol (PEG), to prolong their circulation half-life. The polymer coating of the liposomes can also be engineered to carry a functional group, which can be used for targeting ligand conjugation. Another nanoparticle drug delivery platform, polymer-drug conjugates, has been studied extensively [4]. Small molecule therapeutic agents and proteins usually have two unfavorable properties: short circulation half-life, leading to the need for frequent administration, and non-site-specific targeting, resulting in undesired systemic side effects. The conjugation of drugs to polymeric nanocarriers can reduce these undesirable adverse effects. Polymer-drug conjugates not only prolong the *in vivo* circulation time from several minutes to several hours but also reduce cellular uptake along the endocytic route. This characteristic enhances the passive delivery of drugs to tissues with leaky blood vessels, such as tumors and atherosclerotic plaques.

Many polymers have been proposed as drug delivery carriers but only a few, with linear architecture, have been used in clinic. The major challenges of most polymer-drug conjugates include polymer toxicity, immunogenicity, nonspecific biodistribution, *in vivo* circulation instability, low drug carrying capacity, rapid drug release and manufacturing challenges. PEG, which was first introduced into clinical use in the early 1990s, enhances plasma stability and drug solubility while reducing drug immunogenicity. There are currently six examples of PEG-drug conjugates in clinical use. In addition to PEG, other linear polymers such as polyglutamic acid, *N*-(2-hydroxypropyl) methacrylamide (HPMA), polysaccharide and poly (allylamine hydrochloride), have been harnessed as polymeric drug delivery carriers [5-8]. The polymers in the polymer-drug conjugates can be used for conjugation to a targeting ligand, in turn creating biologically targeted therapeutics. Dendrimers are well defined, regularly branched macromolecules that are 2.5 – 10 nm in size. They are synthesized from either synthetic or natural building blocks such as amino acids, sugars and nucleotides. The core of a dendrimer is denoted generation zero and each additional level of branching adds another generation. Dendrimers' characteristics as carriers of therapeutics include nanoscale spherical architecture, narrow polydispersity, multifunctional surface chemistry and large surface area. Many dendrimer families have been reported. Among the families, the polyamidoamine (PAMAM) and poly (propylenemine) (PPI) families have been most widely used for biomedical applications [9]. The specific molecular structure of dendrimers enables them to carry various drugs through their multivalent surfaces by covalent conjugation or electrostatic adsorption. Alternatively, dendrimers can be loaded with drugs, by using the cavities in their cores through hydrophobic interaction, hydrogen bonding or chemical linkage. Their surface can be engineered to provide precise spacing of surface molecules and to conjugate targeting molecules. Biodegradable polymer nanoparticles have been extensively investigated as therapeutic carriers. They are generally formed by the self-assembly of copolymers consisting of two or more polymer blocks with different hydrophobicity. These copolymers spontaneously assemble into a core-shell micellar structure in an aqueous environment. Specifically, the hydrophobic blocks form the core to minimize their exposure to aqueous surroundings while the hydrophilic blocks form the corona-like shell to stabilize the core. This core-shell structure provides an ideal drug delivery nanocarrier. Its hydrophobic core is capable of carrying therapeutics with varying loading capacity (5 – 25% by weight). The hydrophilic shell not only provides a steric protection for the micelle but also provides functional groups for further particle surface modifications. Polymeric nanoparticles have been formulated to encapsulate either hydrophilic or hydrophobic small drug molecules, as well as macromolecules such as proteins and nucleic acids. The release of encapsulated drugs occurs at a controlled rate in a time or environment dependent manner. Furthermore, the rate of drug release can be controlled by modification of the polymer side chain, development of novel polymers or synthesis of copolymers. In general, these biodegradable polymer systems can provide drug levels at an optimum range over a longer period of time than other drug delivery methods, thus increasing the efficacy of the drug and maximizing patient compliance, while enhancing the ability to use highly toxic, poorly soluble or relatively unstable drugs. Poly (d,l -lactic acid), poly(d,l-glycolic acid), poly(ε-caprolactone), and their copolymers at various molar ratios diblocked or multiblocked with PEG are the most commonly used biodegradable polymers, while PEG is the polymer used most commonly to engineer the polymeric micelle surface.

For example, poly lactide-co-glycolide (PLGA) encapsulated antibiotics have been investigated for the treatment of tuberculosis using murine models [10]. Metallic nanoshells are characterized by a dielectric core coated with a thin metallic shell to improve their biocompatibility and optical absorption. These particles possess a highly tunable plasmon resonance mediated by the size of the core and the thickness of the shell, which in turn determines their absorbing and scattering properties over a broad range of the spectrum from the near-ultraviolet to the mid-infrared.

Gold nanoshells have been developed for *in vivo* photothermal therapy using near infrared light. Similarly, thermally sensitive polymeric hydrogels and optically active nanoshells have been developed for the purpose of photothermally modulated drug delivery. Nanoshell particles with a magnetic core (carbonyl iron) and a biodegradable poly (butylcyanoacrylate) (PBCA) shell have also been developed for the controlled release of 5-fluorouracil [11]. In nucleic acid based nanoparticles, DNA and RNA macromolecules can be used as substrates for developing therapeutic and imaging nanocarriers. By rationally constructing nucleic acid chains that can yield shapes other than the traditional linear or circular shapes, researchers have been able to formulate novel nanoparticles using nucleic acids as building blocks. A multivalent DNA delivery vehicle, with an average size of 100 nm was recently reported for simultaneous targeted drug delivery, imaging and gene therapy [12]. Targeted multifunctional RNA nanoparticles (25 – 40 nm) have also been developed with a trivalent RNA core, RNA aptamers for targeting, and small interfering (si) RNAs for therapeutic effect [13]. Monoclonal antibodies (mAbs) have been the preferred class of targeting molecules for the last several decades. Artificially engineered mAbs have been commonly used for molecular targeting purposes. In order for the engineered antibodies to function in the human body, they have to evade the immune system. Current development of mAbs has thus been focused on chimeric, humanized and fully humanized derivatives to decrease their immunogenicity. The ability of engineered monoclonal antibodies to target disease processes has been demonstrated by the success of several monoclonal antibody therapeutics, including rituximab, trastuzumab, cetuximab and bevacizumab [14]. Despite the vast effort expended on their development, mAbs have their share of limitations. They are large, complex molecules that require significant engineering at the molecular level to be effective. They are expensive to manufacture and there is variation from batch to batch, which limits their efficiency as targeting molecules. Aptamers are small nucleic acid ligands that can bind to targets with high sensitivity and specificity. Aptamers fold by intra-molecular interaction into unique conformations with ligand binding characteristics. For a particular target, aptamers are selected through an *in vitro* process called systemic evolution of ligands by exponential enrichment (SELEX) [15].

This process uses the principles of evolution, where a library of 10 to 15 random oligonucleotides is enriched to identify those aptamers that can bind to the target with the highest affinity and specificity. Aptamers have potential advantages as targeting ligands. They are small in size (15 kDa), and generally have less immunogenicity which leads to better biodistribution. Most importantly, SELEX is a chemical process that can be scaled up with ease, without batch-to-batch variations and with lower costs. More than 200 aptamers have been isolated. For example, RNA aptamers to the VEGF 165 isoform with 2 ′ - *O* - methylpurine and 2 ′ - *F* pyrimidines have been reported [16]. It was found that VEGF aptamers not only can lead to regression of tumor vessels but the aptamers also exhibited a remarkable stability in plasma in monkeys. Pegaptanib, an aptamer targeted against VEGF 165, was approved by the FDA in December 2004 for the treatment of neovascular macular degeneration, underscoring the rapid progress of aptamers from their original conception to clinical application [17]. Aptamers' major shortcomings are their low serum stability and their high production cost. Peptides are an attractive alternative targeting molecule due to their smaller size, lower immunogenicity, higher stability and ease of manufacture. The development of peptide phage libraries (10-11 different peptide sequences), bacterial peptide display libraries, plasmid peptide libraries, and new screening technologies have made their selection much easier, contributing to their popularity as targeting ligands. Peptides can also bind to their targets with high specificity and affinity. For example, cilengitide is a cyclic peptide that binds to integrins, which is currently in Phase II clinical trials for the treatment of non-small cell lung cancer and pancreatic cancer [18].

There are a handful of oligopeptides that are distinct in their characteristics. These include A-domain proteins, AdNectins, and affibodies. A-domain proteins are 40 amino acid oligopeptides that bind to the cell surface through multiple points of attachment. The first A-domain protein was found in the low-density lipoprotein receptor (LDLR) by Tschopp and Mollnes [19]. AdNectins represent another distinct type of peptides. They are thermostable and protease resistant oligopeptides that were initially derived from the 10FN3 domain of human fibronectin. Each AdNectin typically has three distinct loop structures. A large library of AdNectins has been created by introducing diversity into these loops. Recently, an AdNectin for human VEGF receptor 2 (VEGFR2) named Angiocept has been isolated by AdNexux Pharmaceuticals, and entered Phase I clinical trials for treating advanced solid tumors and non-Hodgkin's lymphoma in 2006. Affibodies are small polypeptides derived from an antibody binding domain of staphylococcal protein A. Affibodies targeted against a specific cell marker can be selected using phage-display technology. For example, a 6kDa affibody with selective binding to human EGF receptor 2 (HER2) was found to have subnanomolar affinity [20]. Due to the limitations and challenges of using mAbs discussed earlier, there is increasing interest in using antibody fragments as targeting molecules while retaining the high antigen binding specificity of antibodies. These include the Fab fragments, single chain variable fragments (scFV), minibodies, diabodies and nanobodies [21]. The Fab fragment is composed of one constant and one variable domain of each of the heavy and the light chains. ScFV is a fusion of the variable regions of the heavy and light chains. Minibodies are engineered antibody fragments that are a fusion between scFV and a C H -3 domains that self-assemble into a bivalent dimer [22]. Diabodies are covalently linked dimers or non-covalent dimers of scFvs. Nanobodies, which are the smallest of all fully functional antigen-binding fragments, evolved from the variable domain of heavy-chain antibodies. Nanobodies are typically evolved from single-domain antibodies and antibodies carrying only a functional heavy chain without the light chain. These antibody fragments are engineered to retain a high affinity for target antigens but have less immunogenicity and a smaller size, and thus are better suited for molecular targeting.

Most recently, small antibody mimetics were formulated by Qiu et al. [23]. They fused two complementarity-determining regions that retained the antigen recognition of their parent molecules. These 3-kDa mimetics showed better biodistribution than their parent molecules, suggesting their potential as a new class of targeting ligands. Small molecules have shown great promise as a class of targeting molecules because of their small size and low cost of production. One of the most extensively studied small molecule targeting moieties in targeted drug delivery is folic acid (folate). The high-affinity vitamin folate is a commonly used ligand for cancer targeting because folate receptors (FRs) are frequently overexpressed on tumor cells. Folate specifically binds to FRs with a high affinity (K D = < 10 -9 M), enabling a variety of folate derivatives and conjugates to deliver molecular complexes to cancer cells without causing harm to normal cells [24]. It has been used as a targeting moiety combined with a wide array of drug delivery vehicles including liposomes, protein toxins, polymeric NPs, linear polymers, and dendrimers to selectively deliver drugs into cancer cells using FR-mediated endocytosis. Molecular targeting has been a key concept in recent years. Drugs such as trastuzumab, bevacizumab and rituximab have achieved great results that have eluded conventional therapeutics [25]. Most of the success can be attributed to targeting, as targeted therapeutics can selectively treat diseases without affecting normal tissue. There has been increasing interest in applying molecular targeting to nanoparticle therapeutics and formulating biofunctionalized targeted nanoparticles. Targeted nanoparticles, when compared with non-targeted nanoparticles, have several potential advantages: the ability to partition more of the nanoparticles within target tissue, increased uptake into target cells, higher therapeutic efficacy and lower toxicity. Although there are no clinically approved targeted nanoparticle therapeutics yet, many are in preclinical and clinical development. Almost all the combinations between the nanoparticle platforms and targeting ligands mentioned in the previous sections have been formulated. Data obtained from these targeted nanoparticles so far have supported their theoretical advantages. The most significant effect of functionalizing nanoparticles with targeting ligands is increased intracellular uptake by the target cells. Kim et al. showed that folate-targeted polymeric nanoparticles had more than 6.7 times more cell uptake than non-targeted nanoparticles [26]. Our own data showed that aptamer-targeted nanoparticles displayed a 77-fold increase in intracellular uptake by prostate cancer cells *in vitro* when compared with non-targeted nanoparticles [27]. In a separate study, we demonstrated that by varying the targeting ligand density on the nanoparticle surface (0 – 10%), targeted nanoparticles had more than seven times the intracellular uptake of the non-targeted nanoparticles after two hours of incubation[28]. Oyewumi et al. also studied the uptake of folate targeted polymeric nanoparticles into KB cells. Folate coated nanoparticles showed higher uptake in comparison to PEG-coated nanoparticles. At a nanoparticle concentration of 180 mg/ml, KB cell uptake of folate coated nanoparticles was 20-fold higher than that of non-targeted nanoparticles [29].

Kirpotin et al. formulated anti-HER2 mAb-liposome conjugates to study tumor targeting, where targeted liposomes had a sixfold higher intracellular uptake when compared with non-targeted nanoparticles.

Biofunctionalized targeted nanoparticles also preferentially accumulate in tumors when compared with non-targeted nanoparticles. Kukowska-Latallo et al. demonstrated that folate targeted PAMAM dendritic polymers concentrated in a KB tumor xenograft in SCID mice over 4 days [30]. Our own experience has also shown a higher concentration of targeted nanoparticles in tumors when compared with that of non-targeted nanoparticles. We studied the biodistribution of aptamer targeted polymeric nanoparticles and non-targeted nanoparticles. Tumor accumulation for targeted nanoparticles was 2.5 times higher than that of non-targeted nanoparticles. On the other hand, Kirpotin et al. showed that antibody-targeted lipidic nanoparticles did not increase tumor localization but did increase internationalization in tumor cells (sixfold) [31]. The tumor accumulation of targeted nanoparticles is highly dependent on the characteristics of the targeting ligands and the nanoparticles. As more *in vivo* studies on targeted nanoparticles are reported, we will obtain more information on the factors determining tumor localizations. Targeted delivery of therapeutics has also been shown to achieve greater efficacy. Park et al. studied anti-HER2 immunoliposomes encapsulating doxorubicin in tumor xenograft models. They demonstrated that in four different xenograft models, immunoliposome-doxdoxorubicin was significantly superior to free doxorubicin, liposomal doxorubicin, and anti-HER2 monoclonal antibody [32]. Bartlett et al. showed that transferrin-targeted siRNA nanoparticles are more effective than non-targeted siRNA nanoparticles despite the similar biodistribution and tumor accumulation of the two nanoparticles. Using mouse xenograft tumors expressing luciferase and siRNA against luciferase, they showed that transferring targeted nanoparticles reduced luciferase activity to 50% of the levels with non-targeted nanoparticles. Increased efficacy was also seen in the Kukowska-Latallo study. Folate targeted methotrexate (MTX) lead to statistically slower tumor growth compared with non-targeted MTX [33]. In our own experience, we have developed aptamer-targeted nanoparticles (NP-Apt) that target the prostate specific membrane antigen (PSMA) on prostate cancers. Using NP-Apts encapsulating docetaxel and a murine xenograft model of prostate cancer, it was demonstrated that the targeted nanoparticles effectively decreased tumor size following a single intra-tumor injection while non-targeted nanoparticles did not [34].

9.2. Biomedical Applications of Gold Nanoparticles Functionalized Using Hetero-Bifunctional Poly (Ethylene Glycol) Spacer

Nanotechnology offers unique approaches to probe and control a variety of biological and medical processes that occur at nanometer length scales, and is expected to have a revolutionary impact on biology and medicine. Among the approaches for exploiting nanotechnology in medicine, nanoparticles offer some unique advantages such as sensing, image enhancement, and delivery agents. Several varieties of nanoparticles with biomedical relevance are available including, polymeric nanoparticles, metal nanoparticles, liposomes, micelles, quantum dots, dendrimers, and nanoassemblies. To further the application of nanoparticles in disease diagnosis and therapy, it is important that the systems are biocompatible and capable of being functionalized for recognition of specific target sites in the body after systemic administration.

Nanoparticles based on gold chemistry have attracted significant research and practical attention recently. They are versatile agents with a variety of biomedical applications including use in highly sensitive diagnostic assays, thermal ablation and radiotherapy enhancement, as well as drug and gene delivery. For instance, antibody-modified gold nanoparticles when used for detection of prostate specific antigen, had an almost a million-fold higher sensitivity that a conventional ELISA-based assay. Near-infrared radiation absorbing gold-silica nanoshells have been prepared and evaluated for thermal ablation of tumors after systemic administration [35].

For biomedical applications, surface functionalization of gold nanoparticles is essential in order to target them to specific disease areas and allow them to selectively interact with cells or biomolecules. Surface conjugation of antibodies and other targeting moieties is usually achieved by the adsorption of the ligand to the gold surface. Surface adsorption, however, can denature the proteins or, in some cases, limit the interactions of the ligand with the target on the cell surface due to steric hinderance. Additionally, for systemic applications, long-circulating nanoparticles are desired for passive targeting to tumors and inflammatory sites. Poly (ethylene glycol) (PEG)-modification of nanoparticles affords a long circulating property by evading macrophage-mediated uptake and removal from the systemic circulation [36]. Surface modification of gold nanoparticles through the PEG spacer would, therefore, allow the modified nanoparticles to remain in the systemic circulation for a prolonged period and provide flexibility to the attached ligand for efficient interaction with its target.

The research describes the synthesis of hetero-bifunctional PEG with a thiol group on one terminus and a reactive functional group on the other for conjugation to a biologically-relevant targeting moiety. Using thiol-PEG-coumarin, a model fluorescent dye, describe the "proof of concept" for surface functionalization of gold nanoparticle, cellular cytotoxicity evaluations, and fluorescence confocal analysis of cell uptake and nanoparticle distribution.

9.3. Magnetic Nanoparticles for Biomedical NMR-Based Diagnostics

Rapid and sensitive measurement of clinically relevant biomarkers, pathogens and cells in biological samples would be invaluable for disease diagnosis, monitoring of malignancy, and for evaluating therapy efficacy in personalized medicine.

To translate such molecular measurements into clinical settings, however, an assay would need to 1) provide high sensitivity and specificity, 2) minimize sample preparation and sample volume, and 3) ideally allow concurrent detection of diverse target moieties through multiplexed measurements. Biosensing strategies based on magnetic nanoparticles (MNPs) have recently received considerable attention, since they offer unique advantages over traditional detection methods. Specifically, because biological samples exhibit negligible magnetic background, MNPs can be used to obtain highly sensitive measurements in turbid samples with reduced sample preparation. By contrast, traditional detection strategies based on optical techniques, for example, are often affected by scattering, absorption, autofluorescence, and require extensive sample purification before measurements can be made. To detect biomarkers using MNPs, several technologies have been developed. These include techniques that use magnetometers, such as a superconducting quantum interference device (SQUID), magnetoresistive sensors, and Hall sensors, which directly measure the magnetic fields from magnetically-labeled biological targets. Another technology that has achieved considerable success is diagnostic magnetic resonance (DMR). Based on nuclear magnetic resonance (NMR) as the detection mechanism, DMR exploits MNPs as proximity sensors, which modulate the spin–spin relaxation time of water molecules adjacent to the molecularly-targeted MNPs. The latter creates a local magnetic field and induces a change in the proton relaxation rate in billions of neighboring water molecules. Direct detection of magnetic moments with magnetometers requires MNP-labeled targets to be closely positioned to the sensing elements. DMR assays, however, are faster and simpler since the analytical signal is generated from the entire sample volume.

By developing optimized MNPs, DMR detection sensitivities for various target moieties have been considerably improved. To date, numerous magnetic biosensors have been designed to identify and quantify a wide range of targets including DNA/mRNA, proteins, small molecules/drugs, bacteria, and tumor cells [37]. More recently, the development of miniaturized, chip based NMR detector systems has served to further enhance DMR technology. Such detectors can perform highly sensitive measurements on microliter sample volumes and in a multiplexed format. With the integration of key components (i.e., microcoils, microfluidic networks, NMR electronics, and a portable magnet), the DMR systems have now demonstrated their potential for portable, sensitive and rapid operation in a point-of-care setting and have tremendous potential in the field of biomedical applications, primarily due to their similar size comparable to biological molecules, and because their properties can be fine-tuned during chemical synthesis. In particular, MNPs can be synthesized in such a way as to possess unique superparamagnetic properties, to be biocompatible, and to remain inert with respect to cells and molecules of interest. As the size of magnetic objects shrinks to the nanometer scale, it becomes energetically more favorable for them to have a single magnetic domain than to form domain walls and a consequent multidomain structure. The upper limit for a single domain [$\sim(A/2K)^{1/2}$] is determined by the material properties: the exchange stiffness (A) and the anisotropy constant (K). For most magnetic materials (e.g., ferrite and iron), MNPs with a diameter <20 nm will have a single domain with magnetic moments aligned in a particular direction defined by magnetic anisotropy. At sufficiently high temperatures (above blocking temperature), thermal energy can induce free rotation of the magnetic moment. Thus, when MNPs are grouped together, they display a form of paramagnetic behavior, known as superparamagnetism: MNPs assume overall magnetic moments when placed in an external magnetic field but lose their moments when the field is removed. Distinct from paramagnetism, which arises from individual spins at the atomic or molecular level, superparamagnetism applies to magnetic elements that already assume a magnetically-ordered spin state (typically ferromagnetic or ferrimagnetic). This superparamagnetic property enables MNPs to avoid spontaneous aggregation in solution, a feature that makes them suitable for many biomedical applications. In its simplest form, an MNP is comprised of an inorganic magnetic core and a biocompatible surface coating that stabilizes the particle in physiological conditions. By applying suitable surface chemistry, functional ligands can be integrated and confer the MNP with molecular specificity.

Synthetic methods for MNPs have been recently reviewed [38]. A variety of chemical methods, ranging from traditional wet chemistry to high-temperature thermal decomposition, have been employed to synthesize MNPs. Colloidal iron oxide nanoparticles, which are used as clinical magnetic resonance imaging (MRI) contrast agents, are generally prepared via an aqueous co-precipitation method. During these hydrolytic processes, control of solution pH and the addition of suitable coating surfactants are critical for regulating the nanoparticle size as well as the magnetic properties [39]. Unfortunately, depending on the synthesis procedure used, magnetization can vary significantly among nanoparticles of similar sizes. More recently, high quality MNPs have been prepared through thermal decomposition of organometallic precursors, in nonhydrolytic organic solutions containing surfactants. Monomers are generated via high-temperature thermal decomposition of precursors. Above a supersaturation level, these monomers then aggregate to induce nucleation and nanoparticle growth. By tuning the growth conditions during this procedure (such as precursor choice, monomer concentration, growth temperature and time), it is possible to control the size, composition, and crystallinity of the nanoparticles. While high-temperature decomposition markedly improves size control, size distribution and crystallinity of MNPs, the resulting particles are encased in a hydrophobic coating. In order to achieve nanoparticle stability in aqueous media, this approach requires additional modifications. Techniques, such as the addition of an amphiphilic polymer or surfactant exchange strategies, have been examined for their ability to transfer the hydrophobic MNPs into the aqueous phase [40].

When placed in an external field, each MNP creates a local magnetic field, which increases the field in homogeneity. When water molecules diffuse within the periphery of the MNPs, the coherent precessions of water proton spins are perturbed. The net effect is a change in the magnetic resonance signal, which is measured as a shortening of the longitudinal (T_1, spin–lattice) and transverse (T_2, spin–spin) relaxation times. The capacities of MNPs to decrease T_2 and T_1 are respectively defined as the transverse (r_2) and the longitudinal (r_1) relaxivities. Typically, because the transverse relaxivities (r_2) of MNPs are greater than their longitudinal relaxivities (r_1), T_2 is used for NMR-based biosensing applications. With a higher r_2 relaxivity, fewer numbers of nanoparticles are required to produce detectable T_2 changes.

Within an ensemble of MNPs, magnetic relaxation properties depend on more than simply the particles' relaxivities. The organizational state of the ensemble is also important. Unlike evenly dispersed MNPs, aggregates of nanoparticles (self-assembled magnetic clusters) have been shown to enhance the net rate of transverse relaxation. This unique phenomenon, known as magnetic relaxation switching (MRSw), is a cooperative process in which the interacting nanoparticles become more efficient at dephasing the spins of neighboring water protons, leading to a decrease in T_2 relaxation time. The phenomenon can be explained by the outer-sphere theory. For a given volume fraction of MNPs in solution, T_2 of the sample is inversely proportional to the cross-sectional area of the particles. Thus, the same amount of magnetized material is much more effective when dispersed as fewer large nanoparticles than as a greater number of smaller ones. In MRSw, nanoparticles aggregate to form self-assembled clusters, and the consequent increase in cross-sectional area of the particles shortens T_2 relaxation times.

Analogous to an MRI, DMR exploits targeted MNPs to modulate the spin–spin T_2 relaxation time of biological samples. Depending on the size of the target biomarker, DMR assays can take two forms. For small molecular targets with sizes less than or comparable to that of the MNPs, MRSw assays can be used effectively for their detection and quantification. Small molecular analytes, such as drugs, metabolites, oligonucleotides, and proteins, can cross-link MNPs to promote relaxation switching. As indicated in Figure 1a, MRSw assays can be designed to cause forward switching: a process whereby molecular targets are used as cross-linking agents to assemble MNPs into clusters, thus causing a corresponding decrease in T_2. Alternatively, the assays can cause reverse switching, where enzymatic cleavage or competitive binding of molecular targets disassembles preformed clusters to cause an increase in T_2. Note that MRSw assays are carried out without removing excess unbound MNPs. For larger biological structures such as bacteria, entire mammalian cells or cellular components, targeted MNPs can be used to tag cell surface markers to impart a magnetic moment (Figure 1b). The change of $1/T_2$ is proportional to the number of MNPs bound, and also indicative of the abundance of relevant surface biomarkers. Unlike MRSw assays, this magnetic tagging strategy requires washing steps to remove excess unbound MNPs from the tagged biological targets. To enhance DMR detection sensitivity, MNPs should possess the following characteristics: 1) exhibit superparamagnetic properties; 2) have high stability in aqueous media to avoid spontaneous aggregation, which could mimic target-induced clustering; 3) have high magnetization and transverse relaxivity (r_2) to induce pronounced T_2 changes; and 4) have good surface chemistry to simplify conjugational procedures for attaching affinity molecules, such as antibodies and peptides. The MNPs and their representative strategies described below have been shown to be uniquely suited for DMR applications.

Cross-linked iron oxide (CLIO) nanoparticles have been widely used for DMR applications on account of their excellent stability and biocompatibility. CLIO nanoparticles contain a superparamagnetic iron oxide core (3–5 nm monocrystalline iron oxide) composed of ferromagnetic magnetite (Fe_3O_4) and/or maghemite (γ-Fe_2O_3). The metallic core is subsequently coated with biocompatible dextran, before being cross-linked with epichlorohydrin and activated by ammonia to provide primary amine group functionality. The amine groups can then be easily reacted with various agents containing anhydride, hydroxyl, carboxyl, thiol, or epoxide groups, to confer molecular specificity to the nanoparticle through bioconjugation. Amine-terminated CLIO nanoparticles have an average hydrodynamic diameter of 25–40 nm, approximately 40–80 amines per nanoparticle for bioconjugation, and a r_2 of ~50 $s^{-1}\cdot mM^{-1}$ [Fe]. Despite their relatively low r_2, their unique coating makes CLIO nanoparticles exceedingly robust for use in biological applications [41].

The magnetization of ferrite nanoparticles can be further enhanced by doping the ferrite with ferromagnetic elements such as manganese (Mn), cobalt (Co) or nickel (Ni). Among the singly-doped ferrite MNPs, $MnFe_2O_4$ nanoparticles were found to exhibit the highest magnetization and r_2 value, on account of their electron spin configurations, followed by FeFe2O4, $CoFe_2O_4$, and $NiFe_2O_4$. More recently, it has been demonstrated that magnetization can be further enhanced via additional Zn^{2+} dopant control in $MnFe_2O_4$ nanoparticles.

In addition, nanoparticle magnetization is known to increase with particle size. Ideally, each magnetic spin within a bulk magnetic material would be aligned parallel to the external magnetic field. However, in the nanoscale regime, surface spins tend to be tilted, a feature that reduces the overall magnetic moment. By increasing the MNP size, this surface effect is decreased, which in turn increases the magnetization. It has also been noted that transverse relaxivity r_2 is proportional to the cross-sectional area of the magnetic core. Thus, increasing MNP size is an efficient method for enhancing r_2, since this strategy increases both the magnetization as well as the particle cross-sectional area. Both magnetic doping and sizing strategies were recently employed to produce $MnFe_2O_4$ nanoparticles with superior r_2 relaxivity, for DMR biosensing applications. These particles were synthesized by reacting iron (III) acetylacetonate [Fe $(acac)_3$], manganese(II) acetylacetonate [Mn$(acac)_2$] and 1,2-hexadecanediol at high temperature (300 °C). A seed-mediated growth approach was used to increase the size of the magnetic core from 10 nm to 12, 16, or 22 nm. $MnFe_2O_4$ nanoparticles with a diameter ≤16 nm were found to be highly monodisperse and superparamagnetic at 300 K (Figure 2a).The MNPs were subsequently rendered water-soluble using the small molecule, meso-2,3-dimercaptosuccinic acid (DMSA). DMSA has a terminal carboxylic acid group at one end which interacts directly with the magnetic core, and a sulfhydryl group at the other end which cross-links with other DMSA molecules to increase stability [42].

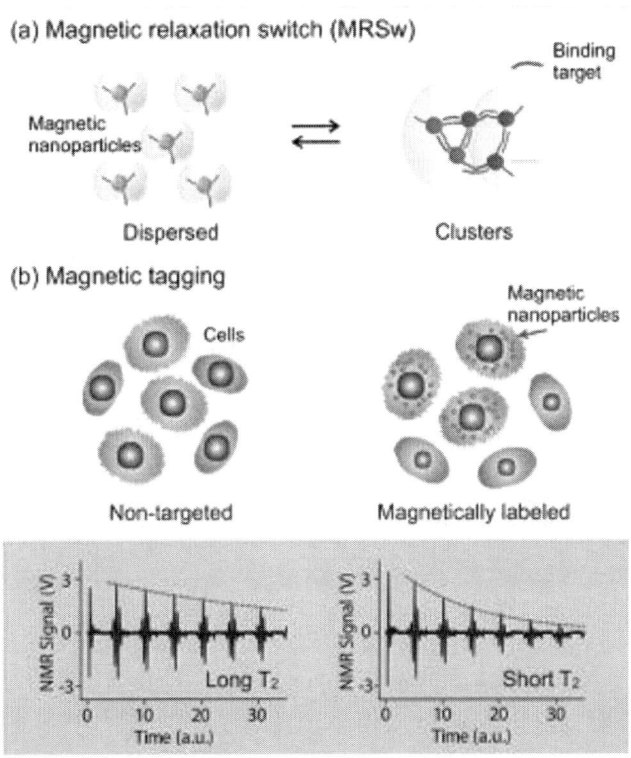

Figure 1. DMR assay configurations with magnetic nanoparticles (MNPs). (a) Magnetic relaxation switching assays detect the clustering of MNPs. (b) Magnetic tagging assays detect the presence of bound MNPs on larger biological entitles.

Due to DMSA's small size, the hydrodynamic diameter of $MnFe_2O_4$ nanoparticles was found to be smaller than that of CLIO nanoparticles, despite their larger magnetic core. More importantly, these $MnFe_2O_4$ nanoparticles possessed superior relaxivities with r_2 values as high as 420 $s^{-1}\cdot mM^{-1}$ [metal] (equal to 6×10^{-11} $s^{-1}\cdot[particle/mL]^{-1}$), more than 8 times greater than CLIO nanoparticles in metal basis (50 $s^{-1}\cdot mM^{-1}$[metal] or 0.7×10^{-12} $s^{-1}\cdot[particle/mL]^{-1}$).

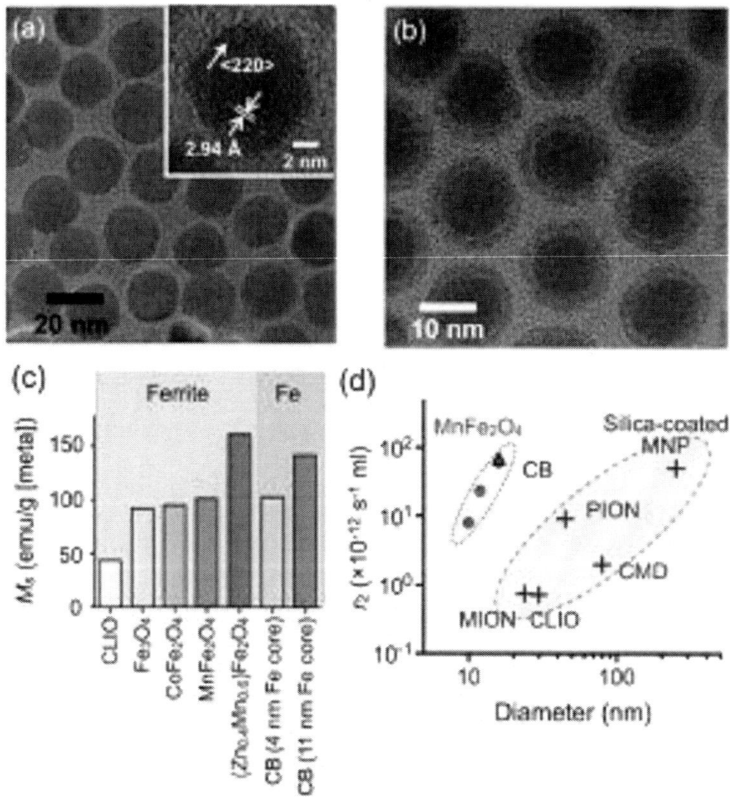

Figure 2. Higher r_2-relaxivity MNPs developed to improve detection sensitivity of in vitro diagnostics.

The synthesis of elemental iron-based nanoparticles (i.e., with elemental iron rather than iron oxides) and their stable dispersion in aqueous media, has remained one of the most attractive goals in magnetic nanomaterial engineering. Elemental iron (Fe) has a higher magnetization than that of metal oxides, which consequently motivates the creation of Fe-core MNPs to achieve high r_2 relaxivities. Because the Fe cores are extremely reactive and subject to rapid oxidation, they need to be encased by a protective shell in order to maintain their magnetic properties. Recently, a 16 nm Fe-core/ferrite shell MNP, known as "cannonball", was developed for DMR applications (Figure 2b). The cannonballs were synthesized by thermal decomposition of iron (0) pentacarbonyl [Fe $(CO)_5$] to form the Fe core.

A protective ferrite shell was formed by controlled oxidation with oxygen gas. This method resulted in a thinner shell than that previously produced by chemical oxidizers, and thus the nanoparticles retained a larger Fe core. The cannonballs were then coated with DMSA as described above. Because of their large Fe core, superparamagnetic cannonballs showed high magnetization (139 emu$\cdot$g^{-1} [Fe]) when compared to other published Fe core–shell structures (Figure 2c). The relaxivity of cannonballs is similar to that of the $MnFe_2O_4$ nanoparticles (6×10^{-11} $s^{-1}\cdot$[particle/mL]$^{-1}$), which is considerably higher than other commercially available or previously reported ferrite nanoparticles (Figure 2d). Bioorthogonal "click" chemistry has emerged as a novel method to label small molecules in complex biological media. Most reported applications, however, rely on either the azide–

alkyne cycloaddition, which requires a copper catalyst, or the strain promoted azide–alkyne cycloaddition, which has relatively slow kinetics. A new bioorthogonal chemistry has been developed based on the Diels–Alder cycloaddition between a 1,2,4,5-tetrazine (Tz) and a *trans*cyclooctene (TCO). The reaction is fast, irreversible (covalent) and can be performed at room temperature without using a copper catalyst. Recently, this chemistry has been successfully adapted to magnetic targeting, so as to improve nanoparticle binding efficiency and detection sensitivity [43].

Termed 'bioorthogonal nanoparticle detection' (BOND), this technique provides a novel targeting platform in which T_z and TCO act as the coupling agents. In a two-step labeling strategy (BOND-2; Figure 3), antibodies against biomarkers of interest are first modified with TCO, which is then used as a target to facilitate the coupling of Tz-modified nanoparticles onto mammalian cells.

Because of the small size of the coupling reagents, their high multiplicity on antibodies/nanoparticles resulted in higher nanoparticle binding to cells. In comparison to alternative standard techniques, such as the avidin/biotin method, BOND-2 not only amplifies the biomarker signals but also significantly improves the detection sensitivity. Moreover, this platform is broadly-applicable and scalable for biomedical use. BOND-2 has already been successfully adapted for molecular profiling of cell samples by DMR, and has now established itself as a major targeting method. Table 1 lists a library of cellular makers tested with BOND-2 and DMR.

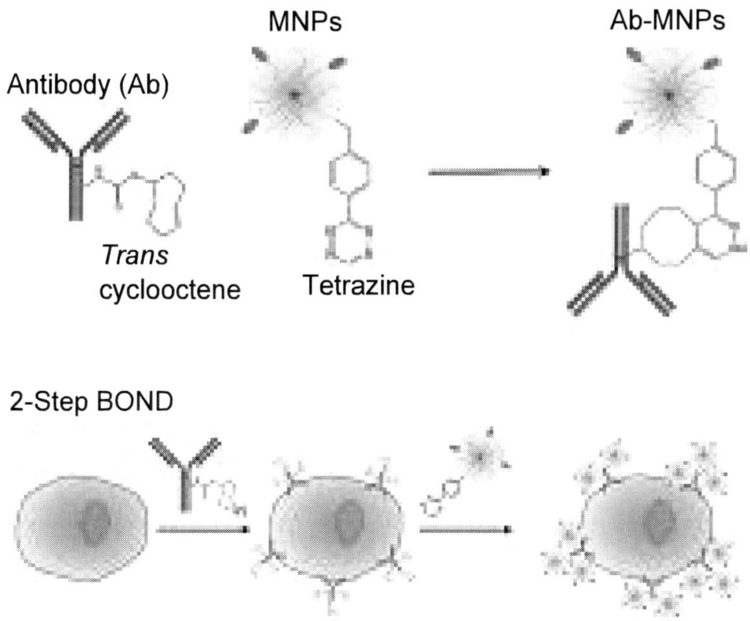

Figure 3. Bioorthogonal nanoparticle detection strategy for DMR detection.

Nuclear magnetic resonance (NMR) can be detected with instruments such as clinical MRI scanners (routinely used for deep tissue whole body imaging), and NMR spectroscopy (used to study proteins and small molecules). Both of these techniques have been used to measure T_2 relaxation time for DMR biosensing. However, because these conventional instruments are bulky and expensive, they remain as specialized equipment in hospitals and laboratories. Benchtop relaxometers, which operate at lower NMR frequencies (100 kHz–50 MHz) with a permanent low-field magnet (<1 T), provide a lower-cost alternative for DMR biosensing [44]. However, these systems lack the capability for performing multiplexed measurements, and require large sample volumes (>100 μL) to achieve accurate measurements.

To overcome the limitations of conventional detectors and to address the need for fast, simple and high-throughput biosensing, a chip-based microNMR (μNMR) device has recently been developed. This miniaturized DMR device consists of an NMR probe containing microcoils for radio-frequency (RF) excitation and NMR signal detection, on-board NMR electronics, a microfluidic network for sample handling, and a small permanent magnet for generating an external magnetic field.

The first μNMR prototype was designed with a 2 × 4 planar microcoil array that was lithographically patterned onto a glass substrate (Figure 4a). This array format enabled the performance of parallel measurements, and each microcoil held 5–10 μl of sample. In the second-generation μNMR, the design was changed to solenoidal coils, since this geometry provides higher signal-to-noise ratio (SNR) by producing more homogeneous radio-frequency magnetic fields for sample excitation. The SNR could be further increased by integrating the coil with a microfluidic channel (Figure 4b). The solenoidal coils were first wound around polyethylene tubes and subsequently immersed into a polymer (polydimethylsiloxan). Following polymer curing, the tubes were retracted to open up fluidic channels. The entire bore of the solenoid thus can be filled with sample to achieve the maximal filling factor (≈1), the fraction of the coil volume occupied by the sample. Due to the larger cross-sectional area of the winding wires, the solenoidal coils also have less electrical resistance than lithographically patterned coils. With these advantages, the sample volume for DMR detection could be reduced by a factor of ~10 (to 1 μL) compared to the previous devices (~10 μL).

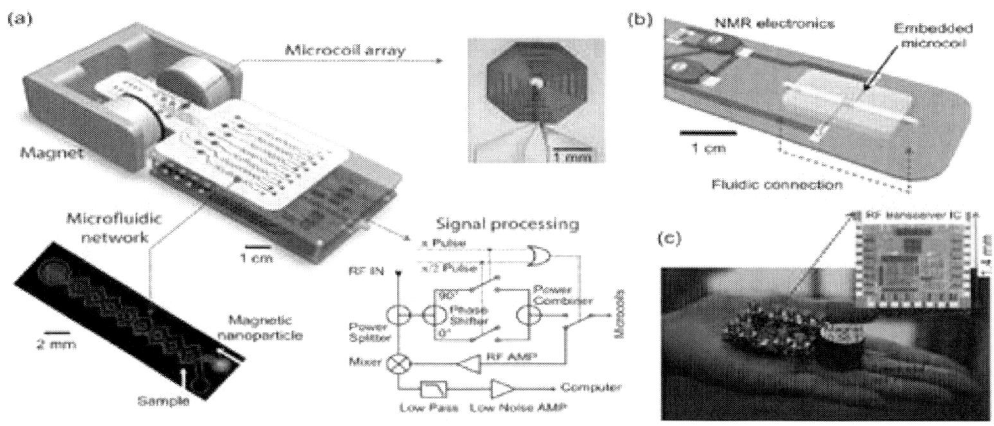

Figure 4. The devices developed for DMR biosensing.

The microfluidic networks in the DMR system facilitate the handling of biological fluids, the effective mixing of MNPs with small sample volumes, and the distribution of small volumes to different coils for parallel sensing. The networks also serve to confine the samples to the most sensitive region of a given microcoil. Furthermore, a membrane filter can be inserted at the outlet of the solenoidal microcoil to retain large biological targets, while removing smaller contaminants such as unbound MNPs. This configuration enables both the concentration of scant samples from large volumes, as well as the performance of on-chip washing steps. The NMR electronics generate versatile RF pulse sequences to measure the longitudinal (T_1) and transverse (T_2) relaxation times, process raw NMR signals (amplification, frequency conversion, filtering) for acquisition by a computer, and handle the multiplexed operation of an array of coils. In the first prototype, the NMR electronics was constructed as a tabletop system using discrete RF chips (e.g., AD9830 for RF generation and AD604 for NMR signal amplification; Analog Devices) and off-the-shelf RF components (e.g., ZAD-1 mixer, ZMSC-2 power splitter, and ZYSWA-2 RF switch; Mini-Circuits) [45]. In newer versions, these functionalities have been integrated onto a single CMOS IC chip. This chip was designed to overcome the adverse effects associated with system miniaturization during NMR measurements, including 1) low NMR signals resulting from small sample volumes, and 2) fast signal decay due to field inhomogeneity produced by the portable magnet. These challenges were addressed by implementing low noise RF amplifiers with high voltage gain, and by developing an on-chip digital pulse generator for various pulse sequences. The latter enabled the accurate measurement of transverse (T_2) relaxation times by generating Carr–Purcell Meiboom–Gill (CPMG) sequenes to compensate for the inhomogeneity of the polarizing magnetic field.

In the latest implementation, the entire DMR system was packaged as a handheld unit for portable operation ("palm" NMR system; Figure 4c). When benchmarked against conventional NMR systems, these miniaturized devices provided both superior detection sensitivities and capabilities for multiplexed measurements on small sample volumes. In view of such advantages, the miniaturized DMR technology is well suited for fast, simple and high-throughput analysis of scant biological samples within a point-of-care setting. DMR applications have been successfully applied to sensitively identify and quantify a wide range of biological targets including DNA/ mRNA, proteins, enzyme activities, small molecules/drugs, bacteria, viruses and mammalian tumor cells. As described previously, the detection mode of DMR depends on the size of its target.

The concept of forward MRSw sensing can be demonstrated with proteins, for example avidin, in model applications. In one early series of experiments, biotinylated MNPs were incubated with varying amounts of avidin. As shown in Figure 5a, the binding of biotin to avidin resulted in the clustering of MNPs and a concomitant avidin concentration-dependent change in T_2. By varying the concentration of MNPs, four orders of dynamic ranges were achieved, indicating that the system has a robust working range. Likewise, specific antibodies can also be used to perform MRSw on target protein molecules [46]. As the second proof-of-principle analysis, green fluorescent protein (GFP)- sensitive nanoparticles were prepared by conjugating CLIO nanoparticles with anti-GFP polyclonal antibodies.

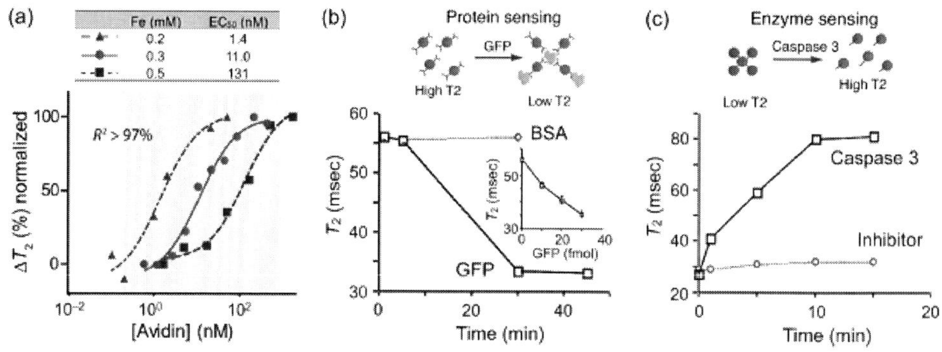

Figure 5. DMR detection of proteins and enzyme activities with MRSw sensors.

Using this system, GFP was rapidly and sensitivity detected in a dose dependent manner while the addition of bovine serum albumin (BSA) protein as a control did not elicit any change in T_2 (Figure 5b). More recently, MRSw biosensors, capable of detecting soluble tumor biomarker proteins (such as CA-125, VEGF, and α-fetoprotein) were described, and used for parallel detection of multiple markers in blood samples with the μNMR device. Finally, using the BOND-2 method, many other cancer proteins have been detected. Reverse sensors have been designed to detect and quantify proteases, endonucleases and methylases. In these assays, the enzyme activity disassembles pre-formed clusters of MNPs. This disintegration translates the enzymatic activity into a detectable T_2 signal. In the first demonstration of this strategy, MNP aggregates were formed with the peptide sequence biotin- GDEVDGC. This sequence served as a linker, binding both an avidin-conjugated CLIO population (via the biotin/avidin interaction) as well as a second CLIO population (via the thiol provided by the terminal cysteine on the peptide). The subsequent addition of caspase-3 disassembled the aggregates by cleaving within the DEVD site, which led to a corresponding increase in the T_2 relaxation time (Figure 5c). This dissociation was not observed when a specific caspase-3 inhibitor was added. A similar reverse switching strategy has been used to detect trypsin, renin, and matrix metalloproteinase 2 activities.

Forward MRSw assays on enzymatic activities have also been demonstrated via the assembly of nanoparticle biosensors (as a result of enzymatic reactions). For example, specific MNPs have been designed to assess human telomerase (hTERT) activity by hybridizing with the 30-base pair telomeric repeat sequences produced by hTERT activity. More recently, myeloperoxidase (MPO) sensors were generated by attaching phenol-containing molecules, such as dopamine or serotonin, to CLIO nanoparticles. In the presence of peroxidase activity, tyroxyl radicals were formed to cross-link the nanoparticles. Using the same assay configuration, leukocyte-derived MPO has been shown to play a critical role in the pathogenesis of atherosclerotic plaques.

Bacteria detection and quantification of large pathogens have been successfully demonstrated using the DMR platform, primarily through magnetic tagging of targets. For example, detection of the bacterium *Staphylococcus aureus* was recently reported with the μNMR device. *S. aureus* were initially incubated with MNPs derivatized with vancomycin, a drug which binds to D-alanyl–D-alanine moieties in the bacterial cell wall to form dense clusters (Figure 6a).

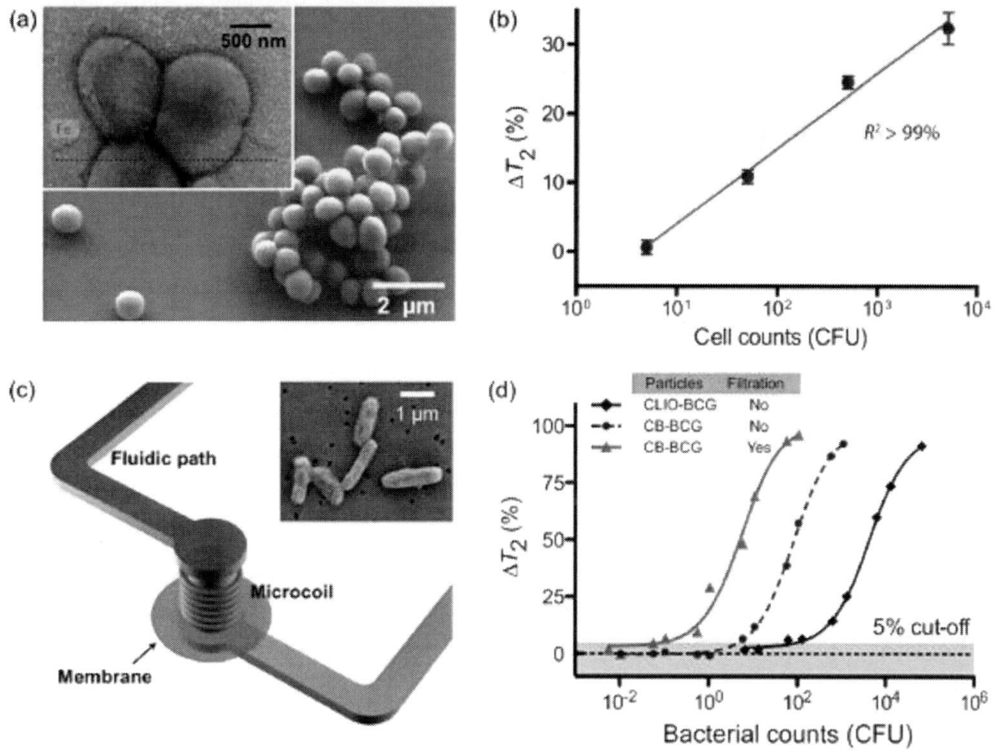

Figure 6. DMR detection of bacteria by tagging the bacterial samples with MNPs.

Due to the low sample volume required by the μNMR device, this first proof-of concept analysis demonstrated a detection sensitivity of only a few colony-forming units (CFUs) per microliter sample (Figure 6b). More recently, tuberculosis (TB) bacteria have been detected using DMR. In one study, the highly magnetic Fe-core/ferrite shell nanoparticles (CB; cannonballs) were used in combination with the second generation DMR device. To evaluate clinical utility of the DMR platform for TB detection, Bacillus Calmette-Guérin (BCG), used as a surrogate for *Mycobacterium tuberculosis*, was spiked into sputum samples. Following liquefaction, the biological samples were incubated with cannonballs conjugated to an anti-BCG monoclonal antibody. Unbound MNPs were then removed via a built-in membrane filter, embedded within the device (Figure 6c). This membrane (~100 nm size cut-off) not only removed excess unbound MNPs but also retained the BCG bacteria. It was effective for both concentrating scant bacteria and removing the background signal. In comparison to standard TB diagnostics, which involve time-consuming culture and acid-fast bacilli (AFB) smear microscopy, the DMR diagnostic technology showed unprecedented detection sensitivity and speed: as few as 20 CFUs could be detected in 1 ml of sputum sample, in less than 30 minutes (Figure 6d). Currently, this detection technology is being adapted to detect infectious pathogens in clinical sputum samples [47].

Sensitive detection and rapid profiling of tumor cell surface markers in unprocessed biological samples will undoubtedly have a significant impact on both the life sciences and clinical practice. DMR molecular profiling of Her2/*neu*, EGFR, and D326 (EpCAM) cancer markers on mammalian cells was first demonstrated using the first-generation DMR device. In these early experiments, CLIO nanoparticles were directly conjugated to monoclonal antibodies. More recently, the use of BOND-2 strategy has further advanced DMR profiling capabilities and Western blot analysis (Figure 7c). Importantly, the DMR detection platform not only required far fewer cells than either of the alternative approaches, but also produced results in a fraction of the time (<15 minutes). The DMR platform has since been shown to be adaptable to rapid multi-target detection, where putative cancer cells can be profiled for multiple biomarkers. DMR is ideally suited to this use since it can perform measurements on a few cells in small sample volumes and in a multiplexed manner. Fine-needle aspirate biopsies from a panel of mouse xenograft tumors have already been successfully analyzed for Her2/*neu*, EGFR, and EpCAM expression [48]. Furthermore, the multiple-marker targeting strategy has been shown to significantly improve the accuracy for correctly diagnosing cancer cells as malignant (Figure 7d).

These, in addition to other advanced refinements to DMR sensing, are currently being applied to clinical trials of cancer cell profiling.

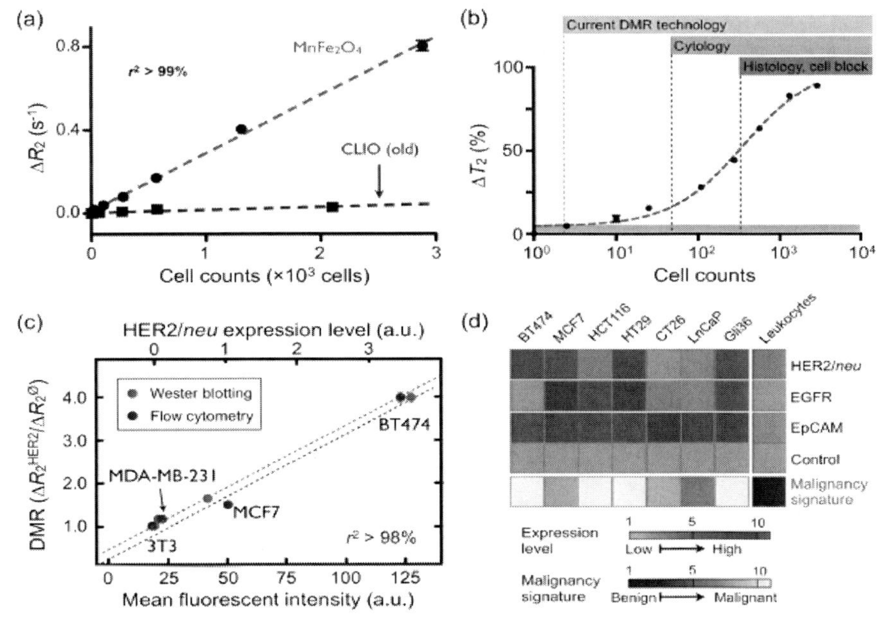

Figure 7. Tumor cell detection and profiling with the μNMR device.

9.4. One-Pot Synthesis of Monodisperse Iron Oxide Nanoparticles for Potential Biomedical Applications

Magnetic nanoparticles with a size up to 20 nm in diameter represent an important class of artificial nanostructured materials. Their magnetic properties change drastically with the sizes as magnetic anisotropic energy, KV, where K is the magnetic anisotropic constant and V is the particle volume, becomes comparable to the thermal energy, *k*T, resulting in moment randomization and superparamagnetism. Such superparamagnetic nanoparticles have great potential for biomedical applications.

Their magnetic signal far exceeds that from any of the known bio-entities, making them readily identified in the ocean of biomolecules. Without an external magnetic field, they show no net magnetic moment, facilitating their long-term stability in various dispersion media. They are smaller than or comparable to a cell (10–100 nm), a virus (20–450 nm), a protein (5–50 nm), or a gene (2 nm wide and 10–100 nm long).

These, plus their capability of being manipulated under an external magnetic field, provide controllable means of magnetically tagging of all biomolecules, leading to highly efficient bioseparation/biodelivery and highly sensitive biolabeling and magnetic resonance imaging (MRI) contrast enhancement.

To apply superparamagnetic nanoparticles for biomedical applications, the nanoparticles should be monodispersed to have uniform physical and chemical properties for controlled biodistribution, bioelimination, and contrast effects. The magnetic nanoparticles should also have high magnetic moment, and can be so modified that they are capable of binding specifically to a biological entity and able to withstand various physiological conditions. Iron oxide nanoparticles, due to their chemical and magnetic stability and low level of toxicity in biological systems, have been widely tested for their use in biomedicine. However, some well-known material problems need to be solved before these nanoparticles can be utilized for any practical applications.

The iron oxide nanoparticles used for the tests are often polydispersed with a large variation not only in size, but also in shape. Consequently, the physical and chemical properties of these particles are not well controlled, and important data on biodistribution/ bioelimination in biological systems, which are essential for in vivo applications, are currently difficult to obtain.

Monodisperse ferrite MFe2O4 nanoparticles were recently made by a high-temperature reduction/ decomposition reaction of metal acetylacetonate. [49]. The size of the particles was controlled up to 8 nm from the one-step reduction/decomposition reaction. Larger size, up to 20 nm, was made possible by seed-mediated growth in which small MFe_2O_4 nanoparticles were used as seeds and more MFe_2O_4 was coated over the seeds. By controlling the heating parameters, the reaction further led to the ferrite nanoparticles with cube- or polyhedron-like shapes.

This reduction/decomposition synthesis is complimentary to other reports in iron oxide nanoparticle syntheses from high-temperature decomposition reactions. The iron precursors used in this new synthesis are commercially readily available and less toxic than the iron pentacarbonyl, $Fe\ (CO)_5$, a common precursor used in thermal decomposition reaction[50]. However, the synthesis of larger nanoparticles (>10 nm) via a seed-mediated growth method involves multiple-step syntheses and as a result, the process leading to 15–20 nm nanoparticles is time-consuming. It is desired that, for practical applications, the iron oxide nanoparticles with a range of sizes can be prepared via a one-step synthesis. The reports show that a one-pot reaction of metal acetylacetonate with polyol in the presence of oleic and oleylamine to prepare Fe_3O_4, $MnFe_2O_4$, and $CoFe_2O_4$ nanoparticles with the size tunable from 5 to 15 nm. The studies demonstrate that these nanoparticles are magnetically stable and can be readily functionalized for potential biomedical applications.

9.5. Advanced Polymeric Materials for Biomedical Applications

The use of polymers as biomaterials has grown tremendously in the past two decades. Currently, various types of polymeric materials are used for a variety of biomedical applications such as sutures, contact and intraocular lenses, drug carriers for sustained delivery, tissue and blood substitutes and artificial organs.

9.5.1. Development of Intelligent Nanoparticles for Cancer Detection and Treatment

Cancer remains one of the leading causes of death in most parts of the world. It remains one of the world's most devastating diseases; with more than 10 million new cases every year. Early diagnosis of cancer is essential to determine the treatment regime. Computerized tomography, positron emission tomography, magnetic resonance imaging and ultrasound imaging are some of the techniques used for cancer detection. However these techniques are expensive, require lab facilities and highly trained technicians, time consuming and cannot be taken to the field condition. Early diagnosis of the disease and better understanding of the pathophysiology of tumor progression has opened many new vistas as therapy options. In most solid tumors, after its surgical removal, the remaining cancer cells are managed with a variety of treatment options including, radiotherapy, chemotherapy, immunotherapy, etc. However, once the cancer is metastasized, the treatment options are limited, and chemotherapy remains the choice of treatment. The main reason for failure of chemotherapy is the poor accessibility of antineoplastic agents to the tumor, requiring higher doses, and the nonselective nature of these agents causes severe toxicity. Nanotechnology is in a unique position to produce a new generation of early cancer detection techniques with higher sensitivity and delivery of the drug to the target site.

9.5.2. Development of Dendrimers/Nanoparticles Based Systems for Bacterial Detection and mitigation

Rapid and sensitive identification of pathogenic bacteria is extremely important for the drinking water supply, in path labs, the veterinary, the food & beverage industry as well as in applications to microbiological threat detection and reduction. Several research groups are working on a DBT sponsored research project entitled "Development of low cost biosensor for rapid detection of pathogenic bacteria" from Aug. 2007. A pathogenic bacterial strain of *Salmonella typhi (S. typhi)* was obtained from Rockland Hospital, Delhi and maintained by repeated subculturing and glycerol stock preparation. Polyclonal antiserum against the *S. typhi* was generated using whole bacterial cells as immunogens in New Zealand White Rabbit and purified using Protein A affinity column. Surface modification of nylon membrane & polyacrylonitrile fiber (PAN) was undertaken with an aim to use them as solid matrices for ELISA/ FIA after immobilization with a bacterial antibody. Modified membranes and fibers were characterized using various techniques. Both the matrices were standardized using

RAG-IgG/GAR-IgG/GAR-HRP as a model antibody system. *S. typhi* bacteria detection in water samples were carried out with the *S. typhi* antibody immobilized nylon membrane (Figure 8). A detection limit of 104 cfu/ml of *S. typhi* could be primarily established as the lower detection limit for the developed Sandwich ELISA method using HRP labeled *S. typhi* antibodies within 2 hours. Research activities are in progress to further decrease the detection limit in the range of 103-102 cfu/ml. Intraday and inter-day variations were also found to be below 10%. Specificity studies for the developed method showed the negligible cross reactivity with other Enterobacteriaceae members (*S. paratyphi A, Klebsiella pneumonia, Shigella sonnei*, *E. coli* and *Pseudomonas aeruginosa*).

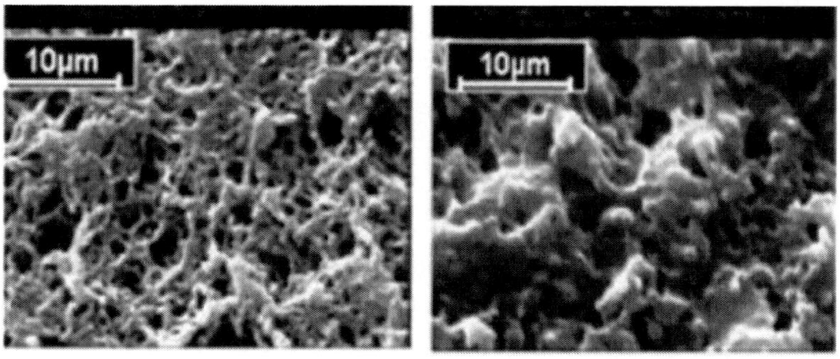

(a) (b)

Figure 8. Scanning electron micrographs of (a) nylon membrane (b) GMA grafted nylon membrane.

Fluorescent core-shell QDs were also synthesized using the hot injection method. Synthesized inorganic QDs were made hydrophilic by the ligand exchange method for the conjugation with biomolecules. Antibody conjugated QDs were used as detector molecules for the detection of *S. typhi.* Preliminary experimental studies were carried out for detection of *S. typhi* in water using a bacteria specific antibody (Sty IgG) bioconjugated to carboxy terminated CdSe/CdS/ ZnS QDs. Bacteria-bioconjugated QDs complex were seen as fluorescent aggregates under a confocal laser scanning fluorescent microscope (Figure 9). Further studies towards the optimization of bacteria detection using bioconjugated QD's and PAN fibers are in progress.

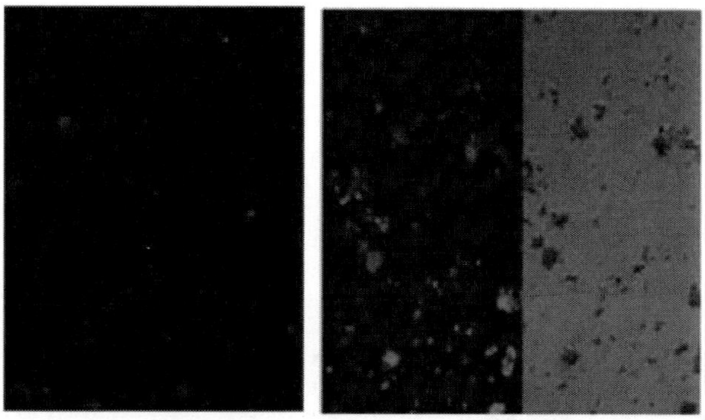

(a) (b)

Figure 9. (a) Confocal Laser Scanning Microscope image (CLSM) of S. typhi antibody conjugated QDs (b) CLSM image (right) and DIC image (left) of conjugated QDs.

9.5.3. Development of Polymeric Micro and Nanoparticle Based Insulin Delivery Systems for Diabetics

The prevalence of diabetes mellitus has been growing by leaps and bounds throughout the world. Major factors behind the increasing diabetic population are hereditary and environmental. India is host to the largest diabetic population in the world with an estimated 35 million people, amounting to 8% of the adult population. Insulin dependent diabetic mellitus is an autoimmune disorder characterized by the destruction of the insulin secreting beta cells of Langerhans of the pancreas and consequent inability to maintain blood glucose homeostasis. Presently the only feasible treatment is administration of insulin through subcutaneous injection. However, the problems encountered with subcutaneous insulin injections are pain, allergic reactions, hyperinsulinemia and chances of infection. The delivery of insulin by nonparenteral routes has gained significant attention over last two decades. The alternate routes explored are ocular, nasal, buccal, rectal, pulmonary and oral. The insulin delivery systems are developing as an alternative to the multiple insulin injections. Following two strategies have been adopted.

Oral insulin delivery system- An oral route is generally not feasible for peptides such as insulin due to rapid enzymatic degradation in the stomach, inactivation and digestion by proteolytic enzymes in the intestinal lumen and poor permeability across the intestinal epithelial because of its high molecular weight. To overcome these problems, the pH sensitive mucoadhesive polymeric micro and nanoparticles of poly (ethylene glycol) dimethacrylate - methacrylic acid have been synthesized and evaluated as a carrier for the oral delivery of insulin. Due to their pH sensitive nature, these particles shrink at low pH in the stomach and protect the insulin from proteolytic enzymes and the harsh acidic environment, while at basic pH in the intestine the particles will swell and release the insulin. Polymeric micro and nanoparticles were synthesized by the free radical suspension and emulsion polymerization techniques. During in vitro insulin release studies, at pH 2.5, the release of insulin from polymeric particles was observed in the range of 5-8% while a significantly higher release (20- 35%) was observed at pH 7.4 during first 15 minutes of in vitro release studies. It was observed that polymeric microparticles were more effective than nanoparticles in reducing the blood glucose level in diabetic rabbits. Biodegradable block copolymers of PLA-PEG with various chain length of PEG (575-4000) were synthesized and characterized by various techniques. Nanoparticles of PEG-PLA were prepared by a double emulsion technique and insulin was encapsulated at the time of particle formation. Loading of insulin increased with an increase in the PEG chain length used during polymerization. Loading efficiency was calculated by HPLC and a maximum loading efficiency of 52% was observed in the PLA-PEG nanoparticles with the PEG chain length of 4000. In-vitro release studies revealed a significant controlled insulin release of up to 8 days. Injected insulin loaded nanoparticles controlled the blood glucose level of diabetic rabbits in the normal range up to 7 days.

9.6. Nanoceramic Matrices: Biomedical Applications

Tissue engineering was defined by Langer and Vacanti as an interdisciplinary field that applies the principles of engineering and life sciences toward the development of biological substitutes that restore, maintain, or improve tissue function [51]. It can also be defined as the use of a combination of cells, engineering materials, and suitable biochemical factors to improve or replace biological functions in an effort to effect the advancement of medicine. The total market for the regeneration and repair of tissues and organs has been estimated to be $25 billion worldwide. Although transplantation of organs has become an established and successful method of therapy, the severe scarcity of donor organs and immune rejection has become a major limitation and has stimulated the tissue engineering field. The most commonly used materials for encapsulating cells or as scaffolds in tissue engineering are synthetic polymers such as poly lactic glycolic acid (PLGA), poly vinyl alcohol (PVA), poly vinyl chloride (PVC), polyacrylonitrile (PAN) etc. and natural polymers like alginates, chitosans or collagen and ceramic matrices like calcium phosphates. Two different approaches were being employed in the assisted self assembly of cells. In the first approach, donor cells and growth factors were seeded *in vitro* into the biodegradable scaffold. After the required period for cell growth and multiplication, the scaffold was surgically implanted into the body for the generation of healthy new tissue. In the second approach, the scaffold was implanted into the damaged area along with the cells and the growth factors to stimulate tissue regeneration. Scaffolding architecture plays an important role on these three dimensional matrix supports composed of cells and extra-cellular matrix similar to natural tissue organization. Mimicking the natural tissue organization contributes significantly to the biological function of the tissue engineered material. Enormous advances have been made in the field of materials science with the advent of nanotechnology. Bone tissue engineering is a specific area in nanotechnology where the development of nanostructured biomaterials may be able to replace hard and soft skeletal tissue, and biocompatible materials for tissue genesis. Other related areas also include creation of nanoporous biocapsules for cellular therapy [52].

Tissue engineering stands to benefit most from the nanotechnology because of the growing ability to fabricate complex nanostructured materials. This section discusses some of the important developments in the utilization of nanoceramic matrices in biomedical technology particularly in tissue engineering.

9.6.1. Bioceramics Used in Biomedical Technology

Calcium phosphate ceramics like hydroxyapatite, tricalcium phosphate and tetracalcium phosphate, alumina, zirconia, silica based glasses or bioactive glasses and also pyrolitic carbons, generally termed as bioceramics, have been used as bone substitutes. These materials range from inert to bioactive based on their biological activity as they can remain unchanged, dissolve or actively take part in physiological processes to enhance bone tissue formation. Bioactive ceramics provide a suitable substrate for the attachment of the cells followed by osteogenic differentiation directly on the surface of the ceramic, which results in bone bonding. Several studies indicated that growth hormones can stimulate bone growth, collagen synthesis and can effect repair *in vivo*. Scaffolds have been developed containing osteoinductive proteins with and without osteogenic cells utilizing various polymers, ceramics and polymer-ceramic composite towards bone tissue engineering. Ceramics are brittle and not suitable for load bearing applications in the case of orthopedic implants. Numerous studies are being reported on polymer-ceramic composites for improving the mechanical properties to match with the natural load bearing tissues [53]. Bioactive ceramic materials like tricalcium phosphates, hydroxyapatite and bioglass gained significant importance as scaffolds for bone tissue engineering in recent times. These scaffolds can either induce the formation of bone from the surrounding tissue or can act as a carrier or guide for enhanced bone regeneration by cell migration, proliferation and differentiation. The main issue regarding the applicability of bioceramics in the field of bone tissue engineering is its degree of biodegradability (resorption) along with its poor mechanical strength. Nanotechnology can play an important role in this regards to develop porous bioceramics with high mechanical strength and enhanced bioactivity and resorbability.

9.6.2. Bone Tissue Engineering

The design of biomaterials with surface properties similar to physiological bone that demonstrates the basic criteria of osteoinduction and osteoconduction, promotes formation of new bone and improved orthopedics. Bone tissue engineering is a comparatively new approach in the repair of bony defects. The recent developments of tissue engineering in the field of orthopedic research make it possible to envisage the association of autologous cells and proteins that promote cell adhesion with osteoconductive material to create osteoinductive materials. The nature of interaction between osteoblast cells and their substrate can influence the ability of these cells to produce an osteoid matrix around an implant, which in turn will determine the fate of the implant. Materials used as a bone tissue engineering scaffold should be biocompatible and porous to be osteoinductive and osteoconductive, and should have the mechanical property compatible with the natural bone tissue. It should also induce cell anchorage and remodel the extracellular matrix in order to integrate with the surrounding tissue. The World Health Organization along with 37 countries and the United Nations have proclaimed the years 2000-2010 as the 'Bone and Joint Decade'. This global initiative is intended to improve the lives of people with musculoskeletal disorders, such as arthritis, and to advance understanding and treatment of musculoskeletal disorders through prevention, education and research.

The chemical composition of the tissue engineered scaffold is crucial for the resorbability and osteoconductive properties together with its internal porous structure for the vascular growth. Porous bioceramics hydroxyapatite (HA) and tricalcium phosphate have been demonstrated to have good osteoconductive properties resulting in a good functional recovery. However, the resorption was poor. A novel polymernano hydroxyapatite porous scaffold has been prepared without the use of organic solvents which exhibited significantly higher cell growth, alkaline phosphatase activity, and mineralization *in vitro* compared to a scaffold prepared with organic solvents [54]. Also on implantation, the histological evaluation showed that bone formation was more extensive on scaffold prepared without using organic solvents. It has been proposed that the use of culture-expanded osteoprogenitor cells in conjunction with hydroxyapatite bioceramics significantly improves the repair of critical size long bone defects. A macroporous hydroxyapatite scaffold coated with PLGA-bioglass demonstrated formation of an apatite layer on sample surfaces immersed in the simulated body fluid (SBF) for 5 days. This also had increased compressive strengths and seems to be a possible scaffold for bone tissue engineering [55]. HA ceramic matrices cultured with mesenchymal stem cells (MSCs) obtained from respective patient's bone marrow cells were forced to differentiate into osteoblasts and bone matrix was formed with hydroxyapatite (HA) ceramic. This tissue-engineered HA was used to fill the patient's bone cavity after tumor curettage and exhibited immediate healing potential. It has been suggested as an alternative to autologous bone grafts. It has also been found effective as 'bone graft' substitute for spinal fusion [56].

9.6.3. Nanotechnology in Bioceramics

Nanosized hydroxyapatite [HA) is the main component of mineral bone in the form of nanometer sized needle-like crystals of approximately 5-20 nm width by 60 nm length. Synthetic HA possesses exceptional biocompatibility and bioactivity properties with respect to bone cells and tissues. They have been widely used clinically in the form of powders, granules, dense and porous blocks and various composites. For drug delivery as well as for tissue engineering applications, the present trend is to develop new formulations of HA with properties closer to those of living bone, such as nanosized and monolithic structures. Nanophase HA properties such as surface grain size, pore size, wettability, etc., could control protein interactions modulating subsequent enhanced osteoblast adhesion and long-term functionality.

Synthetic materials could not be considered as ideal implants as the current average lifetime of an orthopedic implant, such as hip, knee, ankle, etc., is only 15 years. Conventional implants, or those implants constituted with grain size dimensions greater than 1 micron could not invoke natural cellular responses to regenerate tissue that allows longer periods of successful life time [57]. However, since nanophase materials can mimic the dimensions of constituent components of natural tissues, implants developed from nanophase material can be a successful alternative. Several encouraging reports on nanophase materials encourage its use for tissue engineering applications. This has been achieved by the combined effect of its ability to mimic the natural nano dimensions and also the cell responses encouraging high reactivity and in turn helping in regenerating tissues.

NanOss® bone void filer from Angstrom Medica is considered to be the first nanotechnology medical device to receive clearance by the US Food and Drug Administration in 2005. According to the company, NanOss® is an innovative structural biomaterial that is highly osteoconductive and remodels over time into human bone with applications in the sports medicine, trauma, spine and general orthopedics markets[58]. It is engineered synthetic bone developed from nano crystalline calcium phosphate and is the first material that duplicates the microstructure, composition and performance of human bone. Utilizing nanotechnology, calcium and phosphate are manipulated at the molecular level and assembled to produce materials with unique structural and functional properties. It is prepared by precipitating nanoparticles of calcium phosphate in aqueous phase and the resulting white powder is compressed and heated to form a dense, transparent, and nano crystalline material. It is strong and also osteoconductive [59]. Ostim® is an injectable bone matrix in paste form which received CE marking in 2002. It is composed of synthetic nanoparticulate hydroxyapatite which is indicated for metaphyseal fractures and cysts, acetabulum reconstruction and periprosthetic fractures during hip prosthesis exchange operations, osteotomies, filling cages in spinal column surgery, in combination with autogenous and allogenous spongiosa and filling in defects in children etc.

Cell spreading is an essential function of a cell, which has adhered to any surface and precedes the function of cell proliferation. Out of the bone and the ceramic material interactions that take place at the material surface, the interaction of osteoblasts is crucial in determining the tissue response at the biomaterial surface. Attachment and spreading of specific bone forming cells in cell culture has been utilized for predicting the behavior of the calcium phosphate materials *in vivo*. The process of cell interaction on materials is highly dynamic and depends on various parameters influencing the cell responses. It is well known that the size and shape of the cell spreading area, as well as the number, size, shape and distribution of focal adhesion plaques are decisive for further migratory, proliferative and differentiation behavior of anchorage dependent cells. Cells usually do not survive if the attached cells are round and are not spreading. If the cell material contact area is significantly high, the cells tend to skip the proliferation phase and enter the differentiation program sooner. If the adhesion is intermediate, the cells are most active in migration and proliferation. Minerals like zinc and magnesium are known to aid bone growth, calcification and bone density.

Biphasic calcium phosphate ceramic containing zinc also promotes osteoblastic cell activity *in vitro* [60]. The attachment and spreading of osteoblast like cells onto porous ceramic materials made from nanoparticles of zinc phosphate and calcium phosphate containing zinc and magnesium has been evaluated. Furthermore, since the ceramic matrices are made from nanoparticles, it mimics the way nature itself lays down minerals. An important objective of bone tissue engineering is to develop improved scaffold materials or arrangements to control osteoblast behavior significantly affecting its response. Osteoblastic cells on hydroxyapatite exhibited unique attachment and subsequent behavior *in vitro*, which may explain why mineralized tissue formation is better on HA. Divalent cations including Mg^{2+} are known to be active in cell adhesion mechanisms. This investigation demonstrated that the cells were spreading well on a matrix containing an optimum amount of calcium, zinc and magnesium. It seemed that the presence of calcium and magnesium encouraged the spreading and adhesivity of osetoblasts cells onto bioceramic matrices. Cells were attached and spread completely on the ZnCaMgP matrix and this matrix appeared to be comparable to the control group (hydroxyapatite matrix) (Figure 10, 11).

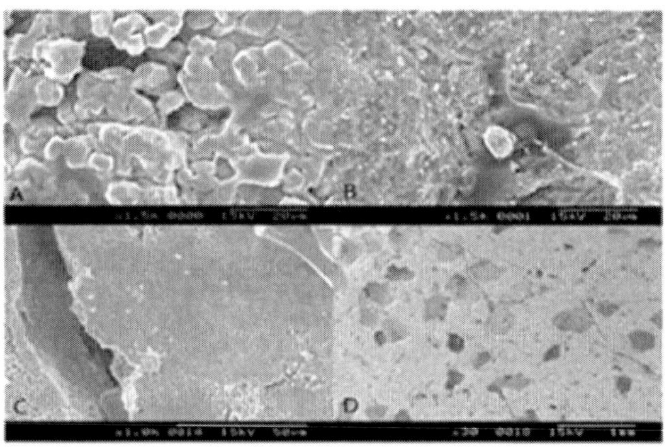

Figure 10. Osteoblast adhesions on hydroxyapatite matrix after 24 hours.

This makes the ZnCaMgP matrix a promising candidate as a bone tissue-engineering scaffold as an alternative to HA. Nanofibers up to a size range of 240nm have been developed based on the sol-gel and electrospinning technique and bioceramic matrices have been developed for possible tissue engineering applications. Attempts have been made to culture osteoblasts onto a nano hydroxyapatite ceramic matrix. This *in vitro* cultured bone may show further bone-forming capability after *in vivo* implantation. This tissue engineering approach is being tried on patients with skeletal problems. Adhesion of osteoblast, synthesis of alkaline phosphatase, and deposition of calcium containing mineral increased on poly lactic-co-glycolic acid (PLGA) scaffolds containing nanophase titania emphasizing the utility of nanotechnology in orthopedic tissue engineering materials [61].

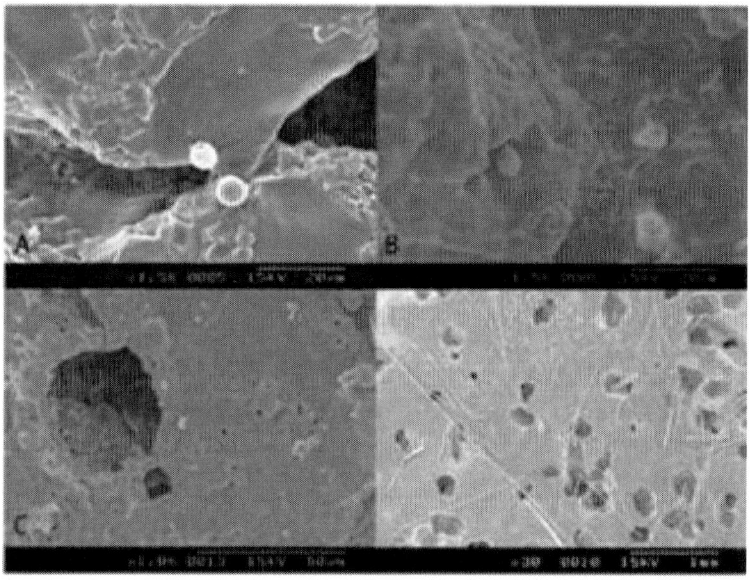

Figure 11. Osteoblast adhesive on CaZnMgP matrix after 24 hours and 5 days.

9.6.4. Oral Insulin Delivery

Type I diabetes is characterized by the inefficiency of pancreatic beta cells to produce insulin. The common form of insulin therapy is by way of twice daily subcutaneous insulin injections. Various attempts have been made to develop a non invasive delivery system for insulin namely, via oral, buccal or transdermal delivery routes etc. with varying levels of success. It is known that divalent metal cations form complexes with helical phosphates of DNA, which stabilizes the structure. Similarly, the two protein subunits in each insulin molecule will be bound to a single zinc atom with a significant increase in insulin stability. Therefore, zinc phosphate nanoparticles could be utilized for attaching insulin and deliver via an oral route with suitable pH sensitive sustained release coating which can protect insulin particles from hydrolysis and enzymatic degradation. The zinc phosphate particles with an average size of 200nm, attach insulin to these particles have been developed and coated with pH sensitive sodium alginate coating. Zinc phosphate nanoparticles were prepared by the precipitation of zinc in a phosphate solution. These particles were ultrasonicated for about 15 minutes at 100W. The particles obtained were washed with distilled water and freeze dried. Insulin (Human insulin 400 IU/ml) was added to the dry particle and kept overnight for insulin binding at 4°C. Insulin bound zinc phosphate nanoparticles were centrifuged to remove excess insulin and dried in a refrigerator. The particle size distribution of zinc phosphate nanoparticles is given in Figure 12. Particle size ranges from 190 to 295nm with a peak diameter of 226nm.

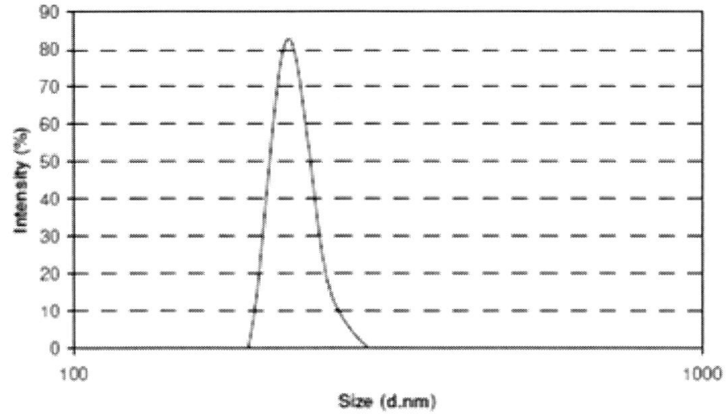

Figure 12. Particle size distributions of zinc phosphate nanoparticles.

Insulin loaded particles were dispersed in 1%, 2% and 4% solutions of sodium alginate for alginate coating. The size of these coated particles was much bigger and was possibly in the range of 1 micron as observed through an optical microscope. Insulin loading in the 3 formulations was evaluated by extracting the loaded insulin into a phosphate buffer pH 7.4 and estimating the protein content by Lowry's method. The activity of the extracted insulin was evaluated by the ELISA technique which was observed as 100%. The drug loading was 23 IU/100mg, 22 IU/100mg and 21 IU/100mg respectively for 1%, 2% and 4% chitosan coating. The release of insulin *in vitro* into simulated gastric as well as intestinal medium is shown in Figure 13.

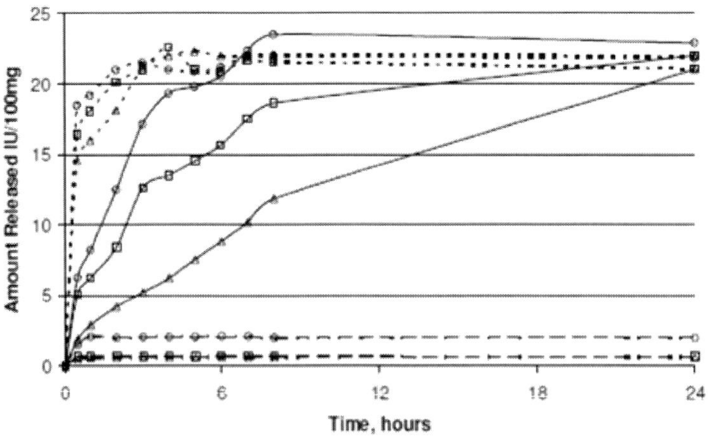

Figure 13. Amount of insulin released into SGF, SIF and SIF, 1%-circle, 2% -square and 4% triangle.

In the gastric medium, the release of insulin was negligible for all coatings and at pH 7.4 the release was almost instantaneous with 100% insulin release in the first 3 hours. However, at pH 6.8 (SIF as per USP) the release of insulin was sustained with significant sustenance for 4% coating. The uptake of particulate carriers by the gastrointestinal epithelium has encouraged a number of research groups to investigate nanoparticulate delivery for insulin. Translocation of alginate coated zinc phosphate particles through the intestinal lymphatic system may be used as a system for delivering insulin via oral route [62]. *In vivo* studies need to be completed to evaluate the uptake of particles through the intestinal epithelium. The pH sensitive ceramic insulin particles from zinc phosphate nanoparticles prepare and demonstrate its release profile *in vitro*. The result was promising towards its possibility in the development of a non-invasive oral insulin delivery system for diabetes. However, the question of degradability of the ceramic particles is yet to be studied. Calcium phosphate particles containing insulin was synthesized in the presence of PEG and coated with casein to obtain the calcium phosphate-PEG-insulincasein (CAPIC) oral insulin delivery system. When tested in NOD mice under fasting or fed conditions, it was observed that the biological activity of insulin is preserved in CAPIC formulation by protecting insulin from degradation while passing through the acidic environment of the GI track and displayed a prolonged hypoglycemic effect after oral administration.

Calcium phosphate nanoparticles can also be represented as a unique class of non-viral vector, which can serve as efficient and alternative DNA carriers for targeted delivery of genes. It has been demonstrated that surface modified calcium phosphate nanoparticles can be used *in vivo* to target genes specifically to the liver. Attachment of galactose moiety onto the particle surface has increased the target ability of the particles to the liver. This surface modification makes it possible for site-specific gene delivery [63].

REFERENCES

[1] Wagner V, Dullaart A, Bock A-K, Zweck A. The emerging nanomedicine landscape. *Nat. Biotech.* 2006 ; 24 : 1211 -7.

[2] Emerich DF, Thanos CG. Targeted nanoparticle-based drug delivery and diagnosis. *J. Drug Target* 2007 ; 15 (3): 163 -83.

[3] Groneberg DA, Giersig M, Welte T, Pison U. Nanoparticle-based diagnosis and therapy. *Curr. Drug Targets* 2006 ; 7 (6): 643 -8.

[4] Harris L, Batist G, Belt R, et al. Liposome-encapsulated doxorubicin compared with conventional doxorubicin in a randomized multicenter trial as first-line therapy of metastatic breast carcinoma. *Cancer* 2002 ; 94 (1): 25 -36.

[5] Uziely B, Jeffers S, Isacson R, et al. Liposomal doxorubicin: antitumor activity and unique toxicities during two complementary phase I studies. *J. Clin. Oncol.* 1995 ; 13 (7): 1777 -85.

[6] Zhang L, Granick S. How to stabilize phospholipid liposome (using nanoparticles). *Nano Lett.* 2006 ; 6 : 694 -8.

[7] Torchilin VP. Recent advances with liposomes as pharmaceutical carriers. *Nat Rev Drug Discov.* 2005 ; 4 : 145 -59.

[8] Duncan R. Polymer conjugates as anticancer nanomedicines. *Nat. Rev. Cancer* 2006; 6: 688 -701.
[9] Tanaka T, Shiramoto S, Miyashita M, et al. Tumor targeting based on the effect of enhanced permeability and retention (EPR) and the mechanism of receptor-mediated endocytosis (RME). *Int. J. Pharm.* 2004; 277 (1-2): 39 -61.
[10] Deguchi J-o, Aikawa M, Tung C-H, et al. Inflammation in atherosclerosis – visualizing matrix metalloproteinase action in macrophages in vivo. Circulation 2006 ; 114 : 55 - 62.
[11] Davis FF. The origin of pegnology. *Adv. Drug Deliv. Rev.* 2002 ; 54 (4): 457 -8.
[12] Svenson S, Tomalia DA. Dendrimers in biomedical applications–reflections on the fi eld. *Adv .Drug Deliv. Rev.* 2005 ; 57 (15): 2106 -29.
[13] Bosman AW, Janssen HM, Meijer EW. About dendrimers: structure, physical properties, and applications. *Chem. Rev.* 1999; 99 (7): 1665 -88.
[14] Moghimi SM, Hunter AC, Murray JC. Long-circulating and target-specifi c nanoparticles: theory to practice. *Pharmacol. Rev.* 2001 ; 53 (2): 283 -318.
[15] Gref R, Minamitake Y, Peracchia MT, et al. Biodegradable long-circulating polymeric nanospheres. *Science* 1994; 263 (5153): 1600 -3.
[16] Torchilin VP. Micellar nanocarriers: pharmaceutical perspectives. *Pharm Res* 2007 ; 24: 1 -16.
[17] Tobio M, Gref R, Sanchez A, et al. Stealth PLA-PEG nanoparticles as protein carriers for nasal administration. *Pharm. Res.* 1998 ; 15 (2): 270 -5.
[18] Perez C, Sanchez A, Putnam D, et al. Poly(lactic acid)-poly(ethylene glycol) nanoparticles as new carriers for the delivery of plasmid DNA. *J. Control Release* 2001; 75 (1-2): 211 -24.
[19] Langer R. New methods of drug delivery. *Science* 1990; 249 (4976): 1527 -33.
[20] Langer R, Tirrell DA. Designing materials for biology and medicine. *Nature* 2004; 428 (6982): 487 -92.
[21] Edelman ER, Mathiowitz E, Langer R, Klagsbrun M. Controlled and modulated release of basic fi broblast growth factor. *Biomaterials* 1991; 12 (7): 619 -26.
[22] Langer R, Folkman J. Polymers for the sustained release of proteins and other macromolecules. *Nature* 1976; 263 (5580): 797 -800.
[23] Langer R. Drug delivery and targeting. *Nature* 1998; 392 (6679 Suppl): 5 -10 24. Farokhzad OC, Cheng J, Teply BA, et al. Targeted nanoparticle-aptamer bioconjugates for cancer chemotherapy in vivo. *Proc. Natl. Acad. Sci. USA* 2006; 103 (16): 6315 -20.
[24] Kabanov AV, Batrakova EV, Alakhov VY. Pluronic block copolymers as novel polymer therapeutics for drug and gene delivery. *J. Control Release* 2002; 82 (2-3): 189 -212.
[25] Pandey R, Khuller GK. Nanotechnology based drug delivery system(s) for the management of tuberculosis. *Indian J. Exp. Biol.* 2006; 44 (5): 357 -66.
[26] Hirsch LR, Gobin AM, Lowery AR, et al. Metal nanoshells. *Ann Biomed Eng* 2006; 34 (1): 15 -22.
[27] Hirsch LR, Stafford RJ, Bankson JA, et al. Nanoshell-mediated near-infrared thermal therapy of tumors under magnetic resonance guidance. *Proc. Natl. Acad. Sci. USA* 2003; 100 (23): 13549 -54.
[28] 2001. Tissue Engineering: Technologies, Markets, and Opportunities, 3rd Edition, *Drug and Market Development Publishing*, MA, USA.

[29] Mastrogiacomo M, S. Scaglione, R. Martinetti, L Dolcini, F. Beltrame, R. Cancedda, R. Quarto, 2006. Role of scaffold internal structure on *in vivo* bone formation in macroporous calcium phosphate bioceramics, *Biomaterials,* 27(17), 3230-3237.

[30] Rush SM. 2005. Bone graft substitutes: osteobiologics, *Clin. Podiatr. Med. Surg. North Am.,* 22(4), 619-630.

[31] Giannoudis PV, H. Dinopoulos, E. Tsiridis, 2005. Bone substitutes: an update, *Injury,* 36 Suppl 3, S20-27.

[32] Hajime O., AI. Caplan, 1999. Stem cell technology and bioceramics: From cell to gene engineering, *J. Biomed. Mater. Res.* B, 48(6), 913 – 927.

[33] Claes O., B. Bengtsson, OGP Isaksson, TT Andreassen, MC Slootweg, 1998. *Growth Hormone and Bone Endocrine Reviews*, 19(1): 55- 79.

[34] Ramay HR., M. Zhang, 2004. Biphasic Calcium Phosphate Nanocomposite Porous Scaffolds for Load-bearing Bone Tissue Engineering, *Biomaterials,* 25(21), 5171-5180.

[35] Ramay HR., M. Zhang, 2003. Preparation of porous hydroxyapatite scaffolds by combination of the gel-casting and polymer sponge methods, *Biomaterials* 24(19), 3293-3302.

[36] Zhang Y, M. Zhang, 2002. Three-dimensional macroporous calcium phosphate bioceramics with nested chitosan sponges for load-bearing bone implants, *J. Biomed. Mater. Res.* 61(1), 1-8.

[37] El-Ghannam A, 2005. Bone reconstruction: from bioceramics to tissue engineering, *Expert Rev Med Devices.*, 2(1), 87-101.

[38] Anselme K., 2000. Osteoblast adhesion on biomaterials, Biomaterials, 21(7), 667-681.

[39] Puleo DA, R. Bizios, 1992. Formation of focal contacts by osteoblasts cultured on orthopedic biomaterials, *J. Biomed. Mater. Res.,* 26, 291.

[40] Scaglione S, A. Braccini, D. Wendt, C. Jaquiery, F. Beltrame, R. Quarto, I. Martin, 2006. Engineering of osteoinductive grafts by isolation and expansion of ovine bone marrow stromal cells directly on 3D ceramic scaffolds, *Biotechnol Bioeng.*, 93(1), 181-187.

[41] Yoshikawa H, A. Myoui, 2005. Bone tissue engineering with porous hydroxyapatite ceramics. *J. Artif. Organs.* 8(3), 131-136.

[42] Mastrogiacomo M, A. Muraglia, V. Komlev, F. Peyrin, F. Rustichelli, A. Crovace, R. Cancedda, 2005. Tissue engineering of bone: search for a better scaffold, *Orthod. Craniofac Res.*, 8(4), 277- 284.

[43] Kim SS, MS Park, O. Jeon, CC Yong, BS Kim, 2006. Poly(lactide-co-glycolide)/hydroxyapatitecomposite scaffolds for bone tissue engineering, *Biomaterials*, 27(8), 1399-1409.

[44] Cancedda R, M. Mastrogiacomo, G. Bianchi, A. Derubeis, A. Muraglia, R. Quarto, 2003. Bone marrow stromal cells and their use in regenerating bone, *Novartis Found Symp.,* 249, 133-143.

[45] Huang X, X. Miao, 2006. Novel Porous Hydroxyapatite Prepared by Combining H2O2 Foaming with PU Sponge and Modified with PLGA and Bioactive Glass, *J. Biomater. Appl.,* [Epub ahead of print].

[46] Morishita T, K. Honoki, H. Ohgushi, N. Kotobuki, A. Matsushima, Y. Takakura, 2006. Tissue engineering approach to the treatment of bone tumors: three cases of cultured bone grafts derived from patients' mesenchymal stem cells, *Artif. Organs*, 30(2),115-118.

[47] Tan KK, GH Tan, BS Shamsul, KH Chua, MH Ng, BH Ruszymah, BS Aminuddin, MY Loqman, 2005. Bone graft substitute using hydroxyapatite scaffold seeded with tissue engineered autologous osteoprogenitor cells in spinal fusion: early result in a sheep model, *Med. J. Malaysia.*, 60 Suppl C:53- 58.
[48] Hunter A, CW Archer, DS Walker, GW Blunn, 1995. Attachment and proliferation of osteoblasts and fibroblasts on biomaterials for orthopedic use, *Biomaterials*, 16(4), 287-294.
[49] Meyer U, A. Büchter, HP Wiesmann, U. Joos, DB Jones, 2005. Basic Reactions of Osteoblasts on Structured Material Surfaces, *European Cells and Materials*, 9, 39-49.
[50] Ba_áková L, E. Filová, F. Rypá_ek, V. Švor_ík, V. Starý, 2004. Cell adhesion on artificial materials for tissue engineering, *Physiol. Res.* 53, s35.
[51] Higashi A, T. Nakamura, S. Nishiyama, M. Matsukura, S. Tomoeda, Y. Futagoshi, M. Shinohara, I. Matsuda, 1993. Zinc kinetics in patients with bone demineralization due to physical immobilization, *J. Am. Coll. Nutr.*, 12, 61.
[52] Tucker KL, MT Hannan, H. Chen, LA Cupples, PWF Wilson, DP Kiel, 1999. Potassium, magnesium and fruit and vegetable intakes are associated with greater bone mineral density in elderly men and women, *Am. J. Clin. Nutr.*, 69, 727.
[53] Sogo Y, A. Ito, K. Fukasawa, T. Sakurai, N. Ichinose, 2004. Zinc containing hydroxyapatite ceramics to promote osteoblastic cell activity, *Materials Science and Technology*, 20, 1079.
[54] Paul W., CP Sharma, 2006. Effect of Calcium, Zinc and Magnesium on the Attachment and Spreading of Osteoblast like Cells onto Ceramic Matrices, *J. Mater. Sci. Mater. Med.,* (in press).
[55] Fitton, JH, 1995. Divalent Cation Modulation of Cell Adhesion to Surface, In: Proceedings 5th Annual Conference of Australian Soc. *Biomater.,* p.A 22.
[56] Hynes, RU, 1992. Integrins: Versatility, Modulation and Signaling in Cell Adhesion, *Cell,* 69, 11.
[57] Kim HW, HE Kim, 2005. Nanofiber generation of hydroxyapatite and fluor-hydroxyapatite bioceramics, *J. Biomed. Mater. Res. B. Appl. Biomater.*, [Epub ahead of print].
[58] 59. Ohgushi H, J. Miyake, T. Tateishi, 2003. Mesenchymal stem cells and bioceramics: strategies to regenerate the skeleton, *Novartis Found Symp*, 249:118-127.
[59] Webster TJ, TA Smith, 2005. Increased osteoblast function on PLGA composites containing nanophase titania, *J. Biomed. Mater. Res. A.*, 74(4), 677-686.
[60] Paul W., CP Sharma, 2005. Bioceramics, Towards Nano-enabled Drug Delivery: A Mini Review, *Trends Biomat. Artif. Org.*, 19(1), 7-10.
[61] Paul W., CP Sharma, 2001. Porous hydroxyapatite nanoparticles for intestinal delivery of insulin, Trends Biomat. *Artif. Org.*, 14(2), 37-38.
[62] Hussain N., V. Jaitley, AT. Florence, 2001. Recent advances in the understanding of uptake of microparticulates across the gastrointestinal lymphatics, *Adv. Drug Deliv. Rev.,* 50, 107-142.

Chapter 10

Nanoparticles for Photodynamic Therapy

10.1. Introduction

Photodynamic therapy (PDT) is a light-activated treatment for malignant or premalignant cancer tumors and other diseases. It has emerged as an important research area in biophotonics. PDT is based on the fact that some photosensitizers (PSs) can be accumulated in a higher concentration in tumor cells than in healthy cells upon systemic administration. By matching the wavelength of the therapeutic light to the absorption peak of the sensitizers, most of the light is absorbed by the PSs, and the excited PS molecules can then transfer their energy to surrounding oxygen molecules, which are normally in their triplet ground state. This results in the formation of reactive oxygen species (ROSs) such as singlet oxygen ($^1O_2$) or free radicals. ROSs are responsible for oxidizing various cellular compartments, resulting in irreversible damage to tumor cells. PSs, oxygen, and light are three crucial components for this photon-induced toxicity effect. Most existing PSs are hydrophobic and may aggregate easily in a biological environment. Even for hydrophilic PSs, the selective accumulation in tumor tissues is not high enough for clinical use [1-6]. To overcome these limitations, colloidal carriers for PSs, such as liposomes and polymeric micelles, have been investigated. The colloidal mesoporous silica nanoparticles with encapsulated protoporphyrin IX (PpIX), which has been widely investigated and officially approved for use in clinical treatments, were synthesized and utilized as drug carriers. This nanoformulation has many advantages over PS encapsulated polymer/liposomes. The PS-encapsulated silica nanoparticles can be easily prepared with the desired size, shape, and porosity, and are very stable [7]. They are more biocompatible than other nanomaterials, and they can effectively protect the doped PSs against denaturation induced by the extreme bioenvironment. Silica encapsulation is relatively transparent for both the activated light and signal light. The mesoporosity property can also ensure oxygen contact with encapsulated PSs so that they can be effectively stimulated and released. The ultra small size (50 nm) and surface charge of silica nanoparticles can help them be uptaken by tumor cells through the "enhanced permeability and retention effect". To accumulate PSs selectively in tumor tissues and reduce the dosage of PSs, their surfaces can be functionalized with various groups and further conjugated with antibodies for specific uptake in an *in vivo* experiment. We report an investigation that used mesoporous silica nanoparticles with encapsulated PpIX for PDT treatment. The nanoparticles were synthesized, characterized, and utilized for *in vitro* imaging of tumor cells.

The effect of photon-induced toxicity of this nanoformulation was also demonstrated after comparison with some control experiments [8-10]. The PpIX-encapsulated mesoporous silica nanoparticles showed good potential applications to, e.g., photodynamic therapy for cancers.

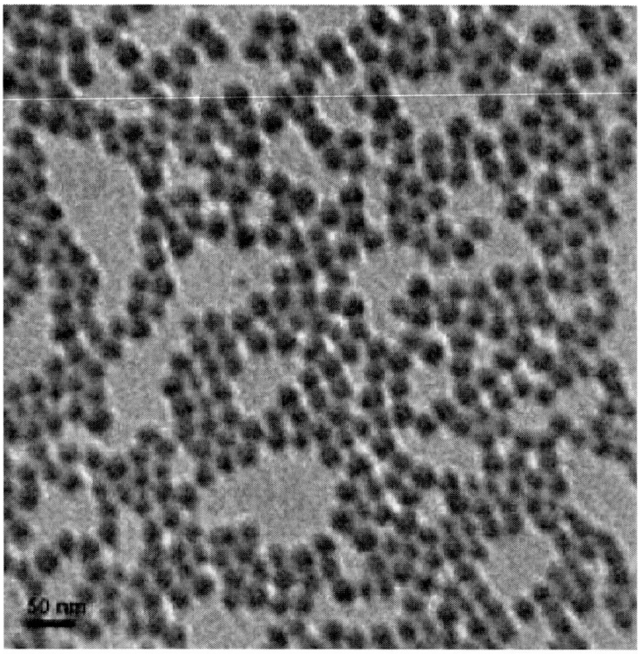

Figure 1. TEM image of some PpIX-encapsulated silica nanoparticles (scale bar is 50 nm).

10.2. Photocytotoxicity of mTHPC (Temoporfin) Loaded Polymeric Micelles Mediated by Lipase Catalyzed Degradation

PDT uses a combination of a photosensitizer (PS) and laser light. Upon illumination of the photosensitizer in the presence of oxygen, reactive oxygen species (ROS) are formed. The anti-tumor effect of ROS is exerted either directly by killing tumor cells, or indirectly by damaging tumor-associated vasculature and inducing an immune response against the tumor cells. PDT has many advantages compared to other cancer treatments, i.e., less invasive than surgery, and more targeted than chemotherapy as a result of selective illumination [11]. Meta-tetra (hydroxyphenyl) chlorine (mTHPC or temoporfin, scheme 1) is a second-generation PS, which is currently clinically used as a formulation in ethanol and propylene glycol (Foscan®). It is registered for the palliative treatment of advanced head and neck squamous cell carcinoma (Biolitec AG: Foscan®, summary of product characteristics), and its clinical efficacy has been studied in PDT of other malignancies as well, such as prostate cancer, oral squamous cell carcinoma, and as adjuvant therapy in combination with surgery in pleural mesothelioma [12].

Although mTHPC performs better than other first- and second-generation PSs, like photofrin and meta-tetra (hydroxyphenyl) porphyrin (mTHPP) in terms of potency, depth of

PDT effect and daylight induced skin toxicity, further improvement is needed with regard to its formulation and biodistribution. Analogous to other PSs, mTHPC is highly hydrophobic, which complicates its formulation and administration. In patients, precipitation and adsorption of mTHPC to the endothelium of the injected vein and adjacent subcutaneous fat tissue have been observed after intravenous administration of Foscan®, and the biodistribution of mTHPC is influenced by its aggregation in the circulation and complexation to serum proteins [13-16]. In addition, skin accumulation obligates patients to prevent exposure to light during the first days after receiving Foscan®. The therapeutic index of mTHPC may be greatly improved by enhancing its tumor accumulation and reducing the skin accumulation. To optimize PDT with mTHPC, alternative, water-soluble formulations were developed. PEG–mTHPC conjugates were designed and their efficacy was shown to depend on the linker used to couple PEG to the PS [17].

However, the therapeutic effect of these conjugates was at best similar to native mTHPC. Encapsulation of mTHPC in bare liposomes (Foslip®) resulted, in vitro, in similar cellular uptake and photocytotoxic characteristics compared to Foscan®. MTHPC has also been incorporated in PEGylated liposomes (Fospeg®). A two to four times higher bioavailability in the tumor, 5.5 times earlier maximal tumor accumulation and four times higher tumor-toskin ratio compared to Foscan® was found, which indicates that PEGylation of the delivery vehicle could be a very interesting approach to formulate mTHPC for in vivo use. However, mTHPC is still rapidly redistributed from the Fospeg® formulation to serum components upon intravenous injection (Biolitec AG, personal communication) [18-21]. Although the amount of released mTHPC is not quantified yet, the binding of mTHPC to serum components could lead to loss in target specificity. Furthermore, although (PEGylated) liposomes are very promising for in vivo application, the release of the contents from these systems can generally not be triggered by environmental factors. Alternatively, polymeric micelles are suitable for the formulation and targeted delivery of PSs to tumors. Their hydrophobic core can accommodate hydrophobic drugs, whereas their hydrophilic shell, which is usually composed of poly(ethylene glycol) (PEG), in combination with their small size (10–100 nm) results in long circulation times, allowing selective accumulation at the tumor site by the so-called enhanced permeability and retention (EPR) effect. To obtain the best result from these favorable features, a high loading capacity, i.e., drug/polymer ratio, is desired and the drug should be stably incorporated in de micellar core until it reaches the target site [22-30]. Several PS-loaded micelles have been studied, such as Pluronic or PEG2000-distearoylphosphatidyl ethanolamine (DSPE) micelles loaded with meso-tetraphenylporphine (mTPP), aluminium phthalocyanine loaded pH-responsive micelles, or thermosensitive micelles loaded with solketal-substituted phthalocyanine. Micelles that can be triggered by environmental factors to release their contents are of particular interest for selective delivery of PSs to the tumor tissue. Until now, there are no formulations available that can stably entrap mTHPC and deliver it directly to the tumor where it is released as a result of an environmental trigger. Therefore, polymeric micelles that degrade at their site of action are attractive systems for the formulation and delivery of mTHPC *in vivo* [31].

In this study we evaluated the suitability of biodegradable mPEG750-b-oligo (ε-caprolactone) based micelles for the formulation of mTHPC. These micelles can be easily prepared by the film-hydration method, and their small size (sub-20 nm) may result in better properties in terms of biodistribution and tumor penetration, compared to the larger micelles that have been studied so far. Moreover, it is expected that stable drug incorporation can be obtained, as a result of derivatization of the terminal hydroxyl groups of mPEG-b-oligo (ε-caprolactone) with an aromatic group, which may improve the compatibility between the mTHPC and the core, thereby stabilizing the drug-loaded micelles. Therefore, the loading of mPEG-b-oligo(ε-caprolactone) based micelles with and without end group modification was studied, followed by examination of the photocytotoxic effect and cellular uptake of the mTHPC loaded micelles, and the influence of lipase, a hydrolytic enzyme, thereon[31-37].

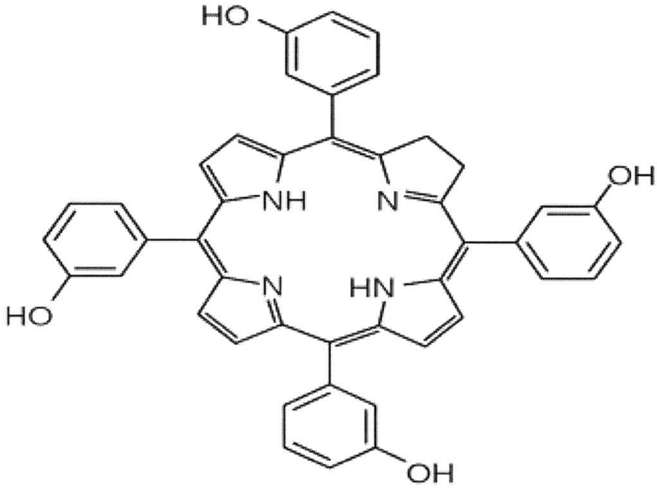

Scheme 1. Structural formula of mTHPc (meta-tetra(hydroxyphenyl) chlorine, temoporfin).

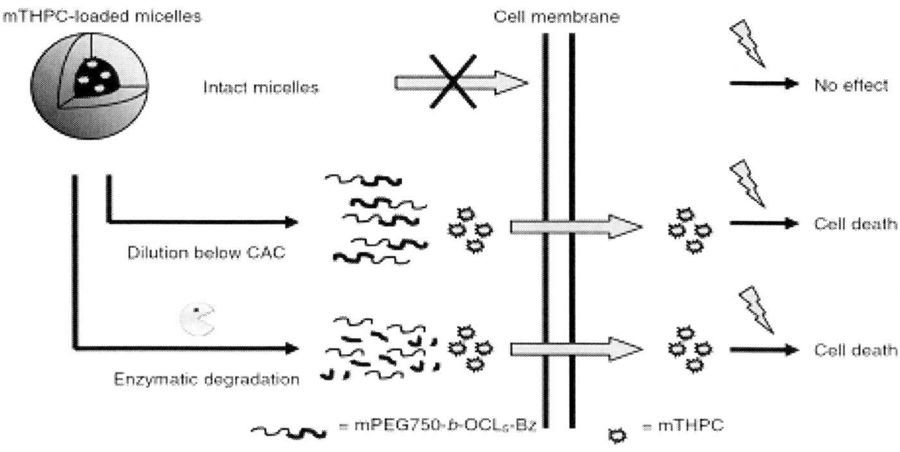

Figure 2. Schematic representation of the processes leading to the photocytotoxicity of mTHpc- loaded mPEG750-b-cCL5-Bz micelles.

Cancer is the second leading cause of death of people around the world. Cancer therapy has been practiced in recent decades and extensive effort has been made to find new harmless and effective solutions. Chemotherapy, renewed surgical techniques to remove tumors, radiotherapy and, newly developed, photodynamic therapy are aimed at prolonging and giving a better standard of life to patients diagnosed with neoplastic disease. Surgery is the oldest (and once the only) way of treating cancer. In the case of micro metastasis, surgery is obsolete and the use of other techniques, such as radiotherapy is mandatory [38].

Like surgery, radiotherapy has its limitations. When dealing with large tumors, there is some risk that the central cells are not affected, which leads to recurrences. That's why chemotherapy has its indications. These methods are sometimes only palliatives, so widespread research is continuing to understand the mechanisms of cancer evolution and to find new ways to effectively deal with it. This research has lead in recent years to photochemotherapy or photodynamic therapy (PDT) [39].

10.3. Photodynamic Therapy

Photodynamic therapy is a method of local treatment of cancer based on the administration of a photosensitizing substances selectively absorbed by the tumor cells which leads, after the exposure to optical radiation, to tumor destruction. The photodynamic action of nonionizing radiation was discovered in 1898 by student Oscar Raab who demonstrated the effect of light and some colorants on paramecia. He noticed that when paramecia were treated with acridine orange dye and exposed to sunlight, a lethal effect has appeared. The paramecia treated with acridine orange dye but kept in the dark and the ones exposed to sunlight without acridine orange had survived. Herman von Tappeiner, in Munich, continued the experiments, and introduced together with Joldbauer in 1904 the term of *photodynamic reaction* containing light, the photosensitizing substance and oxygen [40-45].

Since 1904, many researchers have developed the photodynamic therapy demonstrating its value in the treatment of malignant tumors and have reported many studies and even elaborated some protocols accepted by medical authorities (Table 1) [46].

Table 1. Current status of approved photodynamic therapy

Year	Photosensitizer	Disease
1993	Porfimer sodium	Early bladder tumors
1996	Porfimer sodium	Obstructive tumor of esophagus
1997	Porfimer sodium	Advance lung and esophageal tumors
1997	Porfimer sodium	Advance lung and esophageal tumors
1997	Porfimer sodium	Early lung cancer
1997	Porfimer sodium	Early esophageal, lung, gastric, cervical cancer
1998	Porfimer sodium	Endobronhial cancer
2000	Verteporfin	Age-related macular degeneration

Although in the experimental phase, the phototherapy results obtained in the latest period are remarkable and have led to growing interest in this method. PDT is optimal in endoscopic applications: treatments of carcinomas in GIT, bronchial carcinomas, some carcinomas of the bladder and urinal paths etc. Penetration of light with wavelengths within the range of 630 - 700 nm, which are utilized in PDT for excitation of most of the preparations, seems to be the limiting factor as their penetration is up to 6 mm only. Therefore, PDT is suitable for treatment of early stages of solid carcinomas and/or palliative treatment. The application of photodynamic therapy in current practice raises some problems [47-50]:

- Choosing the "right" photosensitizing substance
- how to prepare and administer this substance
- how much time should pass from the moment of administration of the substance until the exposure to the light radiation
- choosing the type of the radiation source and setting the parameters of the light radiation (wavelength, time exposure, pulse duration, pulse frequency, etc.)
- methods of evaluating the biologic response
- how long the organism is still sensitized to radiation, etc.

There have been many studies aimed at analyzing these factors. The most important results are summarized in the following pages.

10.3.1. The Mechanisms of Photodynamic Therapy

Photodynamic therapy is based on the photooxidation of living matter involving three basic constituents: the photosensitizer (S), the radiation (with a wavelength appropriate to the maximum absorption of the substance in use), and oxygen. Oxygen species, in different toxic forms, represent the most important agents of the cellular death processes during the photodynamic therapy [51]. Active oxygen species can be generated through a type II reaction mechanism (Figure 3) between the photosensitizer molecule, (being in an active triplet state - 3S*) and the biologic environment (B). The sequence of the reactions is:

$$^3S^* + {}^3O_2 \Rightarrow {}^1S + {}^1O_2^*$$
$$^1O_2^* + B \Rightarrow B_{ox}$$

where: 3O2 is the triplet excited state of oxygen, $^1O_2$ is the singlet oxygen molecule, and Box is the target substrate oxidated [52].

Molecular oxygen, as a singlet, ($^1O_2$) acts rapidly and not discriminatingly with electrophyle molecules (B) like unsaturated lipids, proteins, nucleic acids, etc. After transferring the energy to oxygen, the photosensitizer can restore its fundamental state (1S) and the cycle repeats. The molecule of the photosensitizer, as a triplet, can react with a target molecule, other than oxygen, by changing hydrogen or electrons [53]. This type of reaction is called a type I reaction mechanism:

$$^3S^* + RH \Rightarrow SH^{\cdot} + R^{\cdot}$$
$$^3S^* + RH \Rightarrow S^{\cdot -} + RH^{\cdot +}$$

where:, RH is the target substrate molecule, $SH^{\cdot}$ is the radicalic form of photosensitizer, $R^{\cdot}$ is the radical derived from the target substrate molecule, $S^{\cdot}-$ is the radical anion form of photosensitizer, $RH^{\cdot}+$ is the cationic form of target substrate molecule.

Generally speaking, in a given photodynamic system, both types of reactions can appear, their efficiency depending both on the concentration of the photosensitizer that has been accumulated at the level of the tumor and the density of the light radiation energy [53-60].

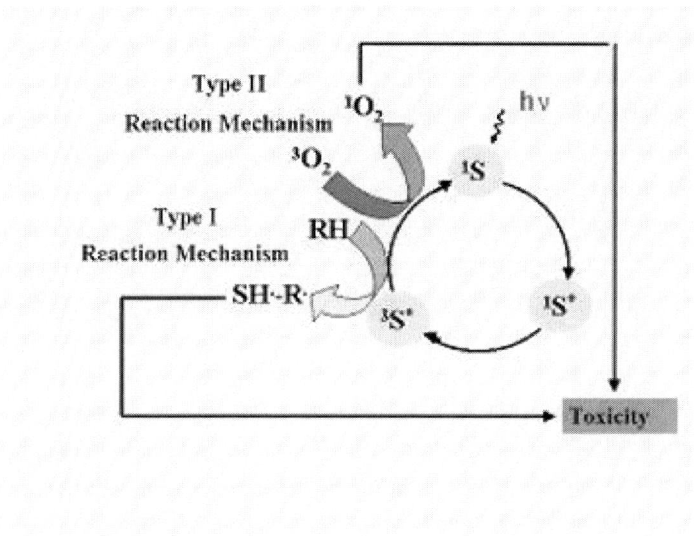

Figure 3. The mechanisms of photodynamic therapy.

10.3.2. Effects of Photodynamic Therapy

There can be direct or indirect effects.

10.3.2.1. Direct Effects

The direct photodynamic effects are mediated, according to the type II reaction mechanism, by the generation of singlet oxygen. Singlet oxygen has a short lifetime (0.01-0.04 μs) and a limited diffusion distance in a biologic environment (0.01-0.02 μm) due to its reactivity, so the cellular deterioration appears in the proximity of the substance in the cell. The location of this molecule is determined by the degree of lipophilicity of the photosensitizing substance [61].

In general, the anionic lipophilic photosensitizers accumulate at the level of the cellular membrane and its organelles (mitochondria, endoplasmic reticulum, lyssosomes, and nuclear membrane). Lipid and protein photooxidation in the cellular membrane generates the activation of the phospholypases from the membrane which leads to: changes in membrane permeability, reduced fluidity and inactivation of enzymatic systems and receptors [62-67].

Inhibiting the mitochondrial enzymes was considered to be the key of cellular death due to photodynamic therapy. Inactivation of transport membrane systems and the depolarization of the plasmatic membrane happen before the inactivation of mitochondrial cytosolic and lysosomic enzymes. Hydrophilic photosensitizers localize preferentially in lysosomes. After the exposure to light radiation, lysosomes disintegrate and the enzymes they contain are spread in the cytoplasm leading to cellular death. The probability of cell inactivation is higher for the lipophilic photosensitizer than for hydrophilic ones for the same dose of light energy, which means that membranous structures are more vulnerable. Photodynamic therapy can also lead to the deterioration of the DNA. In vitro studies have demonstrated that photodynamic therapy can produce: breaks in the DNA chain, changes between twin chromatides and chromosomal aberrations. Lethal, acute, direct effects on tumoral cells and even normal cells depend on: location of the photosensitizer on/or in the cell, the photodynamic efficiency of the substance in the environment, parameters of irradiation and oxygen transport [68-70].

10.3.2.2. Secondary Effects

Due to the direct effect of the photodynamic therapy, most of the tumoral mass is destroyed. In the tumor cells that were not directly affected, secondary effects may appear including: vascular effects, the initiation of apoptosis, inflammation and generation of an immune response.

- Vascular effects

The acute effects of photodynamic therapy most easily observed *in vivo* are on the micro-vasculature which usually leads to stasis [71].

Major determinants for vascular photosensitivity are: the concentration of the circulant photosensitizer and the time from its administration and the tissue irradiation. The demonstrated vascular effects on experimental tumoral systems were: microvascular aggregation followed by transitory vasoconstriction, vasodilatation, blood stasis and hemorrhage. When the skin is the subject of photodynamic therapy, edema and erythema are always the first signs. In clinical situations, these vascular events are documented by whitening and hemorrhage of the treated tissue [72].

- The initiation of apoptosis

Apoptosis (programmed cellular death) occupies an important role (besides direct cell death – necrosis) in photodynamic therapy. During apoptosis, the genetic mechanisms of the cell start a program of cell destruction without involving inflammation. It involves a cytoplasmic condensation with cell destruction in fragments rolled in membrane called apoptotic corpuscles. In the tissue, these corpuscles are destroyed by macrophages. Cancer cells resistant to apoptosis (through genetic mutations) are resistant both to chemotherapy and radiotherapy. Studies concerning photodynamic induced apoptosis were conducted both *in vitro* or *in vivo* and have demonstrated an association between photodegradation of mitochondria and the apoptotic response of the cell [73].

- Inflammation

Photodynamic therapy usually induces a powerful inflammatory response in the treated tissue. Phototoxic effects on the cell membrane cause the release of lipid metabolites which are important activators of inflammation. It also generates an extended immune response which is very important in long term tumor control [74].

The complex nature of tissue response to photodynamic therapy has not yet been discovered.

Table 2. Photosensitizers for photodynamic therapy

Porphyrins	hematoporphyrin derivative
	dihematoporphyrin ether/ester
	porfimer sodium
	tetra-sodium-meso-tetra-phenyl-porphyrin-sulphonate
	metallo-tetra-azoporphyrin
	5,20-Bis(4-sulphophenyl)-10,15-bis(2-metoxi-4-sulphophenyl)-21-tiaporphyrin (2,10tiaporphyrin)
Phthalocyanines	chloroaluminum phthalocyanine tetrasulfonate
	Zn-phthalocyanine
	Si-naphthalocyanine
	cloraluminium phthalocyanine sulphonated
Chlorines	mono-L-aspartil-chlorin e6
	diaspartyl chlorine
	bacteriochlorin a
	benzoporphyrin derivative monoacid ring A
Porphycenes	2,7,12,17-tetraphenylporphycene
	2-hydroxy-ethyl-7,12,17-tris(methoxy-ethyl)porphycene
	21,23-Dithia-3,13-diazaporphycenes
	capronyloxy-tetrakis methyloxyethyl porphycene
Purpurins	metallopurpurin
	tin etiopurpurin
Pheophorbids	pheophorbide a
	bacteriopheophorbide

10.4. Photosensitizing Substances Used in Photodynamic Therapy

Photosensitizing substances can exist naturally in cells and tissues (flavine, pyridine, bilirubine, etc.) or can be administered under control as in photodynamic therapy.

Many substances have been tested both *in vitro* and *in vivo* (Table 2) and none of them are perfect. Current research is focused on discovering new substances. A new concept is the biosynthesis of porfirinic photosensitizers after the administration of 5-aminolevulinic acid, a compound from the hem synthesis pathway.

10.4.1. Hemoporphyrin (Hp)

Hemoporphyrin is composed of several porphyrinic fragments: acetate, biacetate, hydroxy-ethyl, protoporphyrine, vinyl and deuteroporphyrine. Of all these fragments the most active one is biacetate. The results using this fragment are not as good compared the use of the whole mixture. The absorption spectrum of hemoporphyrin shows 5 bands in the visible domain of the electromagnetic spectrum at the following wavelengths: 400 nm, 520 nm, 540 nm, 590 nm and 630 nm. The only one of interest to photodynamic therapy is 630 nm because in this region the biologic tissues show a low absorption so the effect of the photosensitizer is higher. Hp is better fixed by the macrophages and vascular endothelia.

10.4.2. Hemoporphyrin Derivative (HpD)

HpD is a complex mixture of porphirinic monomers (45%) and oligomers (55%) linked by ether or ester bonds. It can be extracted from haemoporphirine mixed with sulphuric and acetic acids.

The absorption spectrum of HpD presents 5 bands in a visible region. The most important one has its maximum at 400 nm (Soret band). The other absorption bands are less intense and have their maximum at 520 nm, 540 nm, 590 nm, 630 nm.

The metabolism and ways of action of the HpD are not fully understood. Some researchers believe that after the i.v. injection, HpD enters all the body cells and is metabolized in the cytoplasm. The concentration of HpD is 2-2.5 times higher in tumoral tissue than normal cells. This great difference permits the application of photodynamic therapy after 48-72 hours without harm on the neighboring normal tissues [75].

10.4.3. Porfimer Sodium (Photofrin)

Photofrin is a mixture of oligomers formed by linking of porphyrine units through ether or ester bonds and the tumoral tissues and, under light radiation produces cititoxic effects by generating active oxygen species, which lead to vascular thrombosis and ischemic necrosis of the tumor.

10.4.4. Phthalocyanines

They are a class of tetra-aza-tetra-benzeneporphyrine which absorbs in the red region of the electromagnetic spectrum. They have 2 important absorption bands, one more intense (650-700) nm with a maximum at $\lambda = 670$ nm and a less intense one in the UV domain with a maximum at $\lambda = 350$ nm. The pyrole groups from phthalocyanines are conjugated with benzene rings through nitrogen bonds. This structure shifts the absorption spectrum to higher wavelengths and the absorption band in the visible spectrum becomes more intense than the one in the near ultraviolet spectrum. For an efficient tissue photosensitization, the lifetime of the triplet state of the substance must be long so that the reaction with singlet oxygen can take place. In the case of phthalocyanines, this is done by incorporating a diamagnetic metal ion like Al or Zn.

The most important substance in this class was the cloraluminium phtalocyanine sulphonated (AlPcS). It is a stable compound, very soluble in water and has a powerful absorption at $\lambda = 675$ nm. It also has a specific absorption in some tumors, a low toxicity in darkness, minimal cutaneous photosensitivity and an excellent photodynamic activity at high wavelength. Another tested compound from this class was chloroaluminum phthalocyanine tetrasulfonate (AlPcS4). Studies have tested its effect on basal cell carcinoma, Kposi tumor and lung cancer.

10.4.5. Chlorins

Chlorins are reduced porphyrins, synthesized by hydrogenating a double bond exo-pyrol of the porphyrinic ring. They show intense absorption in the (640-700) nm spectral domain. The compounds studied as photosensitizers are:

- meso-tetra-hydroxyphenyl-chlorin (m-THPC), known commercially as Foscan. The first study was conducted in 1990 in the treatment of human mesothelioma.
- mono-L-aspartil-chlorin e6 (Npe6 or MACE) shows a maximum of absorption at 654 nm. The *in vitro* and *in vivo* experiments with MACE have demonstrated its efficiency in photodynamic therapy due to its great absorption in the tumor and low skin photosensitivity.
- benzoporphyrin derivative monoacid ring A (BPD) shows a maximum of absorption at $\lambda = 650$ nm. Clinical trials showed great absorption in the tumors and low skin photosensitivity.
- bacteriochlorins have a powerful absorption at $\lambda = 740$ nm and can become good photosensitizing agents, although their stability is under question.

10.4.6. Porphycene

It is formed through reductive coupling of 5,5'diformil-2,2'bipyrol with titanium in lower covalent form. The UV-VIS spectrum of this substance has 5 bands at wavelengths of 358 nm, 370 nm, 558 nm, 596 nm, and 630 nm.

10.4.7. Purpurins and Verdins

Purpurins have an absorption band at 660 nm and can be synthesized at high purity. Histological studies have shown that purpurins and metallopurpurins (octaethylpurpurin and etiopurpurin), exposed to visible light radiation can result in tumor necrosis. *Verdins* have a broad absorption band at 700 nm but with much lower intensity compared to the Soret band.

10.4.8. Pheophorbides

Methyl pheophorbide has an intense absorption band at 650 nm. Studies on biomolecules, human neutrophiles and tumor cells showed that their efficiency is comparable with Hp. Research in this field continues and is directed towards finding new substances with properties closer to ideal.

10.5. Light Sources

For photodynamic therapy of tumors, both coherent (lasers) and non-coherent (lamps and LED) light sources are suitable. Both types have their own advantages and disadvantages.

10.5.1. Lasers

Laser is considered as an ideal light source for photodynamic therapy due to its coherence and monochromaticity. Monochromaticity permits the irradiation with a wavelength near to the maximum absorption of the photosensitizer, and coherence offers the possibility of using optic fibers for the transmission of light to cavities of the body. Lasers that have been used in photodynamic therapy are: argon laser, copper vapors laser, gold vapor laser, dye lasers, and laser diodes such as galliumaluminium- arsenide laser.

10.5.2. Lamps

Lamps have proven useful in photodynamic therapy especially in the cure of skin diseases. Their main advantages are linked to the low cost and reduced dimensions. But they can be used only for superficial diseases. Although the light is filtered, the wavelength bands are wide compared with the absorption bands of the photosensitizing substances which lead to heat release which enhances the tumoricid result. Metal halogen lamps and short arc xenon lamps were used in photodynamic therapy.

10.5.3. LEDs

These are non-coherent light sources that are scarcely used in photodynamic therapy. They generate wavelength bands wider than those from lasers, usually around 25 nm.

10.6. METHODS OF DETERMINING THE BIOLOGIC RESPONSE TO PHOTODYNAMIC THERAPY

The biologic response to this therapeutic method depends on a number of variables among which are: the light flux and its spatial distribution, the concentration of photosensitizer in the target tissue and the oxygen delivery to that tissue. Determining these parameters represents an important step in experimental studies or clinical trials concerning photodynamic therapy. This field of research is nevertheless less developed and has been limited to: radiolabelling of the photosensitizing substance, *in vivo* reflexion spectroscopy and *in vivo* fluorescence spectroscopy in vivo.

10.6.1. Radiolabelling the Photosensitizing Substance

A possible method of determining the amount of the substance and its concentration is to label it with a radioisotope. This method was employed only by B.C.Wilson to determine the concentration of HpD and Photofrin II labeled with $_{64}Cu$. The method has a great sensibility and does not depend on the tissue type but it is expensive and difficult to reproduce in different studies on photodynamic therapy. That is why there is a growing interest in developing optic methods to assess *in vivo* the treatment parameters and the biologic response, methods that are easier to apply for a wide range of photosensitizers. Such methods are fluorescence spectroscopy and diffuse reflectance spectroscopy.

10.6.2. Fluorescence Spectroscopy

The tissue fluorescence can be studied in vivo by point-monitoring of fluorescence or recording the fluorescence images.

10.5.2.1. Point-Monitoring of Fluorescence

Point-monitoring of tissue fluorescence was first employed in 1990 by Profio to determine the concentration of Photofrin II in tumors and normal tissues. The method is based on recording the fluorescence emission spectrum of living tissue for a certain excitation wavelength. The quantification analysis is based on measuring the intensity of the fluorescent bands which is proportional to the concentration and width of the tissue sample through which the incident radiation passes and can lead to real time evaluation of the treatment parameters and helps in choosing the procedure of photodynamic therapy.

10.5.2.2. Imaging Fluorescence Spectroscopy

Imaging fluorescence spectroscopy is based on recording the fluorescent images of the tumoral tissue. The capture, processing and analyzing these images can be accomplished in several ways:

- the incorporation of a device with spectral resolution that measures in points in an image forming instrument and spectroscopic investigation of the regions of points on the image ; - sequentially or parallel image forming on several wavelengths of the excitation or emission wavelength (to subtract self-fluorescence);

- combining several wavelengths of excitation or emission radiations to compensate for the variations in optical properties of the tissues and extracting of linear signals depending on the concentration of the exogenous photosensitizer;
- using a dividing optical device to obtain several images of an object filtered at different wavelengths of the emission wave. These images can be further processed by a computer to obtain an optimal contrast image. Research continues in this field aiming at obtaining better fluorescence images and developing processing software for image analysis.

10.6.3. Diffuse Reflectance Spectroscopy

Like fluorescence spectroscopy, diffuse reflectance spectroscopy can be used in photodynamic therapy both for determining the concentration of the photosensitizer in the tumor and for treatment monitoring. The concentration of the substance is measured by determining the variation of the diffuse reflectance of the tumoral tissue due to the presence of the photosensitizing substance in that tissue. The relation that links the variation of the diffuse reflectance to the absorption coefficient of the tissue (which is directly proportional to the substance concentration $\mu a = \varepsilon C$) can be obtained by solving the radiative transfer equation. The diffuse reflectance can be determined by *steadystate measurements* (using light sources) with continuous or pulse emission (with the pulse duration greater than the time of propagation of the photons in the target volume) or detectors with temporal resolution greater than the photon propagation time or *time-resolved measurements* (using light sources with very short pulses (ps) or detectors with high temporal resolution).

10.6.3.1. Steady-State Measurements

The principle of the method is to send radiation through a optic fiber on the sample to be studied and reception of the reflected radiation through another optic fiber situated at a distance r to the sample. The variation of the local diffuse reflectance (R0(r)/R(r)) of the tissue induced by the photosensitizing substance is determined by measuring the local diffuse reflectance before the substance is administered (R0(r)) and after its administration (R(r)).

Using a calibration curve, Ro(r)/R(r) = f(c), the concentration of the photosensitizing substance can be determined.

This method is not primarily used for quantification, but to establish the time when the concentration of the photosensitizing substance is at its maximum (the ideal moment to irradiate the tissue).

10.6.3.2. Time-Resolved Measurements

Recently, the appearance of light sources with very short emission pulses (ps) has led to the development of time resolved reflectance spectroscopy. The method is based on the irradiation of the living tissue with a pulsed laser and the measurement of the temporal distribution of photons reflected from the tissue. From the dependence of temporal distribution of the photons on the optical properties of the tissue, it is possible to determine the absorption and scattering coefficients of tissue.

10.7. *In Vitro* Studies of Biodegradable Modified Chitosan Nanoparticles for Photodynamic Treatment of Cancer

Photodynamic therapy (PDT) is a two-step treatment used to help treat symptoms caused by cancer or its treatment. It is called two-step since both a drug and laser treatment are used. PDT is most often used to help people who have cancer of the esophagus or lung. This treatment is given in the hospital. Biodegradable nanoparticles have received much attention as possible drug-delivering agents. Their main advantages lie in high drug loading, the possibility of controlling the drug release, and a large variety of materials and manufacturing processes. Biocompatible polymers are predominantly favored over small molecules for drug delivery systems because of the opportunities that polymers offer in achieving optimal medication effects. The chemical composition and architecture of polymers can be readily tailored to accommodate drugs with varying degrees of hydrophobicity, molecular weight, pH, etc. Furthermore, the surface properties, morphologies, and composition of polymer matrices can be easily optimized to achieve the desired biocompatibility, controlled polymer degradation and drug release kinetics. Recently, studies in PDT have focused on the use of polymer-based nanoparticles for PS delivery and release systems, in particular those with biocompatible and biodegradable polymers. For example, the surface of particles can be modified with poly (ethylene oxide) to improve the carrier's biocompatibility and biodistribution.

Different strategies have been developed for the preparation of polymer-based nanomaterials for the delivery of PS molecules in PDT. Among these methodologies, the most common method is the emulsion polymerization or the interfacial polymerization in water-in-oil or oil-in-water microemulsions. The resulting nanomaterials can be either nanospheres or nanocapsules (core-shell nanoparticles) depending upon the synthetic approach. Among the various PSs investigated for PDT, phthalocyanines have been found to be highly promising. However, one of the problems in using phthalocyanines for PDT arises from their intrinsic tendency toward aggregation. This intermolecular association process promotes the non-radiative internal conversion, almost invariably causing a shortening of the triplet lifetime and a drastic reduction in the overall efficiency of singlet oxygen production. As early as 1991, the preparation of tetrasulfonated zinc phthalocyanine (ZnPcS4) or aluminium naphthalocyanine entrapped into poly(isobutylcyanoacrylate) or poly(ethylbutylcyanoacrylate) nanocapsules or nanospheres by interfacial polymerization in water-in-oil emulsion was reported. The prepared nanocapsule sizes ranged from 150 to 250 nm, while the nanospheres had a broader size distribution from 10 to 380 nm. The PS encapsulation efficiency was very high and could reach 100% when prepared under optimal experimental conditions. PS binding in the nanoparticles was between 75 and 80%. In a subsequent study with cultured V-79 hamster cells as a model, it was found that the nanoencapsulation enhanced the cellular localization of the dyes providing a high level of in vitro phototoxic activity. It was found that the formulation of the biodegradable nanoparticle improved the PDT response of the tumor while providing prolonged tumor sensitivity towards PDT. Recently, a similar method has been applied to the preparation of poly (D,L-lactide-co-glycolide) (50:50 and 75:25 PLGA) and poly(D,L-lactide) nanoparticles entrapped with the PS meso-tetra(hydroxyphenyl)porphyrin (p-THPP). The resulting nanoparticles have an average particle size of 130 nm and the PS loading was improved to near 8%. The cellular uptake, localization and phototoxicity of PS-loaded carriers have been studied in macrophage-like J774 cells and HT 29 human adenocarcinoma cells. It was found that the cellular uptake was reduced with mTHPC encapsulated into surface-modified nanocapsules – PLA-PEG and polox PLA – compared with naked PLA. However, the phototoxicity was much less affected and a protecting effect against plasma proteins was observed. As explained above, singlet oxygen is the main cytotoxic species in PDT, and hence a fundamental requirement for a PS carrier is that it should not significantly quench singlet oxygen.

In this research paper, we have synthesized biodegradable and biocompatible polymeric nanoparticles for PS delivery and release system. This system is able to target different organs and control the release of the PS molecules by the incorporation of site-specific moieties. The system is based on a modified chitosan – PS conjugate that uses for PDT. In the present research, we have prepared mTHPP loaded NOCCS /PMAA nanoparticles by the classical method, which involves spreading a uniform layer of polymer dispersion followed by a drying step for removal of the solvent system. Since the methodology of nanoparticles preparation involved a heating step, it may have had a detrimental effect on the chemical stability of drug. Hence, the stability assessment of mTHPP impregnated in nanoparticle was done using the stability-indicating method. For this purpose, mTHPP was extracted from polymer and analyzed by HPLC. A single peak at 14.8 minutes representing mTHPP (with no additional peaks) was detected in the chromatogram, suggesting that the molecule was stable during the preparation of nanoparticles. The composition of the polymer defines its nature as a neutral or ionic network and furthermore, its hydrophilic/hydrophobic characteristics. Ionic hydrogels, which could be cationic, containing basic functional groups or anionic, containing acidic functional groups, have been reported to be very sensitive to changes in the environmental pH. Hydrogels containing basic functional groups are found to increase swelling activity in acidic conditions and reduce it in basic conditions.

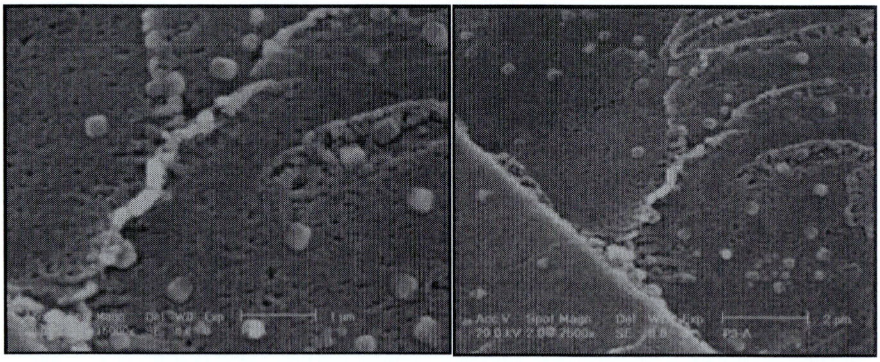

Figure 4. SEM of NOCCS-PMAA-mTHPP nanoparticles.

The pH sensitive anionic hydrogels show low swelling activity in acidic medium and very high activity in basic medium. Figure 1 shows the scanning electron microscope (SEM) of the graft NOCCS copolymer with polymethacrylic acid and nano polymer bonded mTHPP, respectively. As observed from the SEM photomicrographs, the crystals of mTHPP have a different appearance than recrystallized mTHPP. These nanoparticles do not have clearly defined crystal morphological features in the SEM photomicrographs. Hence, it appears that the irregularly shaped particles are surface deposited with poloxamer, which gives them an appearance resembling that of coated particles. The thermal behavior of a polymer is important for controlling the release rate in order to have a suitable drug dosage form. The glass transition temperature (Tg) was determined from the DSC thermograms.

The values are given in Table 1.

Table 3. DSC Data and Composition of Copolymer

Polymer Sample	Molar composition of monomers in the feed				Degree of Substitution(DS)	Tg (°C)
	NOCCS (gr)	MAA (gr)	CA (gr)	IN (gr)		
1	1	3	0.05	0.05	0.53	134
2	1	2	0.05	0.05	0.46	140

The higher Tg values are probably related to the introduction of crosslinks, which would decrease the flexibility of the chains and the ability of the chains to undergo segmental motion, which would increase the Tg values. On the other hand, the introduction of a strongly polar carboxylic acid group can increase the Tg value because of the formation of internal hydrogen bonds between the polymer chains. X-ray diffraction is a proven tool to study crystal lattice arrangements and yields very useful information on the degree of crystallinity. An x-ray diffraction technique is also used to study the degree of crystallinity of pharmaceutical drugs and excipients. A lower °2θ value indicates larger d-spacings, while an increase in the number of high-angle reflections indicates higher molecular state order. In addition, broadness of reflections, high noise, and low peak intensities are characteristic of a poorly crystalline material. A broad hump in the diffraction pattern of chitosan extending over a large range of °2θ suggests that modified chitosan is present in an amorphous state in the nanoparticles. Figure 2 show that the X-ray diffraction pattern of nanoparticles has several high-angle beads at the following 2θ values: 21.20.13, 22.43, 25.21, 26.50, 28.01 and 32.11. The 26.5° 2θ peak had the highest intensity, and the hump in the base line occurred from 7° to 45° 2θ, as observed for modified chitosan nanoparticles.

The loading efficiency and loading capacity of the nanoparticle dispersions are presented in Figure 3a and 3b, respectively. It was shown that mTHPP/polymer feed ratios up to 30%(w/w), the loading efficiency was almost quantitative, resulting in loading capacities up to 30% and an mTHPP concentration in the dispersion of 1.8 mg/mL, which is markedly higher than the aqueous solubility of mTHPP.

At 10% (w/w) loading, the dispersions stayed clear for more than a week. At higher loadings of 20% and 30% (w/w), a precipitate was observed in the dispersions of loaded

NOCCS /PMAA within 1 day after preparation, indicating that the loading capacity of these nanoparticles was exceeded. In contrast, the nanoparticles composed of polymer remained clear for more than a week at 20% and 2 days for 30% loading, which demonstrates that these end groups have a favorable effect on the loading and solubilization capacity of the nanoparticles.

The in vitro photocytotoxicity effect of mTHPP loaded NOCCS /PMAA nanoparticles on 14C cells was determined after 6 hours of incubation and compared with free mTHPP. Either various dilutions of loaded nanoparticles at a fixed mTHPP/polymer ratio of 15%(w/w) were used, or the polymer concentration was fixed at 0.5 mg/mL and the mTHPP loading was varied between 0.004% and 0.32%(w/w) in order to secure a polymer concentration of ≈ 100 times above its critical aggregation concentration. Figure 4 demonstrates that the effect on the viability of 14C cells of mTHPP – polymer nanoparticles at a fixed mTHPP/polymer ratio is comparable to that of free mTHPP. At a fixed high polymer concentration, the nanoparticles with various loadings were hardly toxic to cells even at the highest mTHPP concentration used, as indicated by a cell viability of 80%. Loaded nanoparticles and free mTHPP were diluted in medium to a concentration of 12 μM mTHPP and incubated with 14C cells for 6 hours to determine the cellular uptake of mTHPP [76].

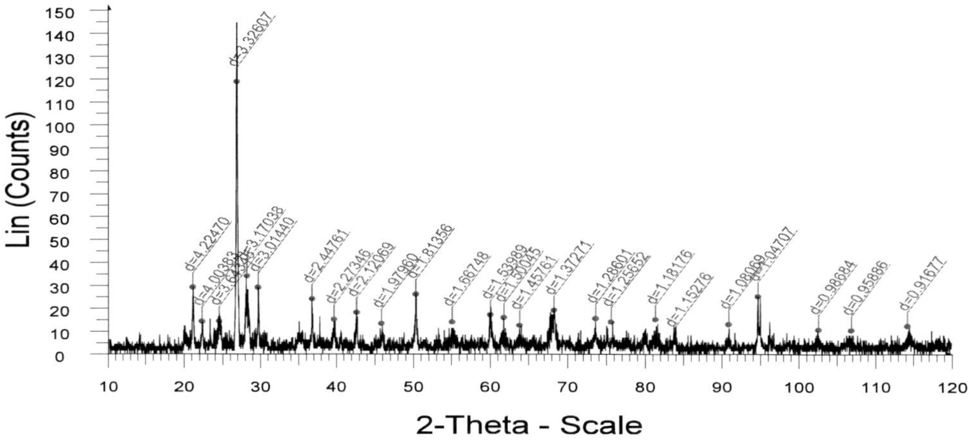

Figure 5. XRD of NOCCS-PMAA-mTHPP nanoparticles.

The 15% (w/w) mTHPP loaded and polymer nanoparticles showed a lower cellular uptake of mTHPP compared to free mTHPP, by a factor of 2.8 and 3.7, respectively. More importantly, hardly any mTHPP uptake was measured after incubation with the nanoparticles yielding a cellular uptake comparable of free mTHPP. The results of the photocytotoxicity and cellular uptake of the different formulations suggest that when the concentration of polymer is kept well above, mTHPP remains stably encapsulated in the nanoparticles even in the presence of serum, and that the intact nanoparticles are not taken up by the cells, which may be ascribed to the presence of a dense polymer-layer. Only after dilution of the mTHPP-loaded NOCCS/PMAA nanoparticles to polymer concentrations below and after nanoparticle degradation, can the released mTHPP be taken up by 14C cells and exert its cytotoxic effect upon illumination. Nano polymer bonded mTHPP (50 mg) was poured into 3 mL of aqueous buffer solution (pH 7.4) (Figure 5).

The mixture was introduced into a cellophane membrane dialysis bag. The bag was closed and transferred to a flask containing 20 mL of the same solution maintained at 37° C. The external solution was continuously stirred, and 3 mL samples were removed at selected intervals. The triplicate samples were analyzed by UV spectrophotometer, and the quantity of mTHPP was determined using a standard calibration curve obtained under the same conditions. The primary mechanisms for release of mTHPP from matrix systems in vitro are swelling, diffusion, and disintegration. In vitro degradation of chitosan beads were prepared by solution casing method occurred less rapidly as the degree 73% deacetylated showed slower biodegradation.

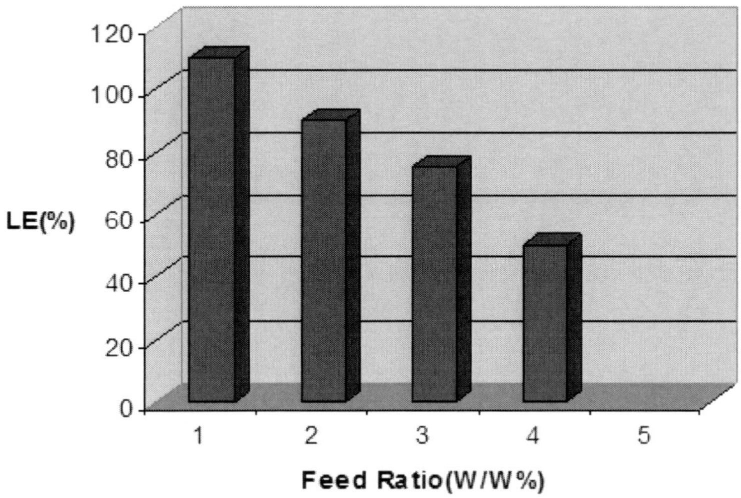

(a)

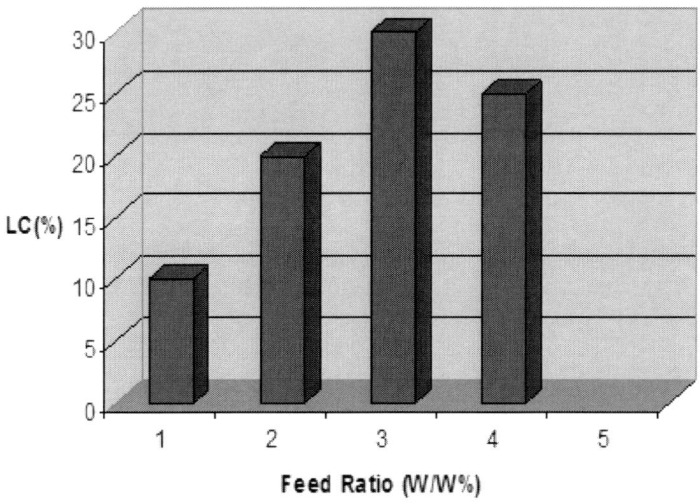

(b)

Figure 6. (a) Loading efficiency and (b) loading capacity of NOCCS /PMAA with mTHPP.

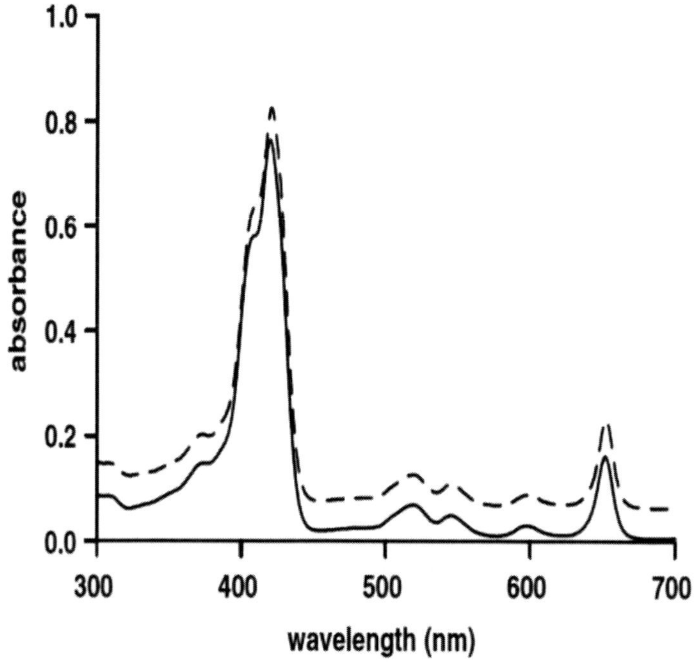

Figure 7. Absorption spectrum of mTHPP (solid line) and mTHPP loaded NOCCS/PMAA copolymer (dash line).

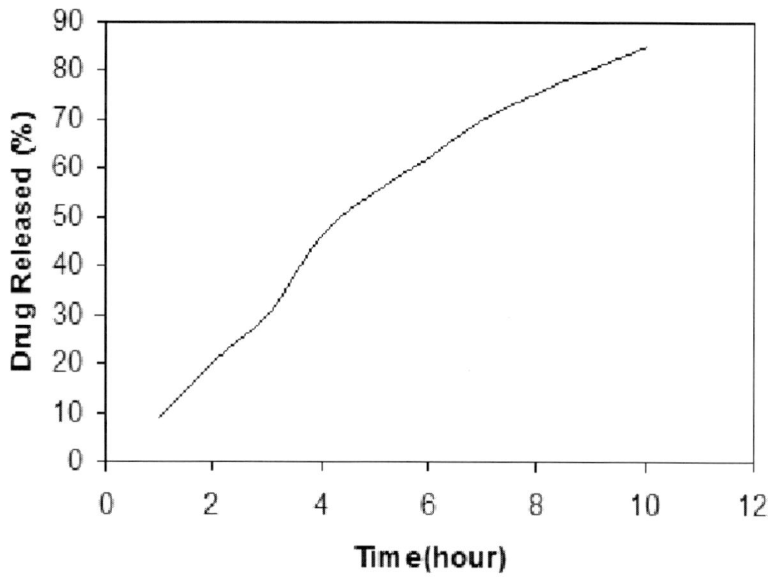

Figure 8. Release of m-THPP from mTHPP loaded NOCCS/PMAA nanoparticles at 37ºC.

Since the grade of chitosan used in the present study was of high molecular weight with a degree of deacetylation ≥85%, significant retardation of release of the mTHPP from bead is attributed to the polymer characteristics. In addition, diffusion of mTHPP may have been hindered by increased tortuosity of polymer accompanied by a swelling mechanism [77].

REFERENCES

[1] Prasad PN, *Introduction to Biophotonics*, Wiley-Interscience, New York, 2004.

[2] Dolmans DEJGJ, Fukumura D, and Jain RK, "Photodynamic therapy for cancer," *Nature* (London) 2003, 5, 380–387.

[3] Pass HI, "Photodynamic therapy in oncology: mechanisms and clinical use," *J. Natl. Cancer Inst.* 1993, 85, 443–456.

[4] Star WM, Marijnissen HP, van den Berg-Blok AE, Versteeg JA, Franken KA, and Reinhold HS, "Destruction of rat mammary tumor and normal tissue microcirculation by hematoporphyrin derivative photoradiation observed *in vivo* in sandwich observation chambers," *Cancer Res.* 1986, 46, 2532–2540.

[5] Wang I, Bendsoe N, Klinteberg CAF, Enejder AMK, Andersson-Engels S, Svanberg S, and Svanberg K, "Photodynamic therapy vs. cryosurgery of basal cell carcinomas: results of a phase III clinical trial," *Brit. J. Dermatol.*2001, 4, 832–840.

[6] Klinteberg CAF, Enejder AMK, Wang I, I Svanberg S, and Svanberg K, "Kinetic fluorescence studies of 5-aminolaevulinic acidinduced protoporphyrin IX accumulation in basal cell carcinomas," *J. Photoch. Photobio.*, B 1999, 49, 120–128.

[7] Konan YN, Gruny R, and Allemann E, "State of the art in the delivery of photosensitizers for photodynamic therapy," *J. Photoch. Photobio.,* B 2001, 66, 89–106.

[8] Taillefer J, Jones MC, Brasseur N, van Lier NJE, and Leroux JC, "Preparation and characterization of pH-responsive polymeric micelles for the delivery of photosensitizing anticancer drugs," *J. Pharm. Sci.* 2000, 89, 52–62.

[9] Wang S, Gao R, Zhou F, and Selke M, "Nanomaterials and singlet oxygen photosensitizers: potential applications in photodynamic therapy," *J. Mater. Chem.* 2004, 14, 487–493.

[10] van Nostrum CF, "Polymeric micelles to deliver photosensitizers for photodynamic therapy," *Adv. Drug Delivery Rev.* 2004, 56, 9–16.

[11] Peng Q, Berg K, Moan J, Kongshaug M, and Nesland JM, "5- aminolevulinic acid-based photodynamic therapy: principles and experimental research," *Photochem. Photobiol.* 1997, 65:2, 235–251.

[12] Peng Q, Warloe T, Berg K, Moan J, Kongshaug M, Giercksky KE, and Nesland JM, "5-aminolevulinic acid-based photodynamic therapy-clinical research and future challenges," *Cancer* 1997, 79_12_, 2282–2308.

[13] Arriagada FJ and Osseo-Asare K, "Synthesis of nanosize silica in aerosol OT reverse microemulsions," *J. Colloid Interface Sci.* 1995, 170, 8–17.

[14] Jain TK, Roy I, De TK, and Maitra AN, "Nanometer silica particles encapsulating active compounds: A novel ceramic drug carrier," *J. Am. Chem. Soc.* 1998, 120, 11092–11095.

[15] Roy I, Ohulchanskyy TY, Bharali DJ, Pudavar HE, Mistretta RA, Kaur N, and Prasad PN, "Optical tracking of organically modified silica nanoparticles as DNA carriers: A nonviral, nanomedicine approach for gene delivery," *Proc. Natl. Acad. Sci. USA*. 2000, 102, 279–284.

[16] Maeda H, Wu J, Sawa T, Matsumura Y, and Hori K, "Tumor vascular permeability and the EPR effect in macromolecular therapeutics. A review," *J. Controlled Release* 2000, 65, 271–284.
[17] Choi K, Burns AA, Williams RM, Zhou Z, Flesken-Nikitin A, Zipfel WR, Wiesner U, and Nikitin AY, "Core-shell silica nanoparticles as fluorescent labels for nanomedicine, " *J. Biomed. Opt.* 2007, 12_6_, 064007.
[18] Spiller W, Kliesch H, Wohrele D, Hackbarth S, Roder B, and Schnurpfeil GJ, "Singlet oxygen quantum yields of different photosensitizers in polar solvents and micellar solutions," *J. Porphyr. Phthalocyanines* 1998, 2, 145.
[19] van de Hulst HC, *Light Scattering by Small Particles*, Wiley- Interscience, New York (2004).
[20] B. N. Khlebtsov, V. A. Khanadeev, and N. G. Khlebtsov, "Determination of the size, concentration, and refractive index of silica nanoparticles from turbidity spectra," *Langmuir* 2008, 24, 8964–8970.
[21] Dougherty TJ, Gomer CJ, Henderson BW, Jori W, Kessel D, Korbelik M, Moan J, and Peng Q. Photodynamic therapy. *J. Natl. Cancer Inst.* 1998, 90:889–905.
[22] Dolmans DE, Fukumura D, and Jain RK. Photodynamic therapy for cancer. Nat. Rev. *Cancer.* 2003, 3:380–387.
[23] Bonnett R. Photosensitizers of the porphyrin and phthalocyanine series for photodynamic therapy. *Chem. Soc. Rev.* 1995, 24:19–33.
[24] Moore CM, Nathan TR, Lees WR, Mosse CA, Freeman A, Emberton M, and Bown SG. Photodynamic therapy using meso tetra hydroxy phenyl chlorin (mTHPC) in early prostate cancer. *Lasers Surg. Med.* 2006, 38:356–363.
[25] Hopper C, Kubler A, Lewis H, Tan IB, and Putnam G. mTHPC-mediated photodynamic therapy for early oral squamous cell carcinoma. *Int. J. Cancer.* 2004, 111:138–146.
[26] Friedberg JS, Mick R, Stevenson J, Metz J, Zhu T, Buyske J, Sterman DH, Pass HI, Glatstein E, and Hahn SM. A phase I study of Foscan-mediated photodynamic therapy and surgery in patients with mesothelioma. *Ann. Thorac. Surg.* 2003, 75:952–959.
[27] Ma L, Moan J, and Berg M. Evaluation of a new photosensitizer, meso-tetra-hydroxyphenyl-chlorin, for use in photodynamic therapy: a comparison of its photobiological properties with those of two other photosensitizers. *Int. J. Cancer.* 1994, 57:883–888.
[28] Kiesslich T, Berlanda J, Plaetzer K, Krammer B, and Berr F. Comparative characterization of the efficiency and cellular pharmacokinetics of Foscan- and Foslip-based photodynamic treatment in human biliary tract cancer cell lines. *Photochem. Photobiol. Sci.* 2007, 6:619–627.
[29] Sasnouski S, Zorin V, Khludeyev I, D'Hallewin MA, Guillemin F, and Bezdetnaya L. Investigation of FoscanÒ interactions with plasma proteins. *Biochim. Biophys. Acta.* 2005, 1725:394–402.
[30] Biolitec Pharma Ltd. Foscan, summary of product characteristics.
[31] Krueger T, Altermatt HJ, Mettler D, Scholl B, Magnusson L, and Ris HB. Experimental photodynamic therapy for malignant pleural mesothelioma with pegylated mTHPC. *Lasers Surg. Med.* 2003, 32:61–68.
[32] Reuther T, Kubler AC, Zillmann U, Flechtenmacher C, and Sinn H. Comparison of the in vivo efficiency of photofrin II-, mTHPC-, mTHPC-PEG- and mTHPCnPEG-

mediated PDT in a human xenografted head and neck carcinoma. *Lasers Surg. Med.* 2001, 29:314–322.
[33] Grahn MF, Giger A, McGuinness A, de Jode ML, Stewart JCM, Ris HB, Altermatt HJ, and Williams NS. mTHPC polymer conjugates: the in vivo photodynamic activity of four candidate compounds. *Lasers Med. Sci.* 1999, 14:40–46.
[34] Buchholz J, Kaser-Hotz B, Khan T, Rohrer Bley C, Melzer K, Schwendener RA, Roos M, and Walt H. Optimizing photodynamic therapy: in vivo pharmacokinetics of liposomal meta-(tetrahydroxyphenyl)chlorin in feline squamous cell carcinoma. *Clin. Cancer Res.* 2005, 11:7538–7544.
[35] Pegaz B, Debefve E, Ballini JP, Wagnieres G, Spaniol S, Albrecht V, Scheglmann DV, Nifantiev NE, van den Bergh H, and Konan-Kouakou YN. Photothrombic activity of m-THPC-loaded liposomal formulations: pre-clinical assessment on chick chorioallantoic membrane model. *Eur. J. Pharm. Sci.* 2006, 28:134–140.
[36] Torchilin VP. Structure and design of polymeric surfactantbased drug delivery systems. *J. Control. Release.* 2001, 73:137–172.
[37] Gaucher G, Dufresne MH, Sant VS, Kang N, Maysinger D, and Leroux JC. Block copolymer micelles: preparation, characterization and application in drug delivery. *J. Control. Release.* 2005, 109:169–188.
[38] Kwon GS, and Okano T. Polymeric micelles as new drug carriers. *Adv. Drug Deliv. Rev.* 1996, 21:107–116.
[39] Jones MC, and Leroux JC. Polymeric micelles—a new generation of colloidal drug carriers. *Eur. J. Pharm. Biopharm.* 1999, 48:101–111.
[40] Kwon GS. Polymeric micelles for delivery of poorly water soluble compounds. *Crit. Rev. Ther. Drug Carrier Syst.* 2003, 20:357– 403.
[41] Torchilin VP. Micellar nanocarriers: pharmaceutical perspectives. *Pharm. Res.* 2007, 24:1–16.
[42] van Nostrum VP. Polymeric micelles to deliver photosensitizers for photodynamic therapy. *Adv. Drug Deliv. Rev.* 2004, 56:9–16.
[43] Rapoport N. Physical stimuli-responsive polymeric micelles for anti-cancer drug delivery. *Prog. Polym. Sci.* 2007, 32:962–990.
[44] Sezgin Z, Yuksel N, and Baykara T. Preparation and characterization of polymeric micelles for solubilization of poorly soluble anticancer drugs. *Eur. J. Pharm. Biopharm.* 2006, 64:261–268.
[45] Taillefer J, Jones MC, Brasseur N, van Lier JE, and Leroux JC. Preparation and characterization of pH-responsive polymeric micelles for the delivery of photosensitizing anticancer drugs. *J. Pharm. Sci.* 2000, 89:52–62.
[46] Le Garrec D, Taillefer J, Van Lier JE, Lenaerts V, and Leroux JC. Optimizing pH-responsive polymeric micelles for drug delivery in a cancer photodynamic therapy model. *J. Drug Target.* 2002, 10:429–437.
[47] Rijcken CJF, Hofman JW, van Zeeland F, van Nostrum CF, and Hennink WE. Photosensitiser-loaded biodegradable polymeric micelles: preparation, characterisation and in vitro PDT efficacy. *J. Control. Release.* 2007, 124:144–153.
[48] Kakizawa Y, Harada A, and Kataoka K. Environment-sensitive stabilization of core–shell structured polyion complex micelle by reversible cross-linking of the core through disulfide bond. *J. Am. Chem. Soc.* 1999, 121:11247–11248.

[49] Katayama Y, Sonoda T, and Maeda M. A polymer micelle responding to the protein kinase a signal. *Macromolecules*. 2001, 34:8569–8573.
[50] Rijcken CJF, Soga O, Hennink WE, and van Nostrum CF. Triggered destabilisation of polymeric micelles and vesicles by changing polymers polarity: an attractive tool for drug delivery. *J. Control. Release*. 2007, 120:131–148.
[51] Carstens MG, Bevernage JJL, van Nostrum CF, van Steenbergen MJ, Flesch FM, Verrijk R, de Leede LGJ, Crommelin DJL, and Hennink WE. Small oligomeric micelles based on end group modified mPEG-oligocaprolactone with monodisperse hydrophobic blocks. *Macromolecules*. 2007, 40:116–122.
[52] Weissig V, Whiteman KR, and Torchilin VP. Accumulation of protein-loaded long-circulating micelles and liposomes in subcutaneous Lewis lung carcinoma in mice. *Pharm. Res*. 1998, 15:1552–1556.
[53] Vonarbourg A, Passirani C, Saulnier P, and Benoit JP. Parameters influencing the stealthiness of colloidal drug delivery systems. *Biomaterials*. 2006, 27:4356–4373.
[54] Sun X, Rossin R, Turner JL, Becker ML, Joralemon MJ, Welch MJ, and Wooley KL. An assessment of the effects of shell cross-linked nanoparticle size, core composition, and surface PEGylation on in vivo biodistribution. *Biomacromolecules*. 2005, 6:2541–2554.
[55] Liu J, Xiao Y, and Allen C. Polymer-drug compatibility: a guide to the development of delivery systems for the anticancer agent, ellipticine. *J. Pharm. Sci*. 2004, 93:132–143.
[56] Yokoyama M, Opanasopit P, Okano T, Kawano K, and Maitani Y. Polymer design and incorporation methods for polymeric micelle carrier system containing water-insoluble anticancer agent camptothecin. *J. Drug Target*. 2004, 12:373–384.
[57] Dougherty TJ, Gomer CJ, Henderson BW, Jori G, Kessel D, Korbelik M, Moan J, and Peng Q. Photodynamic therapy. *J. Natl. Cancer Inst*. 1998, 90, 889–905.
[58] Dolmans DE, Fukumura D, Jain RK. Photodynamic therapy for cancer. *Nat. Rev. Cancer*. 2003, 380–387.
[59] Bonnett R. Photosensitizers of the porphyrin and phthalocyanine series for photodynamic therapy. *Chem. Soc. Rev*. 1995, 24, 19–33.
[60] Moore CM, Nathan TR, Lees WR, Mosse CA, Freeman A, Emberton M, Bown SG. Photodynamic therapy using meso tetra hydroxy phenyl chlorin (mTHPC) in early prostate cancer. *Lasers Surg. Med*. 2006, 38, 356–363.
[61] Hopper C, Kubler A, Lewis H, Tan IB and Putnam G. mTHPC-mediated photodynamic therapy for early oral squamous cell carcinoma. *Int. J. Cancer*. 2004, 111, 138–146.
[62] Friedberg JS, Mick R, Stevenson J, Metz J, Zhu T, Buyske J, Sterman DH, Pass HI, Glatstein E and Hahn SM. A phase I study of Foscan-mediated photodynamic therapy and surgery in patients with mesothelioma. *Ann. Thorac. Surg*. 2003, 75, 952–959.
[63] Ma L, Moan J and Berg K. Evaluation of a new photosensitizer, meso-tetra-hydroxyphenyl-chlorin, for use in photodynamic therapy: a comparison of its photobiological properties with those of two other photosensitizers. *Int. J. Cancer*. 1994, 57, 883–888.
[64] Kiesslich T, Berlanda J, Plaetzer K, Krammer B, and Berr F. Comparative characterization of the efficiency and cellular pharmacokinetics of Foscan- and Foslip-based photodynamic treatment in human biliary tract cancer cell lines. *Photochem. Photobiol. Sci*. 2007, 619–627.

[65] Sasnouski S, Zorin V, Khludeyev I, D'Hallewin MA, Guillemin F and Bezdetnaya L. Investigation of FoscanÒ interactions with plasma proteins. *Biochim. Biophys. Acta.* 2005, 1725, 394–402.

[66] Biolitec Pharma Ltd. Foscan, summary of product characteristics.

[67] Krueger T, Altermatt HJ, Mettler D, Scholl B, Magnusson L and Ris HC. Experimental photodynamic therapy for malignant pleural mesothelioma with pegylated mTHPC. *Lasers Surg. Med. 2003*, 32, 61–68.

[68] Reuther T., Kubler AC, Zillmann, U., Flechtenmacher C. and Sinn H. Comparison of the in vivo efficiency of photofrin II-, mTHPC-, mTHPC-PEG- and mTHPCnPEG-mediated PDT in a human xenografted head and neck carcinoma. *Lasers Surg. Med.* 2001, 29, 314–322.

[69] Grahn MA, Giger A, McGuinness A, de Jode ML, Stewart JCM, Ris HB, Altermatt HJ and Williams NS. mTHPC polymer conjugates: the in vivo photodynamic activity of four candidate compounds. *Lasers Med. Sci. 1999*, 14, 40–46.

[70] Buchholz J, Kaser-Hotz B, Khan T, Rohrer Bley C, Melzer K, Schwendener RA, Roos M, and Walt H. Optimizing photodynamic therapy: in vivo pharmacokinetics of liposomal meta-(tetrahydroxyphenyl)chlorin in feline squamous cell carcinoma. *Clin. Cancer Res.* 2005, 11 7538–7544.

[71] Pegaz B, Debefve E, Ballini JP, Wagnieres G, Spaniol V, Albrecht DV, Scheglmann N, Nifantiev EN, van den Bergh H and Konan-Kouakou YN. Photothrombic activity of m-THPC-loaded liposomal formulations: pre-clinical assessment on chick chorioallantoic membrane model. *Eur. J. Pharm. Sci.* 2006, 28, 134–140.

[72] Torchilin VP. Structure and design of polymeric surfactantbased drug delivery systems. *J. Control. Release*. 2001, 73, 137–172.

[73] Gaucher G, Dufresne MH, Sant VP, Kang N., Maysinger D. and Leroux JC. Block copolymer micelles: preparation, characterization and application in drug delivery. *J. Control. Release*. 2005, 109, 169–188.

[74] Kwon GS and Okano T. Polymeric micelles as new drug carriers. *Adv. Drug Deliv. Rev.*, 1996, 21, 107–116.

[75] Jones MC and Leroux JC. Polymeric micelles—a new generation of colloidal drug carriers. *Eur. J. Pharm. Biopharm.*, 1999, 48, 101–111.

[76] Kwon GS. Polymeric micelles for delivery of poorly water soluble compounds. *Crit. Rev. Ther. Drug Carrier Syst.*, 2003, 20, 357–403.

[77] Saboktakin MR, Tabatabaie RM, Maharramov A, Ramazanov MA. Synthesis and In vitro Studies of Biodegradable Modified Chitosan Nanoparticles for Photodynamic Treatment of Cancer, *Int. J. Bio. Macromol*, 2011.

About the Contributors

Authors

Abel Maharramov - Academician, Professor, Doctor of Science in Chemistry
Baku State University
Academic Zahid Khalilov str.23
Baku City, Azerbaijan 1148
E-mail: rector@bsu.az

Mahammad Ali Ramazanov – Professor, Doctor of Science in Physics
Nanotechnology Department
Baku State University
Academic Zahid Khalilov str.23
Baku City, Azerbaijan 1148
E-mail: mamed_r50@mail.ru

Mohammad Reza Saboktakin, Ph.D.
International Research Institute of Arian Chemie Gostar
1026, Koye 13, Mikhak str., Boustan str.
Tabriz, Iran
E-mail: Sabotakin123@yahoo.com

Editor

Dr. Olgun Guven - Professor
Head of the Radiation Chemistry and Polymer Research Laboratories
and Researcher and Lecturer at Hacettepe University
Chemistry Department
06800, Beytepe, Ankara, Turkey
Telephone and Fax No.: + 90 312 2977977
E-mail: guven@hacettepe.edu.tr

INDEX

A

B

C

D

E

F

G

H

I

J

K

L

M

N

Q

R

S

T

U

V

W

X

Y

Z